OXFORD–WARBURG STUDIES

General Editors

CHARLES HOPE *and* IAN MACLEAN

Oxford–Warburg Studies comprise works of original research on the intellectual and cultural history of Europe, with particular reference to the transmission and reception of ideas and images originating in the ancient world. The emphasis of the series is on elite rather than popular culture, and the underlying aim is to foster an interdisciplinary approach based on primary sources, which may be visual as well as written, and may extend to materials in a wide range of vernaculars and ancient languages. The authors of the series have addressed in particular the relationship between classical scholarship and the Christian tradition, the influence of modes of transmission on the uptake of ideas, the contributions of great scholars to the learning of their day, and the study of the Italian and Northern manifestations of humanism and their aftermath.

OXFORD–WARBURG STUDIES

The Apocryphal Apocalypse
The Reception of the Second Book of
Esdras (4 Ezra) from the Renaissance
to the Enlightenment
ALASTAIR HAMILTON

Children of the Promise
The Confraternity of the Purification and the
Socialization of Youths in Florence
1427–1785
LORENZO POLIZZOTTO

Machiavelli—The First Century
Studies in Enthusiasm, Hostility, and Irrelevance
SYDNEY ANGLO

History of Scholarship
A Selection of Papers from the Seminar on the History of Scholarship held annually
at the Warburg Institute
Edited by CHRISTOPHER LIGOTA and JEAN-LOUIS QUANTIN

Transmitting Knowledge
Words, Images, and Instruments in Early Modern Europe
Edited by SACHIKO KUSUKAWA and IAN MACLEAN

The Copts and the West, 1439–1822
The European Discovery of the Egyptian Church
ALASTAIR HAMILTON

Commonplace Learning
Ramism and its German Ramifications 1543–1630
HOWARD HOTSON

The Church of England and Christian Antiquity
The Construction of a Confessional Identity in the 17th Century
JEAN-LOUIS QUANTIN

John Selden
A Life in Scholarship
G. J. TOOMER

A History of Renaissance Rhetoric 1380–1620
PETER MACK

The Experience of Beauty in the Middles Ages
MARY CARRUTHERS

Johann Heinrich Hottinger
Arabic and Islamic Studies in the Seventeenth Century
JAN LOOP

Criticism and Confession
The Bible in the Seventeenth Century Republic of Letters
NICHOLAS HARDY

Making Mathematical Culture

University and Print in the Circle of Lefèvre d'Étaples

RICHARD J. OOSTERHOFF

OXFORD
UNIVERSITY PRESS

Great Clarendon Street, Oxford, OX2 6DP,
United Kingdom

Oxford University Press is a department of the University of Oxford.
It furthers the University's objective of excellence in research, scholarship,
and education by publishing worldwide. Oxford is a registered trade mark of
Oxford University Press in the UK and in certain other countries

First Edition published in 2018

Impression: 1

Published in the United States of America by Oxford University Press
198 Madison Avenue, New York, NY 10016, United States of America

British Library Cataloguing in Publication Data
Data available

Library of Congress Control Number: 2017964289

ISBN 978–0–19–882352–0

Printed and bound by
CPI Group (UK) Ltd, Croydon, CR0 4YY

Acknowledgements

Among many other things, Elora, *dilectissima mea*, has convinced me that the convention of thanking those dearest last is confusing. No one has given more to this book. So I thank her first.

The practices of authorship in the modern university are quite unlike those I uncover in this book. But we remain dependent on others. Robert Goulding first shepherded the dissertation that has been mostly rewritten as this book—he has seen this work in more states of disarray than anyone else, signs of his endurance and intellectual strength of character. Likewise, Margaret Meserve, Lynn Joy, John van Engen, and Jill Kraye are all exemplars of the well lived scholarly life, and early on nudged me with questions that still haunt me. Isabelle Pantin encouraged some very early thoughts with great kindness, Emmanuel Faye and James Hirstein have graciously welcomed me into the *sodalitas Rhenani*, and Jean-Marc Mandosio has been an enduring source of encouragement, even offering welcome suggestions on a late version. Benjamin Wardhaugh commented in depth and at length, first on a chapter and then on the entire thing, and gave wise and useful advice—most of which I attempted to follow.

Many friends have conspired to keep me closer to the intellectual straight and narrow. Alex Marr continues to inspire as a model of intellectual acuity, sagacity, and insight. Raphaële Garrod and José Ramon Marcaida are congenial spirits ministering to a soul in need of succour. Tim Chesters and Richard Serjeantson offered their innate good sense with their usual eloquence. All read parts of this work in one form or another. Sundar Henny, Kate Isard, Arthur Keefer, and Nathan Ristuccia also commented on sections at key moments, with habitual good judgement and good cheer. Sachiko Kusukawa, Sietske Fransen, and Katie Reinhart are the best of cotravellers. Anthony Ossa-Richardson has been the embodiment of scholarly friendship, reading each chapter with care, saving me from countless solecisms and infelicities, and commiserating to the end. I am profoundly grateful for such friends, and sharply conscious of how little these words of thanks are. The remaining errors are not theirs.

I owe special thanks for the gifts of institutions. Librarians at the Bibliothèque Humaniste de Sélestat have been unfailingly gracious every time I swept through with too many requests. I found librarians of tremendous professionalism at the Huntington Library, the Houghton Library, Duke Humphrey's Library (back in the day), Cambridge University Library, the British Library, and indeed the Bibliotheque nationale de France. I am grateful for fellowships from the Nanovic Institute (Notre Dame) and the Notre Dame Institute for Advanced Study, the Warburg Institute, the Houghton Library, and the Huntington Library. For the last steps on this project, the European Research Council provided wonderful support, in the form of the project 'Genius before Romanticism: Ingenuity in Early Modern Art and Science'. Without Gaenor Moore, who did the big kindness of obtaining many of the final images, I might have forgotten many things, including to say this: the research leading to these results has received funding from the European Research Council under the European Union's Seventh Framework Programme (FP7/2007–2013)/ERC grant agreement no. 617391. Not least, I must thank the Maire of Sélestat, and Mr Laurent Naas of the Bibliothèque Humaniste de Sélestat, for permission to use images of annotations by Beatus Rhenanus. Finally, I am grateful to Kavya Ramu, Howard Emmens, and Andrew Hawkey for so carefully seeing these sometimes recalcitrant pages through the press.

Then there are the clans. I owe so much to the Deliyannides, the Vander Klippes, the Wallaces, and the Oosterhoffs I grew up with. Sydney, Norah, Judah, and (yes) Elora: this is yours.

Contents

List of Illustrations

3.1. Beatus Rhenanus' student notebook: a page from Beatus Rhenanus' *Cahier d'étudiant*, BHS MS 58, fol. 147r, with the mnemonic 'NACAMILUT' in the margin. This acronym labels circles found in Beatus' textbook on natural philosophy (see Figure 3.2). 58

3.2. Beatus Rhenanus' textbook on natural philosophy: Lefèvre and Clichtove, *Totius philosophiae naturalis paraphrases, adiecto commentario* (1502), BHS K 1199, 2v. 59

3.3. Beatus orders the authorities: Lefèvre, *Libri logicorum* (Paris: Hopyl and Estienne, 1503), BHS K 1047, guard page, detail. 66

3.4. The *minor mundus* (man) reflects the *maior mundus* (world): Lefèvre, *Libri logicorum* (1503), BHS K 1047, end page with annotations by Beatus Rhenanus. 76

3.5. Beatus observes analogies between grammar and city: Aristotle, *Politica*, in *Opera* (Venice, 1496), BHS K1276, 260r. 78

4.1. Contents of two Fabrist compendia of mathematics: Lefèvre et al., *Epitome Boetii, etc.* (1503) and Johannes Caesarius (ed.), *Introductio Stapulensis, etc.* (Deventer, 1507). 97

4.2. Aristotle's and Plato's modes of philosophizing: Lefèvre, *Libri logicorum* (1503), BHS K1047, endpage. Beatus contrasts the resolutive and compositive modes of philosophizing, portrayed as variants of the Porphyrian tree. The translation does not include the diagrams of substances, which fill in layers of colour for different composites: black for a substance with body, or bodily nature; blue represents the addition of life, or animate nature; red for the addition of soul, or sensible nature; and white for rational nature. The descent from universals is identified with Plato and the order of nature (left, moving from substance to the individual Plato and friends), and the ascent from particulars with Aristotle and the order of cognition (right, moving from individuals towards substance). 99

4.3. The table (*formula*) introducing Lefèvre's *Epitome Boetii* (1496), listing all the key terms of number theory from broadest headings down to narrowest subcategories. Note how each term is next to an example, so that a *numerus multiplex triplus* is shown as a ratio of 3:1. Lefèvre, *Elementa arithmetica, etc.* (1496), h8r. By kind permission of the Syndics of Cambridge University Library, Shelfmark: Inc.3.D.1.21. 110

4.4. The *figura introductionis* which outlines the basic structure of sublunar physics in Aristotelian natural philosophy, with major headings and minor subdivisions. Lefèvre, *Paraphrases philosophiae naturalis* (1492), b2r. By permission of the British Library, Shelfmark IA.40121. 112

4.5. Typography of a textbook: list of theses, linked by marginal numbers to discursive arguments in paraphrases of Aristotle (see Figure 4.6). Lefèvre, *Paraphrases philosophiae naturalis* (1492), a2r. By permission of the British Library, Shelfmark IA.40121. 114

4.6. Typography of a textbook: a numbered thesis. Lefèvre, *Paraphrases philosophiae naturalis* (1492), b5r. By permission of the British Library, Shelfmark IA.40121. 115

4.7. Epitome of arithmetic. The *formula* of core propositions collated from Boethius and Jordanus shares a goal with the table from Ramus shown in Figure 4.8: to epitomize a discipline. Lefèvre, *Elementa arithmetica, etc.* (1496), i5r. By permission of the British Library, Shelfmark C.106.a.8. 118

4.8. Epitome of dialectic: Peter Ramus, *Dialecticae institutiones* (Paris: Jacques Bogard, 1543), 57r. By permission of the British Library, Shelfmark C.106.a.8. 119

5.1. The 'studious athlete' in Bovelles, *Liber de intellectu, etc.* (1511), 60v. Note the lines between *lectio, scriptio,* and vision, which together with hearing and speaking fill the scholar's starry imagination. By kind permission of the Syndics of Cambridge University Library, Shelfmark: Acton.b.sel.32. 126

5.2. A typical page in Lefèvre's *Textus de sphera*: table, text, commentary (in smaller type), with diagrams and section numbers both printed outside the forme. Lefèvre, *Textus de sphera* (1495 [here 1516]), a5v. © Fitzwilliam Museum, University of Cambridge. 136

5.3. Frontispiece from Sacrobosco, *Sphaera mundi* (Venice: Ottaviano Scoto, 1490). © Fitzwilliam Museum, University of Cambridge. 138

5.4. Frontispiece from Lefèvre, *Textus de sphera* (Paris: Estienne, 1516), a3v, reusing the woodblock from the first edition of 1495. © Fitzwilliam Museum, University of Cambridge. 139

5.5. The exaggerated realism of Sacrobosco, *Sphaera mundi* (1490), a8r. © Fitzwilliam Museum, University of Cambridge. 140

5.6. The *introductoria additio* in Lefèvre, *Textus de sphera* (Paris: Estienne, 1500), BHS K 1046c, a2v–a3r. Lefèvre's primer explains the geometrical objects needed to understand astronomy, as contemporary editions of Sacrobosco sometimes also did (*right*). Here Lefèvre goes further, explaining how to perform sexagesimal arithmetic, which is the subject of Beatus Rhenanus' extensive notes on the facing page (*left*). 142–3

5.7. Illustration of a sphere on a lathe, with a semi-circular cutting tool. Lefèvre, *Textus de sphera* (Paris: Estienne, 1516), a4r. © Fitzwilliam Museum, University of Cambridge. 145

5.8. Lefèvre shows how to divide any interval equally, 'without a certain or fixed ratio of number'. From his *Elementa arithmetica, musicalia, etc.* (1496), G6v, proposition III.35. By kind permission of the Syndics of Cambridge University Library, Shelfmark: Inc.3.D.1.21. 158

5.9. Beatus Rhenanus records the performance of calculations of the distances between the various planetary spheres: on the bottom of the page, he records various 'modern' opinions on the number of these heavenly spheres—the omission of what is evidently 'Regiomontanus' in the first line suggests he missed the name during dictation and intended to fill it in later. Detail from Beatus Rhenanus' copy of Lefèvre, *Textus de sphera* (1500), BHS K 1046c, a1v. 166

Conventions and Abbreviations

CONVENTIONS

All translations, unless indicated differently, are my own. Titles of Latin books are left in the original, unless they seemed more commonly familiar in English translation.

For the sake of readability, I have regularized Latin transcriptions, notably capitalization, i/j, u/v, punctuation, and I have silently expanded abbreviations. I have counted signatures in arabic numerals, so that signature e, folio IV, verso is e4v.

Dates have been normalized to modern standards. Before the gradual adoption of the Gregorian calendar (in France in 1582, in England not fully until 1752) the legal year normally began on 25 March. Therefore a date given as February 1494 I record as February 1495.

For names, I have tried to follow the English conventions most often found in modern scholarship. Therefore, the full name is 'Nicholas of Cusa', while the single cognomen is 'Cusanus'; 'Peter Ramus', not the more accurate 'Pierre Ramus'. Lefèvre, *Faber* in Latin, permits the pun 'Fabrist' as an adjective. On Oronce Fine, I follow his close friend Antoine Mizauld, who made it rhyme with *doctrine* (Emmanuel Poulle, 'Oronce Fine', *Dictionary of Scientific Biography*, ed. Charles C. Gillespie, vol. 15 (New York: Charles Scribner's Sons, 1978), 156).

In the bibliography, printed editions of classical and medieval works are listed by editor.

ABBREVIATIONS

BHS	Bibliothèque Humaniste de Sélestat
CE	Peter G. Bietenholz and Thomas B. Deutscher, *Contemporaries of Erasmus: A Biographical Register of the Renaissance and Reformation*, 3 vols (Toronto: University of Toronto Press, 1985)
Cusanus	*Opera omnia*, in the edition of the Heidelberger Akademie der Wissenschaften (Hamburg: Felix Meiner, 1927–2005), available at http://www.cusanus-portal.de/
Elementa	Jacques Lefèvre d'Étaples, *Elementa arithmetica; Elementa musicalia; Epitome in libros arithmeticos divi Severini Boetii; Rithmimachie ludus que et pugna numerorum appellatur* (Paris: Johannes Higman and Wolfgang Hopyl, 1496)
Epitome	Jacques Lefèvre d'Étaples, Josse Clichtove, and Charles de Bovelles, *Epitome compendiosaque introductio in libros arithmeticos divi Severini Boetii, adiecto familiari [Clichtovei] commentario dilucidata. Praxis numerandi certis quibusdam regulis (auctore Clichtoveo). Introductio in geometriam Caroli Bovilli. Astronomicon Stapulensis* (Paris: Wolfgang Hopyl and Henri Estienne, 1503)
Euclidis Elementa	Jacques Lefèvre d'Étaples (ed.), *Euclidis Megarensis mathematici clarissimi Elementorum geometricorum libri xv, Campani Galli transalpini in eosdem commentariorum libri xv, Theonis Alexandrini*

	Bartholamaeo Zamberto Veneto interprete, in tredecim priores, commentariorum libri xiii, Hypsiclis Alexandrini in duos posteriores, eodem Bartholomaeo Zamberto Veneto interprete, commentariorum libri ii (Paris: Henri Estienne, 1517)
Liber de intellectu etc.	Charles de Bovelles, *Que hoc volumine continentur: Liber de intellectu; Liber de sensu; Liber de nichilo; Ars oppositorum; Liber de generatione; Liber de sapiente; Liber de duodecim numeris; Epistole complures. Insuper mathematicum opus quadripartitum: De numeris perfectis; De mathematicis rosis; De geometricis corporibus; De geometricis supplementis* (Paris: Henri Estienne, 1511)
Libri logicorum	Jacques Lefèvre d'Étaples, *Libri logicorum ad archteypos recogniti cum novis ad litteram commentariis ad felices primum Parhisiorum et communiter aliorum studiorum successus in lucem prodeant ferantque litteris opem* (Paris: Wolfgang Hopyl and Henri Estienne, 1503)
PE	Eugene F. Rice Jr. (ed.), *The Prefatory Epistles of Jacques Lefèvre d'Étaples and Related Texts* (New York: Columbia University Press, 1972)
Politica etc.	Jacques Lefèvre d'Étaples, *Contenta. Politicorum libro octo. Commentarii. Economicorum duo. Commentarii. Hecatonomiarum septem. Economiarum publ. unus. Explanationis Leonardi [Bruni] in Oeconomica duo* (Paris: Henri Estienne, 1506)
Renaudet	Augustin Renaudet, *Préréforme et humanisme à Paris pendant les premières guerres d'Italie, 1494–1517* (1916; 2nd edn, Paris: Édouard Champion, 1953)
Textus de sphera	Jacques Lefèvre d'Étaples, *Textus de sphera Johannis de Sacrobosco, cum additione (quantum necessarium est) adiecta: novo commentario nuper edito ad utilitatem studentium philosophice parisiensis academie: illustratus* (Paris: Wolfgang Hopyl, 1495)

1
Introduction

It was a Sunday afternoon in Rome, sometime in 1507. The celebrated Paris master Jacques Lefèvre d'Étaples (*c.*1455–1536) had been here twice before, but it was the first time for his close friend and colleague Charles de Bovelles (1479–*c.*1567). Walking through the Jewish quarter, a companion recognized a face in the crowd: Bonetus de Latteis, the Jewish astrologer and physician to the pope. Bovelles knew Bonetus from his brief book on an astronomical 'ring', a marvel of precious craft and ingenuity, a fully functioning astrolabe in miniature. In fact, Bovelles had published the book in Paris several years before as a supplement to Lefèvre's astronomical handbook. He called out in greeting, and soon the men were talking of the ring, which Bonetus promptly displayed in its case. The venerable Jew welcomed the travelling Frenchmen in from the street and their conversation filled the afternoon. As they entered a synagogue, their talk progressed from the ring to the heavens, and from there to the Jewish and Christian scriptures and the teaching of the Trinity.[1]

This exchange sets out what these men loved best: books; scholarly friendship and conversation; the relationship of natural knowledge to divine truth; and the marvels of mathematics. In this book, I hope to orient the discussion of Renaissance culture toward this last term, mathematics. A familiar story sets the emergence of mathematical culture in late sixteenth-century Italian courts, inexorably aimed at the seventeenth century of Galileo and Descartes.[2] By contrast, my argument, taken at its narrowest, is that mathematics became the engine for transforming knowledge a century earlier, in the most surprising of places: the scholastic stronghold of Europe, the University of Paris, where this band of scholars experimented with new print to renovate the structure of university knowledge.

To elaborate this mathematical culture, I scrutinize sources often scorned: university books, from manuscript lecture notes and cheap short-cut manuals to thick printed treatises heavy with student notes. Historians of science and humanism alike have been seduced by a polemic common from Petrarch to Bacon, that university books were full of pedantic obscurantism rather than new knowledge. Many were (and are). But novelty emerges in the practices of recycling what 'everyone' knows. Consider the book Bovelles and Lefèvre enriched

[1] Bovelles later reconstructed the conversation as a dialogue *De trinitate*, appended to his *Quaestionum theologicarum libri septem* (Paris: Josse Bade, 1513), 53r ff.

[2] The classic account is Paul Lawrence Rose, *The Italian Renaissance of Mathematics* (Geneva: Droz, 1975).

with the treatise on the astronomical ring, the *Textus de sphera*.[3] Ostensibly it was the *Sphere* of Sacrobosco, the medieval university's omnipresent introduction to astronomy, the very symbol of the supposedly static knowledge eclipsed by the new sciences of the seventeenth century. In fact, Lefèvre had transformed it into a compendium of up-to-date astronomical knowledge. Previous printed editions of Sacrobosco supplied only the text; Lefèvre's version of 1495 tripled the text with commentary.[4] Previous printed editions, if illustrated, simply replicated figures in the manuscript tradition. Lefèvre's edition collated these old images with new images, whether borrowed from other genres or restyled old ones. The old diagrams, illustrations, and tables were redesigned, incorporating new tables of planetary distances, mapping coordinates of cities, and astrological tables. These were intended not merely to train students to theorize the movements of the heavens, but to calculate them and put them to use—the book included a short primer on how to do the sexagesimal arithmetic of astronomy, and exercises for students to use the tables. The very apparatus of the text—another first for Sacrobosco, whether in manuscript or print—advertised the fact that this was no medieval textbook: a table of the book's contents helped readers navigate not Sacrobosco, but Lefèvre's commentary. No wonder Bonetus' miniature astrolabe seemed an appropriate addition. Materially and visually, this printed university book did not simply replicate a text, but was the site of knowledge in the making.

This book and others like it are doubly important because they bear witness to a pivotal moment between 1490 and 1510, a moment historians know better for two other reasons. First, it was during these years that Lefèvre transformed intellectual culture at the Collège du Cardinal Lemoine in Paris. He travelled to Italy, sealing relations with Italian humanists such as Marsilio Ficino, Angelo Poliziano, and Giovanni Pico della Mirandola, in their last years, on the eve of Charles VIII's invasion of Italy in 1494. While exchange can be traced long before and after, Lefèvre's Italian travels have been mythologized as the moment when the swell of humanism moved north, already in his own lifetime and still now in standard histories of the French Renaissance.[5] Lefèvre's books therefore have come to emblematize the new humanist Aristotle, the first wave of the rising tide of northern humanism in France.[6] Second, during these years Lefèvre's circle

[3] Jacques Lefèvre d'Étaples, *Textus de Sphera Johannis de Sacrobosco, Cum Additione (quantum necessarium est) adiecta: Nouo commentario nuper edito ad utilitatem studentium Philosophice Parisiensis Academie: illustratus. Cum compositione Anuli Astronomici Boni Latensis. Et Geometria Euclidis Megarensis* (Paris: Henri Estienne, 1500).

[4] Paris: Wolfgang Hopyl, 1495.

[5] Symphorien Champier, *Duellum epistolare: Gallie & Etalie antiquitates summatim complectens* (Lyons: Froben et al., 1519), a3r ff. Representative accounts include Walter Mönch, *Die italienische Platonrenaissance und ihre Bedeutung für Frankreichs Literatur- und Geistesgeschichte (1450–1550)* (Berlin: Matthiesen Verlags, 1936); Philippe de Lajarte, *L'Humanisme en France au XVIe siècle* (Paris: Honoré Champion, 2009).

[6] Most influentially, Charles B. Schmitt, *Aristotle and the Renaissance* (Cambridge, MA: Harvard University Press, 1983), 34, 47, 80. See also Eugene F. Rice, 'Humanist Aristotelianism in France: Jacques Lefèvre d'Étaples and His Circle', in *Humanism in France*, ed. A. H. T. Levi (Manchester: Manchester University Press, 1970), 132–49.

incubated the leaders of movements that would define France, and indeed Europe, in the century ending with the Wars of Religion.[7] Some of his students, such as Josse Clichtove (*c.*1472–1543), would play central roles in the conservative reform movements at the University of Paris, notably as a leading member of the theology faculty that unsuccessfully tried to limit Luther and Henry VIII. Other students would follow Lefèvre in exploring new approaches to religious reform, the French Reformation; one of these, Guillaume Farel, would convince Calvin to stay in Geneva in 1532. This circle occupied a remarkable space between Italian learning and French reform; in fact, they are just as remarkable for their contributions to mathematical culture.

Indeed, the very richness of this pregnant moment in Europe's history makes it imperative not to read the end into the beginning. For those living in this moment, the epochal consequences of religious reform, humanism, and the printed page were far from set, much less foreseen. Lefèvre's circle is an opportunity for reconsidering a period marked by transformations so profound that they have worn deep ruts into our histories: medieval or Renaissance, manuscript or print, scholasticism or humanism, mathematics or erudition. At the cusp of these transformations, but not party to any one of these categories, Lefèvre's circle shows the poverty of such dichotomies. Therefore this book strives not to dwell on any one of these terms at the expense of the others. And therefore my broader argument: that this group inhabits the tensions of this moment in European culture.

MATHEMATICAL CULTURE

This book addresses a signal episode in the long-range transmission of ancient and medieval mathematical culture into early modern Europe. Historians have sometimes defined the rise of modernity as a shift from words to number, capturing the remarkable transformation of the mathematical arts from the neglected domain of low-prestige practitioners in the fifteenth century to the new 'queen of the sciences' by the eighteenth, with countless practical applications.[8] Books such as Lefèvre's *Textus de sphera* were once deliberately left out of this account because mathematics and the history of science in general seemed like a pure succession of theories that floated free of institutions, especially conservative ones such as universities.[9] In this

[7] Note that Lefèvre is the central character for the foundational modern study of the French Reformation: see Lucien Febvre's 1929 essay 'The Origins of the French Reformation: A Badly-Put Question?', translated in *A New Kind of History: From the Writings of Lucien Febvre* (London: Routledge and Kegan Paul, 1973), 44–107.

[8] e.g. Michael E. Hobart and Zachary S. Schiffman, *Information Ages: Literacy, Numeracy, and the Computer Revolution* (Baltimore, MD: Johns Hopkins University Press, 2000); Mary Poovey, *A History of the Modern Fact: Problems of Knowledge in the Sciences of Wealth and Society* (Chicago: University of Chicago Press, 1998).

[9] Lefèvre's works were known; they just were not read: e.g. Moritz Cantor, *Vorlesungen über Geschichte der Mathematik*, vol. II (Leipzig: Teubner, 1892), 364–7.

version of the 'mathematization of the world picture' Lefèvre seemed only a null node on a line from Nicholas of Cusa to Descartes:

> No one, not even Lefèvre d'Étaples who edited [Cusanus'] works, seems to have paid much attention to them, and it was only after Copernicus... that Nicholas of Cusa achieved fame as a forerunner of Copernicus, and even of Kepler, and could be quoted by Descartes as an advocate of the infinity of the world.[10]

By abstracting a couple of mathematizing ideas (namely, that space is relative and that motion can be relativized against infinity) Koyré brilliantly revealed what Cusanus shared with Albert Einstein. But the fact that Cusanus and Einstein were right obscures any possibility of explaining how we get from one to the other. To say Galileo, Descartes, and Newton succeeded because they were right ignores the time-bound practices and materials by which rightness and facticity are assembled.[11]

I find 'mathematical culture' a term usefully capacious for keeping these practices and materials in view.[12] For a culture involves more than a set of disciplines and their uses; culture involves practices of caring, nourishing, and cultivating. Culture only exists through cultivation. In recent decades historians have uncovered just how much early modern sciences depended on school learning and its manuscript, material, and visual cultures. Bacon and his followers borrowed from the textual practices of erudites to make his new natural history;[13] Galileo and Descartes borrowed logical tools and topics from the very Jesuits' textbooks they ridiculed;[14] they were debated by a growing demographic of university-trained readers.[15] In each case, the 'new' sciences bear deep continuities with the bookish practices and school disciplines they eclipsed. Scientific habits and priorities were cultivated.

[10] Alexandre Koyré, *From the Closed World to the Infinite Universe* (New York: Harper Torch, 1958), 18–19.

[11] Michael Bycroft offers more nuanced reflection on this issue in 'How to Save the Symmetry Principle', in *The Philosophy of Historical Case Studies*, ed. Tilman Sauer and Raphael Scholl (Dordrecht: Springer, 2016), 11–29.

[12] Oronce Fine once refers to a *cultura Mathematicarum*, in a prefatory letter to Juan Martínez Silíceo, *Arithmetica in theoricen, et praxim scissa*, ed. Oronce Fine (Paris: Henri Estienne, 1519), a2r. For more on the term, see Alexander Marr, *Between Raphael and Galileo: Mutio Oddi and the Mathematical Culture of Late Renaissance Italy* (Chicago: University of Chicago Press, 2011), introduction.

[13] Ann Blair, 'Humanist Methods in Natural Philosophy: The Commonplace Book', *Journal for the History of Ideas* 53, no. 4 (1992): 541–51; Fabian Krämer, *Ein Zentaur in London: Lektüre und Beobachtung in der frühneuzeitlichen Naturforschung* (Affalterbach: Didymos-Verlag, 2014); Richard Yeo, *Notebooks, English Virtuosi, and Early Modern Science* (Chicago: University of Chicago Press, 2014).

[14] Peter Dear, 'Mersenne's Suggestion: Cartesian Meditation and the Mathematical Model of Knowledge in the Seventeenth Century', in *Descartes and His Contemporaries: Meditations, Objections, and Replies*, ed. Roger Ariew and Marjorie Grene (Chicago: University of Chicago Press, 1995), 44–62; Garrod, '"La politesse de l'esprit": The Jesuit Pedagogical Background to the *Regulae*' (forthcoming); Roger Ariew, *Descartes among the Scholastics*, History of Science and Medicine Library 20 (Leiden: Brill, 2011).

[15] This argument is one theme in Jed Z. Buchwald and Mordechai Feingold, *Newton and the Origin of Civilization* (Princeton, NJ: Princeton University Press, 2012); Renée Raphael, *Reading Galileo: Scribal Technologies and the Two New Sciences* (Baltimore, MD: Johns Hopkins University Press, 2017). A particularly fine example is by Richard Serjeantson, 'The Education of Francis Willughby', in *Virtuoso by Nature: The Scientific Worlds of Francis Willughby FRS (1635–1672)*, ed. Tim Birkhead (Leiden: Brill, 2016), 44–98.

Similarly, recent studies show how early modern mathematics became authoritative *because* embedded in specific cultural spaces. We have learned how, in libraries from Urbino to Oxford, scholars such as Federico Commandino and Henry Savile matched humanist scholarship with mathematical acumen, using Proclus's commentary on Euclid, Archimedes and other rediscovered ancient texts to construct the rising prestige of mathematical disciplines.[16] We have also learned how a new sort of intellectual moved between workshop, university, and court, epitomized by Oronce Fine in Paris, Kepler in Prague, Galileo in Venice and Florence, or more typically Mutio Oddi in Milan, who built their personae around the mathematical practices such as drawing architectural plans, calibrating horoscopes, surveying, fabricating instruments, taking observations, and teaching students. These mathematical practitioners established their authority through courtly networks of friends and masterfully devised books and instruments.[17] Figures such as Savile, Kepler, and Oddi did not make this mathematical culture in heroic opposition to the universities, but depended on them as their cultural substratum. Savile offered one of the age's most sophisticated accounts of mathematical history in his ordinary lectures on astronomy;[18] Kepler drew deeply on Melanchthonian universities in his conviction that mathematics would resolve heaven's mysteries and earth's conflicts;[19] Oddi developed his mathematical friendships out of his teaching in the *studium* at Milan.[20] These suggest how important university practices—authorship, friendship, teaching and learning, making and appropriating books and instruments—were for legitimating mathematical culture.[21] Indeed, we have come to see lively intellectual battles waged over mathematics at later-sixteenth-century institutions: Wittenberg, the Royal College of Paris (est. 1530) under Oronce Fine and Peter Ramus, Oxford and Cambridge, and the Collegio Romano.[22] In Lefèvre's circle, we see the first steps in the formation of this mathematical culture—the first

[16] More generally on the circulation of new texts, see Rose, *The Italian Renaissance of Mathematics*. On new genealogies, see e.g. Domenico Bertoloni Meli, 'Guidobaldo Dal Monte and the Archimedean Revival', *Nuncius* 7, no. 1 (1992): 3–34; Robert Goulding, *Defending Hypatia: Ramus, Savile, and the Renaissance Rediscovery of Mathematical History* (New York: Springer, 2010).

[17] Stephen Johnston, 'The Identity of the Mathematical Practitioner in 16th-Century England', in *Der 'Mathematicus': Zur Entwicklung und Bedeutung einer neuen Berufsgruppe in der Zeit Gerhard Mercators*, ed. Irmgarde Hantsche (Bochum: Brockmeyer, 1996), 93–120; Lesley B. Cormack, Steven A. Walton, and John A. Schuster (eds), *Mathematical Practitioners and the Transformation of Natural Knowledge in Early Modern Europe* (New York: Springer, 2017).

[18] Besides Goulding, *Defending Hypatia*, 75–115.

[19] Robert S. Westman, 'The Melanchthon Circle, Rheticus, and the Wittenberg Interpretation of the Copernican Theory', *Isis* 66, no. 2 (1975): 164–93; Charlotte Methuen, *Kepler's Tübingen: Stimulus to a Theological Mathematics* (Aldershot: Ashgate, 1998).

[20] Marr, *Oddi*, chapter 2.

[21] An exemplary effort to take such practices seriously, for a later period, is Andrew Warwick, *Masters of Theory: Cambridge and the Rise of Mathematical Physics* (Chicago: University of Chicago Press, 2003). Historians of modern science have, belatedly following Kuhn, begun to consider the foundational role of textbooks: e.g. Marga Vicedo, 'Introduction [to Focus Section]: The Secret Lives of Textbooks', *Isis* 103, no. 1 (2012): 83–7.

[22] The fundamental study is Mordechai Feingold, *The Mathematician's Apprenticeship: Science, Universities and Society in England, 1560–1640* (Cambridge: Cambridge University Press, 1984). The Jesuits based their collegiate model roughly on Paris; on their mathematical interests, see Peter Dear, *Discipline and Experience: The Mathematical Way in the Scientific Revolution* (Chicago: University of Chicago Press, 1995); Antonella Romano, *La contre-reforme mathematique: Constitution et diffusion*

printed textbooks serving a new programme of teaching the disciplines, beginning with mathematics. In a real sense, therefore, in Lefèvre's circle we see the cultivation of a renovated mathematics: a new mathematical culture.

This mathematical culture built on a conceptual core supplied by medieval sources. The books made by Lefèvre's circle, taught in their classrooms and annotated by their students, supply the primordial constituents: the quadrivium. Latin scholars had defined the quadrivium since at least Boethius and Martianus Capella in late antiquity as the four disciplines of arithmetic and music, geometry and astronomy. Through commonplaces and mangled etymologies of *mathemata* (literally, 'things learned') stretching back at least to Plato, these four disciplines were privileged as 'liberal arts', to be set alongside the trivium of grammar, logic, and rhetoric. Arithmetic dealt purely with discrete magnitude or numbers; geometry dealt with continuous magnitude or spatial extension in points, lines, and surfaces. In a disciplinary idiom that held into the seventeenth century, arithmetic and geometry also served as the theoretical basis for other applications—what Aristotelians called 'mixed sciences'. Music and astronomy were considered mixed because they combined pure magnitude with matter: music dealt with number applied to sound, while astronomy considered the moving geometry of the heavens. The list of other such sciences was growing. Aristotle mentioned optics as a mixed science, and medievals also looked to cosmography, the *scientia practica* of measuring weights, distances, or casting horoscopes—all instances where quantities were measured out over physical spans of earth or heavens. Nevertheless, despite the ancient conceptual grouping of the quadrivium, Renaissance scholars had many reasons not to give mathematics authority: it was rumoured to be too difficult and obscure, making wits unsociably sharp; it was too abstract, and so useless for explaining physical phenomena (as Aristotle said); or it was too practical, because linked to 'mechanical' arts of measuring (as humanists sometimes hinted)—not to mention impiety.

It may be necessary to stress how strange premodern mathematics can seem. Take the title *mathematicus*, which often, since antiquity, referred to a divinatory astrologer; and even though some philosophers took the time to discriminate between the theoretical discipline of *astronomia* and its practical application in *astrologia*, the terms were often used as synonyms. In Renaissance culture, mathematical practitioners used modes of reasoning properly delineated by arithmetic and geometry as well as their close neighbours of music (or harmonics), astronomy, optics, and so on. Nevertheless, contemporaries always heard with the word 'mathematics' a heterogeneous constellation of connotations and practices.[23] Lefèvre and his students

d'une culture mathematique jesuite a la Renaissance (1540–1640) (Rome: Ecole française de Rome, 1999). For more on Wittenberg and Paris, see notes 72, 74, and 78 below.

[23] e.g. J. Peter Zetterberg, 'The Mistaking of "the Mathematicks" for Magic in Tudor and Stuart England', *Sixteenth Century Journal* 11, no. 1 (1980): 83–97; Katherine Neal, 'The Rhetoric of Utility: Avoiding Occult Associations for Mathematics Through Profitability and Pleasure', *History of Science* 37 (1999): 151–78.

struggled to balance these broader meanings with the specific texts and practices at the core of the university's curriculum.

Authorities fixed university mathematics within an ambiguous conceptual place. Lefèvre and his students revered 'our Boethius', seeing the early sixth-century senator and philosopher as a representative of Christian antiquity.[24] Not only had he coined the term *quadrivium*, but he had provided medieval students with the fundamental Latin texts for its study: his free paraphrases of Pythagorean arithmetic and music were set texts until the sixteenth century, and some fragments of a translation of Euclid travelled under his name.[25] In the opening lines of *De arithmetica* he presented number as flowing directly from God's mind, a straightforwardly Pythagorean view of number underlying all things. His influence extended beyond mathematical texts, and at points in his theological works Boethius' quadrivium held the potential to spill over into other parts of philosophy. In the *Republic* and the *Timaeus*, Plato had not clarified whether mathematics itself reasons about pure ideas—as the Pythagoreans taught—or simply pointed to a deeper dialectic. Aristotle shared this ambiguity in *Metaphysics* E, where he placed mathematics between the theoretical or speculative sciences of physics and metaphysics. Here mathematics occupied a hazy middle ground: when it dealt with moving bodies, it shared ground with physics; when it looked to its objects as if wholly abstracted from matter, it shared ground with metaphysics. Boethius took up this convergence of Plato and Aristotle in *De trinitate*, book 2, embedding the threefold division of physics, mathematics, and metaphysics deep into medieval classifications of the theoretical disciplines. On rare occasions, therefore, Boethian mathematics could slip down into physics or up into theology, becoming a kind of universal philosophy already in the Middle Ages.[26]

As its profile rose between the fifteenth and eighteenth centuries, the quadrivium disintegrated in at least three major conceptual shifts. A first was internal to mathematics: the erosion of the ancient distinction between discrete magnitude (integers 1, 2, 3, . . .) and continuous magnitude (points, lines, curves). Collapsing this distinction would render unstable the division of geometry and arithmetic, a key element in the rise of algebra and the development of calculus. A second shift was the gradual collapse of ratio theory as a distinct subject. Ratios (sometimes labelled *proportiones*, or in certain cases *analogiae*) were presented by Boethius

[24] Michael Masi, 'The Liberal Arts and Gerardus Ruffus' Commentary on the Boethian De Arithmetica', *Sixteenth Century Journal* 10, no. 2 (1979): 23–41; Ann E. Moyer, 'The Quadrivium and the Decline of Boethian Influence', in *A Companion to Boethius in the Middle Ages*, ed. Noel Harold Kaylor and Philip Edward Phillips (Leiden: Brill, 2012), 479–517.

[25] Cassiodorus reported that Boethius wrote for all four disciplines in *Variae* i.45.4. The lost works are discussed by David Pingree, 'Boethius' Geometry and Astronomy', in *Boethius: His Life, Thought and Influence*, ed. Margaret Gibson (Oxford: Oxford University Press, 1981), 155–61. On Boethius' early significance as a textbook, see Gillian Rosemary Evans, 'Introductions to Boethius's "Arithmetica" of the Tenth to the Fourteenth Century', *History of Science*, 16, no. 1 (1978): 22–41. He first names the quadrivium in the Proemium to the *Arithmetica*. After 1500, Lefèvre included the Boethian fragments of geometry with his *Textus de sphaera*, alongside de Lattes's treatise on the astronomical ring.

[26] David Albertson, *Mathematical Theologies: Nicholas of Cusa and the Legacy of Thierry of Chartres* (Oxford: Oxford University Press, 2014).

and Euclid as a third class of mathematical objects, distinct from number and magnitude; just as magnitude eventually expanded to absorb number, it also absorbed ratios.[27] Later stages of the collapse of ratio theory are related to the rise of algebraic equations and the notion of function. But first there was a great deal of controversy within ratio theory. Medieval theorists experimented with practices of manipulating ratios, often with the goal of applying ratio theory to real-world quantities, especially music and natural philosophy—Galileo and Newton first expressed their physics of time and velocity in terms of this older ratio theory. Such goals required a third shift, namely the increasing dependence of other disciplines on mathematical principles. This eventually turned natural philosophy inside out by replacing final causes with mechanical descriptions of nature's operations. It also extended the reach of mathematics deep into modern art and engineering, making modern culture mathematical in a profound sense.

A long-range perspective helps us notice these shifts beginning in those books of Lefèvre's circle designed to cultivate general learning. Most often, these shifts have been traced to the *mathesis universalis* that emerged in the reception of Proclus' commentary on Euclid, available in Greek in 1533, in Latin in 1570, and paraphrased by Ramus, Descartes, and Leibniz.[28] That story is important too. But this circle raised the status of Boethian mathematics and developed its logic within practices of university teaching, collaborative authorship, and making books. When we recognize textbooks as a dynamic site of the craft of knowledge, we can see these intellectual shifts gestating within the legacy of the medieval curriculum.

LEFÈVRE AND FRIENDS

This book proceeds not as a biography, but pursues Lefèvre's and his students' mathematical books through contexts of late medieval reform, classrooms, print culture, and philosophy. A biography of Lefèvre and his circle would compete with the monumental scholarship of Augustin Renaudet and Eugene Rice, both based on decades in archives.[29] Instead, I pick out a thread that runs through this circle's

[27] Early accounts of this shift include Carl B. Boyer, 'Proportion, Equation, Function: Three Steps in the Development of a Concept', *Scripta Mathematica* 12 (1946): 5–13; Edith Dudley Sylla, 'Compounding Ratios. Bradwardine, Oresme, and the First Edition of Newton's Principia', in *Transformation and Tradition in the Sciences: Essays in Honour of I. Bernard Cohen*, ed. Everett Mendelsohn (Cambridge: Cambridge University Press, 1984), 11–43. See also Sabine Rommevaux, Philippe Vendrix, and Vasco Zara (eds.), *Proportions: Science, musique, peinture et architecture* (Turnhout: Brepols, 2011).

[28] Giovanni Crapulli, *Mathesis universalis: Genesi di un'idea nel XVI secolo* (Rome: Ataneo, 1969); David Rabouin, *Mathesis Universalis: L'idée de 'mathématique universelle' d'Aristote à Descartes* (Paris: Épiméthée, 2009); Guy Claessens, 'Het denken verbeeld: De vroegmoderne receptie (1533–1650) van Proclus' Commentaar op het eerste boek van Euclides' Elementen', PhD dissertation, University of Leuven, 2011.

[29] Augustin Renaudet, *Préréforme et humanisme à Paris pendant les premières guerres d'Italie, 1494–1517* (1916; 2nd edn, Paris: Édouard Champion, 1953); Eugene F. Rice (ed.), *The Prefatory Epistles of Jacques Lefèvre d'Étaples and Related Texts* (New York: Columbia University Press, 1972) (hereafter PE). I reference other works throughout this book; but some key works include Charles-Henri Graf, *Essai sur la vie et les écrits de Jacques Lefèvre d'Étaples* (Strasbourg: G.L. Schüler, 1842); Guy Bedouelle, *Lefèvre d'Étaples at l'intelligence des Écritures* (Genève: Droz, 1976); Philip Edgcumbe

collective biography and yet has remained largely overlooked: their enduring interest in mathematics. Still, something of that biography will be useful here at the outset.

This circle's profile in early modern learning centred on Lefèvre d'Étaples. In his own day, Lefèvre was counted among the illustrious men of the Renaissance, comparable to Giovanni Pico della Mirandola in Florence a generation earlier, or to Erasmus in Basel a generation later. In 1512, when the German schoolmaster Johannes Bugenhagen asked the famous Dutch teacher Johannes Murmellius which contemporary thinkers he thought equalled Augustine, Albert the Great, and Bonaventure, the answer was that only two living philosophers approached those ancients: Pico and Lefèvre.[30] This is the authoritative Lefèvre we find in the sixteenth-century discourse of great lives, from Theodore Beza's *Icones* to Paolo Giovio's *Elogia virorum illustrium*, which constructed bookish halls of heroes for the Republic of Letters.[31]

Historians do note Lefèvre as a mathematical author, if only in passing.[32] His contemporaries are more forthcoming. A source from 1512 illuminates Lefèvre's authority: the 'who's who' of Renaissance scholars, the *De scriptoribus ecclesiasticis* of Trithemius was updated that year by an anonymous Paris editor who called Lefèvre the 'one glory of all France' (*totius Galliae unicum decus*).[33] The list begins with his commentaries on Aristotle's logic, ethics, natural philosophy, and metaphysics. But nearly a third of the list comprises Lefèvre's books on arithmetic, music, and astronomy—that is, his contribution to the mathematical arts of the quadrivium. Lefèvre's mathematics made him a respected authority in early modern mathematical erudition. Much later, Bernardino Baldi, the historian of mathematics from Urbino, included an entry for Lefèvre.[34] Giuseppe Biancani, a Jesuit who recounted Galileo's telescope observations in his own astronomy textbook, listed Lefèvre among the mathematicians of modern times.[35] Erudites

Hughes, *Lefèvre: Pioneer of Ecclesiastical Renewal in France* (Grand Rapids, MI: Eerdmans, 1984). Jean-Marie Flamand is preparing a large edition of documents related to Lefèvre's circle.

[30] Next to them, Murmellius sets Charles de Bovelles and Johann Reuchlin: Otto Vogt (ed.), *Dr. Johannes Bugenhagens Briefwechsel* (Stettin: Leon Saunier, 1888), 6: 'Duo hac aetate clarissimi philosophi theologique et qui proxime ad veteres accedunt meo judicio sunt Joannes Franciscus Picus comes Mirandulanus, qui variae doctrinae multa scripsit opera, inter quae tres hymnos heroicos cum eruditionis reconditissimae commentariis: et Jacobus Faber Stapulensis, qui in Aristotelis plerosque libros, carmina Davidis et Pauli Tarsensis epistolas commentarios scripsit. Hic addo Carolum Bovillum et Capnion Phorcensem.' Note that Murmellius was first a student of Hegius, then of Johannes Caesarius, himself a student of Lefèvre.

[31] Theodore Beza, *Icones* (Geneva: Ioannes Laonius, 1580), X4r; Paolo Giovio, *Elogia doctorum virorum ab avorum memoria publicatis ingenii monumentis illustrium* (Antwerp: Ioannes Bellerus, 1557), 248–9. Various *éloges* are also edited in PE.

[32] Besides Renaudet, PE, and Cantor, see e.g. Jens Høyrup, *In Measure, Number, and Weight: Studies in Mathematics and Culture* (Buffalo, NY: State University of New York Press, 1994), 215–17.

[33] Johannes Trithemius, *De scriptoribus ecclesiasticis*, ed. anon. (Paris: Bertold Rembolt and Jean Petit, 1512), 215v–16r.

[34] Bernardino Baldi, *Cronica de Matematici* (Urbino: A.A. Monticelli, 1707), 108–9.

[35] Giuseppe Biancani, *De mathematicarum natura dissertatio* (Bologna: Bartholomaeus Cochius, 1615), 59r. See also Lefèvre listed among modern mathematical authorities in Edward Sherburne, *The Sphere of Marcus Manilius Made an English Poem: With Annotations and an Astronomical Appendix* (London: Nathanael Brooke, 1675), 44.

in the world of natural philosophy and classical scholarship such as Marin Mersenne and Gerhard Vossius cited Lefèvre as an important modern music theorist.[36]

This authority was accrued during a life as university professor, biblical commentator, theological controversialist, diocesan reformer, and court chaplain. Born to a family of modest means in the port town of Étaples, Picardy, Lefèvre matriculated at the University of Paris in 1475, probably graduating a Master of Arts by 1480.[37] For the following decade, he disappears from record (likely he was teaching arts students at the Collège Boncour) but we find him again in 1491, journeying to Italy to do the sorts of things humanists were supposed to do: hunting for manuscripts in Paduan monasteries, meeting Ermolao Barbaro in Rome, and Marsilio Ficino and Giovanni Pico della Mirandola in Florence. After this first trip to Italy, Lefèvre returned to the Collège du Cardinal Lemoine in Paris, where he taught for at least fifteen years. During his tenure as the longest-standing regent master at Lemoine, he made the college's reputation as a centre for humanist education in Paris by writing on the entire range of the arts curriculum, from natural philosophy and mathematics to logic and moral philosophy. These interests still marked his scholarship after 1508, when he retired to the abbey of Saint-Germain-des-Près just outside the walls of Paris. Although acclaimed as an editor and commentator especially on the Psalms (1509) and the Letters of Saint Paul (1512), he also was hard at work on editions of mathematics and philosophy, including works of Nicholas of Cusa (1514), Aristotle (Bessarion's translation of the *Metaphysics*, 1515) and Euclid (1516). For the next decade, Lefèvre was embroiled in controversies that reverberated throughout Europe, defending the Hebrew interests of Johannes Reuchlin, disputing the judgement of Desiderius Erasmus on the Greek New Testament, and, in what was known as the matter of the 'Three Marys', querying the Theology Faculty's authority. These controversies resonated with scholars and theologians around Europe, and figures such as Thomas More, Agrippa of Nettesheim, and Nikolaus Ellenbog eagerly kept abreast of the gossip these conflicts generated.[38] Then, from 1521, just as Luther transformed Wittenberg into an epicentre of Protestant reform, Lefèvre piloted a *réforme evangelique* in the diocese at Meaux. When the Meaux experiment disintegrated in 1525 under the critique of the Paris Faculty of Theology, Lefèvre fled, first to Strassburg and then under royal

[36] Marin Mersenne, *Traité de l'harmonie universelle* (Paris: Fayard, 2003), 32; Gerardus Joannes Vossius and Franciscus Junius, *De quatuor artibus popularibus, de philologia, et scientiis mathematicis, cui operi subjungitur, chronologia mathematicorum, libri tres* (Amsterdam: Ioannis Blaeu, 1660), 95–6. See also L. Lucács (ed.), *Monumenta Paedagogica Societatis Jesu* (Rome, 1965–92), 2:148–50 (for the mathematics course designed by Jeronimo Nadal, 1552); Antonio Possevino, *Bibliotheca selecta: qua agitur de ratione studiorum* (Rome: Typographia Apostolica Vaticana, 1593), 2:184.

[37] V. Carrière, 'Lefèvre d'Étaples à l'Université de Paris (1475–1520)', in *Etudes historiques dediées à la mémoire de M. Roger Rodière* (Arras, 1947), 109–20.

[38] Some sense of the debates can be gathered from Sheila M. Porrer, *Jacques Lefèvre d'Etaples and the Three Maries Debates* (Geneva: Droz, 2009); Erika Rummel, *Erasmus and His Catholic Critics* (Nieuwkoop: De Graaf, 1989); Ann Moss, *Renaissance Truth and the Latin Language Turn* (Oxford: Oxford University Press, 2003), 95–111; Marc van der Poel, *Cornelius Agrippa, the Humanist Theologian and His Declamations* (Leiden: Brill, 1997), 32–6.

protection to the court of the king's sister Marguerite de Navarre, where he died in 1536.[39]

A paradox runs through Lefèvre's life. He became one of the great authorities of the Renaissance, yet he never wrote a single magnum opus, whether philosophical treatise or literary work. Unlike humanist poets such as Sebastian Brandt, encyclopaedists such as Giorgio Valla, or even Erasmus, whose *Adagiae*, *Colloquia*, and satires stand on their own, Lefèvre's eminence rested purely on an accumulated *oeuvre* of commentaries, paraphrases, and translations—genres that serve other authorities, rather than claiming authority directly for themselves. Lefèvre's authority reflects a period in which authorship depended on the practices of commentary.

But the paradox goes deeper, for Lefèvre's published authority also rested on his circle of students, a relationship which will be a main theme of this book. If we turn to the steady stream of printed editions that allows us to track Lefèvre's life, we find that the material and social construction of Lefèvre's authority is inextricable from the names of students. Lefèvre's own name first appears on a book in 1492, as author of a paraphrase of Aristotle's natural philosophy: *Iacobi Fabri Stapulensis in Aristotelis octo Physicos libros Paraphrasis*.[40] His name is not on the first page—it would be a decade before title pages became common—or even the first gathering of the book's pages, which is filled with indexes, tables of theses, and an introductory figure (more on these paratexts later). Rather, Lefèvre's name is found at the beginning of the first major section that makes up the volume. And when we turn to the last page, we find a poem by Josse Clichtove, who later became a major figure in the Faculty of theology at Paris, but began his career as Lefèvre's student.[41] Clichtove praises Johann Higman, the printer, for venturing the publication 'at his own expense', and records his own role in seeing the work through the press, along with a certain 'Bohemian'. In other words, the book neither begins nor ends with Lefèvre, but with editorial matter. Anthony Grafton has explored the manifold responsibilities of correctors from minor emendations to whole-cloth ghost writing.[42] This book's paratexts—the opening tables, figures, and closing poem—were elements that students such as Clichtove likely prepared for the book as their first foray into authorship. For many years Clichtove was a regent at the Collège du Cardinal Lemoine with Lefèvre, where he taught from Lefèvre's various introductions to the arts; most later editions of Lefèvre's epitomes and paraphrases included Clichtove's detailed commentary.[43] Lefèvre recognized Clichtove's intellectual friendship by

[39] On Lefèvre's central place in this context, see Jonathan A. Reid, *King's Sister – Queen of Dissent: Marguerite of Navarre (1492–1549) and Her Evangelical Network*, 2 vols (Leiden: Brill, 2009).

[40] Paris: [Johann Higman], 1492, b5r.

[41] Jean-Pierre Massaut, *Josse Clichtove, l'humanisme et la réforme du clergé*, 2 vols (Paris: Société d'Edition 'Les Belles Lettres', 1968); James K. Farge, *Biographical Register of Paris Doctors of Theology, 1500–1536* (Toronto: Pontifical Institute of Mediaeval Studies, 1980), 90–104.

[42] Anthony T. Grafton, *The Culture of Correction in Renaissance Europe* (London: British Library, 2011).

[43] This edition first appeared as Jacques Lefèvre d'Étaples and Josse Clichtove, *Totius philosophiae naturalis paraphrases, adiecto commentario* (Paris: Henri Estienne, 1502). Some later editions also included commentaries by Lefèvre's student François Vatable, later professor of Hebrew at the Collège Royal (see bibliography in PE).

making him a main interlocutor in book 2 of his *De magia naturali*, composed in the mid-1490s.[44] Clichtove's career was intertwined with Lefèvre's pedagogical works from the beginning.

Student collaboration especially permeates mathematical works, which are rich in diagrams and tables—paratexts likely originating with students. The first edition of the *Textus de sphera* mentioned at the beginning of this chapter was published in 1495 by Wolfgang Hopyl, an early partner of Henri Estienne (the Elder), founder of the great print dynasty which would enjoy a long association with Lefèvre and his circle.[45] Recall that this large folio edition was filled with updated commentary, diagrams, and tables.[46] Who added these elements? One trace is a colophon in which the book's makers identify themselves as 'lovers of mathematics':

> Printed at Paris in the *Rue St-Jacques*, near the sign of Saint George, on the twelfth of February in the year 1494 [i.e. 1495] of Christ, creator of the stars, by the ingenious printer Wolfgang Hopyl. This thought is always firmly in his mind: great deeds are done not by strength or speed or physical agility,[47] but by planning, judgement, and authority. [Done with the help of] the most diligent correctors Luc Walter Conitiensis, Guillaume Gontier, Jean Grietan, Pierre Griselle, lovers of mathematics.[48]

In what sense was Lefèvre the author of this book? Certainly, he was responsible for most of the text in the volume—but he authored the *commentary*, not the main text. And Lefèvre's students helped him at a couple of stages. In the prefatory letter, he thanked his 'domestic' Grietan: 'he is very studious in the skill of abacus and arithmetic, and knowledgeable in the rest of the mathematical arts. He wrote the work and, like Atlas, offered his shoulder to an exhausted man.'[49] Lefèvre also depended on his students to see the book through the press; we must see this work as the collective effort of a community. Their hands are visible here, in the colophon of this first edition alone, and very likely in the many other paratexts that make up the book. These paratexts—tables, diagrams, headings, indexes, letters, with annotations and reference marks printed in the margins—would have made

[44] *De magia naturali*, MS Olomouc M.I.119, book 2.

[45] Mark Pattison, 'Classical Learning in France: The Great Printers Stephens', *Quarterly Review* 117 (1865): 323–64 (responding to the bibliographical scholarship of Philippe Renouard); Elizabeth Armstrong Tyler, 'Jacques Lefèvre d'Etaples and Henri Estienne the Elder, 1502–1520', in *The French Mind: Studies in Honour of Gustave Rudler*, ed. W. Grayburn Moore (Oxford: Sutherland and Starkis, 1952), 17–33; Jeanne Veyrin-Forrer, 'Simon de Colines, imprimeur de Lefèvre d'Etaples', in *Jacques Lefèvre d'Etaples (1450?–1536)*, ed. Jean-François Pernot (Paris: Honoré Champion Éditeur, 1995), 97–117.

[46] This work is set in context in my 'A Book, a Pen, and the *Sphere*: Reading Sacrobosco in the Renaissance', *History of Universities* 28, no. 2 (2015): 1–54, at 5–8. A full bibliography of early printed editions is given by Jürgen Hamel, *Studien zur 'Sphaera' des Johannes de Sacrobosco* (Leipzig: Akademische Verlagsanstalt, 2014).

[47] Cicero, *De senectute* 17. Thanks to Anthony Ossa-Richardson for this reference.

[48] Lefèvre, *Textus de sphera*, colophon: 'Impressum Parisij in pago diui Jacobi ad insigne sancti Georgij Anno Christi siderum conditoris 1494 duodecima februarij Per ingeniosum impressorem Wolfgangum hopyl. Cui hec sententia semper firma mente sedet: Non viribus aut velocitatibus aut celeritate corporum res magne geruntur: sed Consilio, Sententia, et Auctoritate Recognitoribus diligentissimis: Luca Uualtero Conitiensi, Guillermo Gonterio, Johanne Griettano, et Petro Grisele: Matheseos amatoribus.'

[49] Lefèvre, *Textus de sphera*, sig. a1v (PE, 27).

the book difficult to print, so it was for good reason that Hopyl relied on four correctors. Grietan, Walter, Gontier, and Griselle, therefore, identify themselves as the community who produced this book. In such paratexts, students not only framed Lefèvre as an authority, but also declared themselves part of the community.

In these practices of collaborative authorship, we see Lefèvre's community take form, especially in the mathematical works. Charles de Bovelles is the name most closely linked to Lefèvre and Clichtove. The one book all three authored together is an introduction to arithmetic and geometry, the *Epitome Boetii, etc.* (1503), comprising Lefèvre's introduction to medieval number theory (with Clichtove's extensive commentary), Clichtove's treatise on practical arithmetic, and Bovelles's works on geometry and optics. Bovelles especially shared Lefèvre's vision for mathematics, and Bovelles traced his first steps in 'sound learning' (*sanis disciplinis*) to a chance meeting with Lefèvre in 1495, out in the countryside where scholars had fled from a plague in Paris. Lefèvre had shown him, he said, the 'shining sun of the disciplines. In the Pythagorean style, using introductions to numbers and preludes to arithmetical studies, you became the cause of all my philosophical and literary studies.'[50] By *praeludia* Bovelles no doubt meant Lefèvre's distinctive epitome to Boethius' *Arithmetica*, which he published in 1496 with the *Elementa arithmetica* of Jordanus de Nemore and his own *Elementa musicalia*.[51] With Clichtove, Bovelles quickly became one of Lefèvre's most intimate students and collaborators at the Collège du Cardinal Lemoine, an exuberant reader of Pseudo-Dionysius the Areopagite, Ramon Lull, and Nicholas of Cusa. As Ernst Cassirer suggested and Emmanuel Faye has shown in fine detail, Bovelles was a creative philosopher, deserving a prominent place in the history of Renaissance anthropologies.[52] But underpinning his synthetic philosophical programme filled with visual symbols and psychological analogies, however, was a fascination with numbers. Like Cusanus and Lefèvre, he found numbers good to think with, and numerology pervaded his reconstructions of natural philosophy, metaphysics, and theology.[53] He was also a creative mathematician, writing textbooks on geometry and optics, contributions to number theory, and practical geometries for vernacular readers.

[50] Charles de Bovelles, *Que hoc volumine continentur: Liber de intellectu; Liber de sensu; Liber de nichilo; Ars oppositorum; Liber de generatione; Liber de sapiente; Liber de duodecim numeris; Epistole complures. Insuper mathematicum opus quadripartitum: De numeris perfectis; De mathematicis rosis; De geometricis corporibus; De geometricis supplementis* (Paris: Henri Estienne, 1511), 168v: 'Parhisiis, quod anno 1495 peste affecti sunt... te ruri illustrem disciplinarum solem ostenderi. Tu nempe per introductiones numerorum, per arithmetice discipline preludia, Pythagorico more, totius mei philosophici profectus ac litterarii studii extitisti causa.'

[51] On these introductory genres, see Chapter 4.

[52] The first modern edition of Bovelles's work was of *De sapiente* in Ernst Cassirer, *Individuum und Kosmos in der Philosophie der Renaissance* (Leipzig: Teubner, 1927). See now Emmanuel Faye, *Philosophie et perfection de l'homme: De la Renaissance à Descartes* (Paris: Vrin, 1998).

[53] This can be sampled in the useful work of Joseph M. Victor, *Charles de Bovelles, 1479–1553: An Intellectual Biography* (Genève: Droz, 1978). N.B. Bovelles's date of death was probably 1567. See now Jean-Claude Margolin, *Lettres et poèmes de Charles de Bovelles* (Paris: Champion, 2002); Anne-Hélène Klinger-Dollé, *Le* De sensu *de Charles de Bovelles (1511): conception philosophique des sens et figuration de la pensée* (Geneva: Droz, 2016); and Michel Ferrari and Tamara Albertini (eds.), *Charles de Bovelles' Liber de Sapiente, or Book of the Wise*, Special Issue of *Intellectual History Review* (London, 2011).

In these, Bovelles reimagined the whole world through a mathematical lens—even promoting a mathematical physics, as a later chapter in this book will argue.

The intellectual preoccupations of Lefèvre, Clichtove, and Bovelles can best be seen in the extant library of their student Beatus Rhenanus (1485–1547). He later codified new humanist practices of emendation and Germanic historiography, while serving Erasmus as editor and biographer.[54] Beatus apprenticed in these skills, however, as a student at the Collège du Cardinal Lemoine, where he first helped Lefèvre publish editions of Ramon Lull and Aristotle's *Politics*.[55] The rich library he began to collect in Paris survives, and his editions of textbooks by Lefèvre, Clichtove, and Bovelles are filled with student notes and marginalia. This nearly complete record of a Renaissance student's university studies is, to my knowledge, wholly unique, and throughout this book I will draw on this precious witness to the actual practice of Fabrist pedagogy.[56] (I have given a handlist of his annotated books in the Appendix.)

Lefèvre, Clichtove, and Bovelles anchored a much larger community. Beatus represented a whole generation of Alsatian students who came to Paris to study with Lefèvre around 1500, including Jerome Gebwiler, Michael Hummelberg, Johannes Sapidus, and the sons of Erasmus' printer Johannes Amerbach.[57] In Paris, their circle included the scholar-printers Henri Estienne and Josse Bade, and a network of students who became famous in their own right, such as François Vatable, the first royal lecturer of Hebrew in the Collège Royal.[58] Symphorien Champier may never have studied with Lefèvre, but he nevertheless considered him an intellectual father.[59] Likewise, major humanists in the Low Countries such as Johannes Caesarius claimed Lefèvre as their teacher. Visitors came from England

[54] John F. D'Amico, *Theory and Practice in Renaissance Textual Criticism: Beatus Rhenanus Between Conjecture and History* (Berkeley, CA: University of California Press, 1988); James Hirstein (ed.), *Epistulae Beati Rhenani: La Correspondance latine et grecque de Beatus Rhenanus de Sélestat. Edition critique raisonnée avec traduction et commentaire, Volume 1 (1506–1517)* (Turnhout: Brepols, 2013).

[55] Prefatory poems record that Beatus corrected: Ramon Lull, *Contenta. Primum volumen Contemplationum Remundi duos libros continens. Libellus Blaquerne de amico et amato*, ed. Jacques Lefèvre d'Étaples (Paris: Guy Marchant for Jean Petit, 1505); Jacques Lefèvre d'Étaples, *Contenta. Politicorum libro octo. Commentarii. Economicorum duo. Commentarii. Hecatonomiarum septem. Economiarum publ. unus. Explanationis Leonardi [Bruni] in Oeconomica duo* (Paris: Henri Estienne, 1506).

[56] Gustav Knod, *Aus der Bibliothek des Beatus Rhenanus: ein Beitrag zur Geschichte der Humanismus* (Leipzig, 1889); Emmanuel Faye, 'Beatus Rhenanus lecteur et étudiant de Charles de Bovelles', *Annuaire des Amis de la Bibliothèque Humaniste de Sélestat* 45 (1995): 119–38.

[57] Peter G. Bietenholz, *Basle and France in the Sixteenth Century* (Geneva: Droz, 1971), *passim* but esp. 181–6. See also Richard J. Oosterhoff, 'The Fabrist Origins of Erasmian Science: Mathematical Erudition in Erasmus' Basle', *Journal of Interdisciplinary History of Ideas* 3, no. 6 (2014): 3–37.

[58] Lefèvre's influence in Paris is visible in a list of his students who became the regent masters at Cardinal Lemoine, besides Clichtove and Bovelles: Jérome Clichtove (Josse's cousin), Thomas Doullet (chaplain at Cardinal Lemoine), Jean Drouin, Robert Fortunat, Jean Lagrène, David Laux, Pierre de Gorris, Nicolas de Grambus, Salmon Macrin, Thibault Petit, Gilles of Lille, Jean Molinar, Philippe Prévost, Jean Pelletier, Gérard Roussel, Louis Savel, Alain de Varennes. This list compiled from James K. Farge (ed.), *Students and Teachers at the University of Paris: The Generation of 1500. A Critical Edition of Bibliothèque de l'Université de Paris (Sorbonne), Archives, Registres 89 and 90* (Leiden: Brill, 2006); Nathalie Gorochov, 'Le collège du Cardinal Lemoine au XVIe siècle', *Mémoires de Paris et l'Ile-de-France* 42 (1991): 219–59.

[59] Brian P. Copenhaver, *Symphorien Champier and the Reception of the Occultist Tradition in Renaissance France* (The Hague: Mouton Publishers, 1978), 56–7.

and Italy to meet Lefèvre in Paris.[60] Admirers consistently praised Lefèvre's learning, and especially his kindly demeanour. At Erasmus' suggestion, the Swiss polymath Heinrich Glarean spent five years in Paris and boasted to his compatriot Ulrich Zwingli that 'Lefèvre d'Étaples is now often my close companion. Above all, this wholly honest and eminent man sings, plays, disputes, and laughs with me, especially at this foolish world, this man so wonderfully humane and kind.'[61]

The sociability of friendship fills the letters, poems, excipits, and dedications of this circle's books.[62] Lefèvre frequently portrayed his own texts as a response to conversations with students, including his trademark genre of dialogues.[63] He published a mathematical game at a student's request, in the hope that such serious play would both teach and refresh young students.[64] Sociability between master and student was governed by an ethos of friendship. In 1492, in the preface to dialogues on Aristotle's *Physics*, Lefèvre addressed a certain Stephanus.[65] Lefèvre referred to their old friendship and allowed that 'outsiders may marvel at how great friendship [*benevolentia*] is among those who cultivate the liberal arts in our Paris studium, where this experience is well known.'[66] The 'holy oath of friendship' was as if born from Minerva, 'once thought' the goddess of wisdom and peace. Lefèvre went on to emphasize the close relationship between philosophy and the bonds of friendship, for 'what is philosophy but love of wisdom? And what is a philosopher but a true lover of the same? Thus it behooves philosophers (as they rightly think) to be friends.'[67] Lefèvre proceeded to demote the 'envious, malevolent men who cut each other up with their teeth' to the rank of mere dogs, no longer philosophers.

[60] On the visit of the Cambridge theologian Robert Ridley, see John Monfasani, 'A Tale of Two Books: Bessarion's *In Calumniatorem Platonis* and George of Trebizond's *Comparatio Philosophorum Platonis et Aristotelis*', *Renaissance Studies* 22, no. 1 (2008): 1–15. Notes from visitors are found e.g. in Rice, PE, 126–7, 146–7.

[61] To Zwingli, 29 August 1517 (*Zwinglis sämtliche Werke*, vol. VII [= Corpus Reformatorum 94] (Leipzig: Heinsius, 1911), ep. 26, 59): 'Faber Stapulensis, qui saepe iam comui meae fuit. Is supra modum me amat, totus integer et candidus, mecum cantillat, ludit, disputant, ridet mecum stultum praecipue hunc mundum, vir humanissimus atque ita benignus, ut nonnunquam videatur—quamquam id revera minime facit—fideatur tamen suae gravitates oblitus'.

[62] On this currency of the republic of letters, see Kristine Louise Haugen, 'Academic Charisma and the Old Regime', *History of Universities* 22, no. 1 (2007): 203–9.

[63] e.g. Jacques Lefèvre d'Étaples, *Introductio in metaphysicorum libros Aristotelis*, ed. Josse Clichtove (Paris: Higman, 1494), b1v (PE, 22).

[64] Jacques Lefèvre d'Étaples, *Elementa arithmetica; Elementa musicalia; Epitome in libros arithmeticos divi Severini Boetii; Rithmimachie ludus que et pugna numerorum appellatur* (Paris: Johannes Higman and Wolfgang Hopyl, 1496), i6v (PE, 37).

[65] Rice (PE, 15) suggests that this may be Stephanus Martini de Tyne, of the diocese of Prague, who took the MA at Paris in 1481, and would have been in the medical faculty at Paris, as a bachelor in 1494 and a licenciate in April 1496.

[66] Lefèvre, *Totius Aristotelis philosophiae naturalis paraphrases*, sig. J8r. (PE, 15): 'Mutua nos multos annos astrinximus benevolentia. Carissime Stephane, quanta sit animorum benevolentia inter liberalium artium cultores in hoc nostro Parisio studio (ubi res cognita esset) exteri mirarentur.' For Aristotle, *benevolentia* is the basis of true friendship. Lefèvre simplified: 'Amicitia est mutua aliquorum benevolentia cum eos non lateat' (Lefèvre and Clichtove, *Artificialis introductio per modum Epitomatis in decem libros Ethicorum Aristotelis adiectis elucidata [Clichtovei] commentariis* (Paris: Wolfgang Hopyl and Henri Estienne, 1502), 31v).

[67] Lefèvre, *Totius Aristotelis* (PE, 16): 'Quid enim philosophia nisi sapientiae amor? Quid philosophus nisi verus eiusdem amator? Iure decet itaque (ut recte sentiunt) ipsos esse amicos.'

By drawing on their friendship, Lefèvre was evoking both an image of philosophical harmony as well as acknowledging a debt: Stephen had demonstrated his love for Lefèvre by serving as a corrector for the book in which Lefèvre had published these dialogues, for in print books 'often deviate from the original, unless an attentive corrector is present'.[68] Lefèvre provided acknowledgments regularly—not standard practice in early modern Europe. Such acknowledgments cemented a culture of *amicitia*, in a moment of high ideals that would come under heavy strain later in the decades of the French Reformation.[69]

The circle's culture of friendship and shared authorship is especially evident in their mathematical works. Early on Clichtove and Bovelles joined Lefèvre in his mathematical interests, but continued to write on these topics long after all of them had ceased to teach the arts course. Lefèvre assembled the works of Nicholas of Cusa (1514), especially praising his mathematical acumen, and then edited Euclid's *Elements* (1517). Even when he was a prominent doctor of theology, Clichtove published his commentary on Lefèvre's more advanced astronomy, the *Astronomicon* (1517). Clichtove wrote a great deal independently, though many of his works bear the imprint of his training with Lefèvre—for example his short treatise *De mystica numerorum* (1513), a handbook on the meanings of numbers in Scripture. Just before beginning evangelical reform at Meaux, Lefèvre's student Gérard Roussel extensively commented on the classic *Arithmetica* of Boethius (1521). Bovelles maintained the longest trajectory of influence, not only writing mathematically inflected philosophy for several decades, but also collaborating with Oronce Fine, the first professor of mathematics at Francis I's new royal college, in the various versions of his *Geometrie practique* (1511, 1542, 1547)—the first geometry to be printed in French.[70]

The circle's works echoed through the sixteenth century, especially in the university reform movements that gave new prominence to mathematics. In Spain, when Pedro Ciruelo returned from Paris to set up a new university at Alcala, he reprinted Lefèvre and Clichtove's works alongside his own new perspective treatise to inaugurate the mathematical learning of Renaissance Spain.[71] In the German-speaking

[68] Lefèvre, *Totius Aristotelis* (PE, 16): 'Tu curasti ut nostrae Paraphrases castigatae ab ipsorum efformatorum manibus prodeant in lucem, qui saepe ab archetypo deviant nisi vigilans castigator affuerit.'

[69] One of the longest sections of Lefèvre and Clichtove's introduction to Aristotelian moral philosophy was devoted to *amicitia*, and in particular to Aristotle's highest form of friendship: that pursued for the sake of the other, rather than selfish utility. Lefèvre and Clichtove, *Artificialis introductio*, 31v–38v. The Fabrist language of friendship is explored more fully below, Chapter 6, and in Richard J. Oosterhoff, 'Lovers in Paratexts: Oronce Fine's Republic of Mathematics', *Nuncius* 31, no. 3 (2016): 549–83. Historians have stressed the 'interested' nature of friendship in the later Republic of Letters: e.g. Peter N. Miller, 'Friendship and Conversation in Seventeenth-Century Venice', *Journal of Modern History* 73, no. 1 (2001): 1–31; Vera Keller, 'Painted Friends: Political Interest and the Transformation of International Learned Sociability', in *Friendship in the Middle Ages and Early Modern Culture*, ed. Marilyn Sandidge and Albrecht Classen (Berlin: de Gruyter, 2011), 661–92.

[70] It was translated into Dutch and Latin too. See Richard J. Oosterhoff, '"Secrets of Industry" for "Common Men": Charles de Bovelles and Early French Readerships of Technical Print', in *Translating Early Modern Science*, ed. Sietske Fransen and Niall Hodson (Leiden: Brill, 2017), 207–29.

[71] Pedro Ciruelo, *Cursus quatuor mathematicarum artium liberalium* (Alcalá: A.G. de Brocar, 1516). On Ciruelo and mathematics, see the passing remarks of María Portuondo, *Secret Science: Spanish*

lands, the universities reinvented on Melanchthon's Wittenberg model became noted for their particular attention to mathematics.[72] Here too we find traces of the Fabrists intertwined with the heritage of Johannes Regiomontanus. For instance, the author of the first handbook to be called a 'cosmography' may have been influenced by Lefèvre's *Textus de sphera*; certainly, Ringmann's innovative effort to teach grammar using card games claimed inspiration from Lefèvre's version of the mathematical game *Rithmomachia*.[73] Lefèvre's edition of Euclid, which set medieval and humanist translations side by side, was republished in Basel, where it seems to have informed Simon Grynaeus and Conrad Dasypodius, both important figures in the reintegration of Euclid into schoolrooms. In France and especially Paris, Lefèvre's shadow loomed longest.[74] Oronce Fine took up Lefèvre's mantle as restorer of mathematics within the University of Paris,[75] but he did so by republishing Lefèvre's works, first an edition of Lefèvre's *Astronomicon* in 1515, then the *Textus de sphaera* in 1521. He continued with various works from Lefèvre, Clichtove, and Bovelles that he excerpted and appended to his own edition of Gregor Reisch's *Margarita philosophica* (1535)—this version of the *Margarita* enjoyed remarkable longevity, last republished in 1608.[76] Fine's mission to set mathematics among the liberal arts was carried out in the most prestigious mathematical post of Europe, the mathematical chair in Francis I's Collège Royal. A similar vision would fuel Peter Ramus' own bid for that chair.[77] Ramus' followers would hold the post throughout the late sixteenth century, beginning a new wave of mathematical publishing in the 1550s and 1560s.[78]

Cosmography and the New World (Chicago: University of Chicago Press, 2009), 39; on the relation of these to the University of Paris, see Ricardo García Villoslada, *La universidad de Paris durante los estudios de Francisco de Vitoria (1507–1522)* (Rome: Gregoriana, 1938), 220ff.

[72] Peter Ramus praised German accomplishments in mathematics to Catharine de Medici after his travels through Basel: *Proemium mathematicum* (Paris: Andreas Wechelus, 1567). On mathematics in the German schools, see Westman, 'The Melanchthon Circle'; Sachiko Kusukawa, *The Transformation of Natural Philosophy: The Case of Philip Melanchthon* (Cambridge: Cambridge University Press, 1995), 134–60; Methuen, *Kepler's Tübingen*.

[73] Matthias Ringmann, *Grammatica figurata* [St Die, 1509], 2r; Matthias Ringmann, *Cosmographiae introductio cum quisbusdam geometriae ac astronomiae principiis ad eam rem necessariis insuper quatuor Americi Vespucii navigationes universalis Chosmographiae descriptio* (St. Dié: [Vautrin and Nicolas Lud], 1507). On Lefèvre's game, see Ann E. Moyer, *The Philosophers' Game: Rithmomachia in Medieval and Renaissance Europe* (Ann Arbor, MI: University of Michigan Press, 2001).

[74] Jean-Claude Margolin, 'L'Enseignement des mathématiques en France (1540–70): Charles de Bovelles, Fine, Peletier, Ramus', in *French Renaissance Studies, 1540–70: Humanism and the Encyclopedia*, ed. Peter Sharratt (Edinburgh: Edinburgh University Press, 1976), 109–55; Timothy J. Reiss, *Knowledge, Discovery and Imagination in Early Modern Europe* (Cambridge: Cambridge University Press, 1997).

[75] Isabelle Pantin, 'Oronce Fine's Role as Royal Lecturer', in *The Worlds of Oronce Fine: Mathematics, Instruments and Print in Renaissance France*, ed. Alexander Marr (Donington: Shaun Tyas, 2009), 13–30; Angela Axworthy, *Le Mathématicien renaissant et son savoir: le statut des mathématiques selon Oronce Fine* (Paris: Classiques Garnier, 2016), 14, 28.

[76] On these editions, Andrew Cunningham and Sachiko Kusukawa (eds), *Natural Philosophy Epitomised: Books 8–11 of Gregor Reisch's Philosophical Pearl (1503)* (Farnham: Ashgate, 2010), xxviii–xxx.

[77] Robert Goulding, 'Method and Mathematics: Peter Ramus' Histories of the Sciences', *Journal of the History of Ideas* 67, no. 1 (2006): 63–85; Goulding, *Defending Hypatia*.

[78] François Loget, 'L'algèbre en France au XVIe siècle: Individus et réseaux', in *Pluralité de l'algèbre à la Renaissance*, ed. Sabine Rommevaux, Maryvonne Spiesser, and Maria Rosa Massa

MAKING IN UNIVERSITIES AND PRINT SHOPS

This book approaches Lefèvre's circle by considering how a community *made* a culture: it deals with craft skill, localism, and collective production. I share the premise of Mary Carruthers' account of literary invention—like mathematical practice, often portrayed only conceptually—that premodern thought was a matter of craft, sometimes explicitly patterned on artisanal techniques.[79] Carruthers based her account on monastic practices of reading, especially those of the twelfth-century Victorines whom Lefèvre and Clichtove admired and often published. Their greatest hero was Nicholas of Cusa, who explicitly modelled intellectual insight on the craft of a spoon maker.[80] Friends punned on Lefèvre's name of 'maker' (*faber*), praising him 'as another craftsman Daedalus' who used universities to shape the chariot of the soul, making it a fit vehicle for philosophy.[81] The Fabrist circle brought monastic forms of mental making to new contexts: universities and print shops. These two contexts will intersect throughout this book, so it is worth sketching them now.

First, universities, and specifically the University of Paris. If thought was a craft, then by the fifteenth century its workshops were increasingly university classrooms. When Lefèvre turned to mathematics in the early 1490s, he was a regent master of the Collège du Cardinal Lemoine, a college of 'artists' or scholars of the arts course, situated near the city gate which led to the grand old Augustinian convent of St Victor. Lemoine was one of the older arts colleges of the University of Paris and deeply embedded in the city and region, just as the University was in Europe.

The fifteenth century saw a profound transformation of Europe's universities. In 1400, there were thirty-four universities in Europe. In 1500 that number had doubled to sixty-six.[82] North of the Alps, most of these were modelled on Paris, as

Esteve (Paris: Honoré Champion, 2012), 69–101; Giovanna Cifoletti, 'Mathematics and Rhetoric: Peletier and Gosselin and the Making of the French Algebraic Tradition', PhD dissertation, Princeton University, 1993; Isabelle Pantin, 'Les problèmes spécifiques de l'édition des livres scientifiques à la Renaissance: l'exemple de Guillaume Cavellat', in *Le Livre dans l'Europe de la Renaissance* (Paris: Promodis, 1988), 240–52.

[79] Mary Carruthers, *The Craft of Thought: Meditation, Rhetoric, and the Making of Images, 400–1200* (Cambridge: Cambridge University Press, 1998).

[80] This is the central conceit of Cusanus's three books *De idiota*. The profile of Cusanus is discussed in the section 'Northern Conversions' in Chapter 2.

[81] *Divini Gregorii Nyssae Episcopi qui fuit frater Basilii Magni libri*, ed. Johannes Cuno and Beatus Rhenanus (Basel: Mathias Schurer, 1512), A2r, Cuno to Beatus Rhenanus. Having noted Beatus' gifts of *ingenium* and *fortuna*, Cuno praises his university formation under Lefèvre, 'who, having artfully crafted an ornate chariot as if another artisan Daedalus, brought in Aristotelian philosophy in a cloak of elegance and beauty of phrase to be seen by all' (*qui ut alter Dedalus faber carpento ornato affabre fabrefacto, philosophiam Aristotelicam eleganti stola et phrasi decorata, cunctis aspiciendam invexit*). Cuno then suggests that Beatus was thereby prepared for the 'Platonic, nay, Christian philosophy of Gregory of Nyssa' (*verum etiam ultra hanc ad Platonicam philosophiam immo Christianam Gregorii Nysseni* [...] *animum vertas*). The chariot is evidently a reference to the soul as developed in Plato's *Phaedrus*; on turning the soul, cf. the role of conversion in the next chapter.

[82] These numbers are from Jacques Verger, 'Patterns', in *A History of the University in Europe*, Vol. 1: *Universities in the Middle Ages*, ed. H. de Ridder-Symoens and Jacques Verger (Cambridge: Cambridge University Press, 2003), 35–68, at 58–9.

rulers lured Parisian masters to set up new *studia generalia* in the wake of the Great Schism of 1478.[83] Such rulers hoped to counterbalance the authority of the University of Paris, which uniquely retained links to the pope, who had first granted Paris masters the 'right to teach anywhere' (*ius ubique docendi*) and still supplied benefices directly to Paris arts masters, a privilege no other university enjoyed.[84] Thus the University of Paris was woven by a thousand threads into the fabric of Europe, as the theological *magistra* of Europe and the de facto model for university life in Northern Europe—especially through its far-flung college connections. During the fourteenth and fifteenth centuries, colleges became the main focus of university life, growing out of housing arrangements (*hospites* and *bursae*) and grammar schools (preparatory *pedagogies*); perhaps seventy were founded in Paris alone.[85] By 1500, the number of arts students inhabiting Paris was around ten thousand.[86] Many lived in private residences and shared halls throughout the city, but the centre of university life had gradually been moving into the colleges.[87] With their endowments of lands and external revenue, colleges gave students and masters stability; Lemoine owned lands in several dioceses around Paris and Picardy. Each college was associated with one of the four 'Nations' which made up the Faculty of Arts, and so were tied to linguistic and regional constituencies; they were instruments of political power, founded by kings and churchmen, and frequent litigants with the Paris *parlement*.[88] Many of Lefèvre's patrons and students were part of a new *noblesse du robe* involved in both diocesan and Parisian politics.

Given this astounding growth of universities in the wake of the Black Death, the question of social mobility is pressing. Colleges not only drew their students from the wealthy, but also gave stipends to those they designated as 'poor' students, who often came from the new city grammar schools in outlying dioceses. Firm numbers are hard to come by and harder to interpret, but the period yields outstanding examples such as Jean Gerson, who came from a modest village in Champagne to become the University's Chancellor and a leading public figure of all Europe; similarly, a Rhineland merchant's son, Nicholas of Cusa, drew on his university learning to become one of Rome's most influential cardinals. Lefèvre, Clichtove, and Bovelles came from reasonably well-off families in small towns in the Low Countries and Picardy. The university supplied a structure in which men of

[83] R. N. Swanson, *Universities, Academics and the Great Schism* (Cambridge: Cambridge University Press, 1979); a key example is Vienna, as discussed by Michael H. Shank, *'Unless You Believe, You Shall Not Understand': Logic, University, and Society in Late Medieval Vienna* (Princeton, NJ: Princeton University Press, 1988).

[84] Donald E. R. Watt, 'University Clerks and Rolls of Petitions for Benefices', *Speculum* 34, no. 2 (1959): 213–29.

[85] On their emergence in Paris, see Astrik L. Gabriel, *Student Life in Ave Maria College, Mediaeval Paris* (Notre Dame, IN: University of Notre Dame Press, 1955). A great many of the early colleges quickly disappeared. See Jacques Verger, 'Patterns'; Marie-Madeleine Compère, *Les Collèges français—16e–18e siècles* (Paris: INRP, 2002).

[86] L. W. B. Brockliss, 'Patterns of Attendance at the University of Paris, 1400–1800', *Historical Journal* 21, no. 3 (1978): 503–44.

[87] See chapter 3(a).

[88] Legal conflict between town and gown is a constant theme in Cécile Fabris, *Étudier et vivre à Paris au moyen âge: le Collège de Laon, XIVe–XVe siècles* (Paris: École nationale des chartes, 2005).

modest beginnings could aspire to become public intellectuals, to steer Europe through political quagmires and ecclesiastical minefields.[89]

Lefèvre's students also identified him with a rising class of university critics: Renaissance humanists. It is important to see that such critique was largely pursued *within* the universities. Princes of the pen such as Juan Luis Vives and Erasmus aimed their barbs at the very education that legitimated their own personae.[90] Even in Italy the majority of humanists started out in universities.[91] This was all the truer in Paris, where the university attracted humanist visitors and patrons alike. A major figure in the Law Faculty, the Mathurin Robert Gaguin was a hub for humanist endeavours in the second half of the fifteenth century. His circle included the founders of the first Paris press, the Sorbonne theologians Johannes Heynlin and Guillaume Fichet, as well as the early Greek émigrés who came to Paris, including Gregory Tifernas and George Hermonymus—who would teach Greek with varying success to Johannes Reuchlin, Lefèvre, Erasmus, Budé, and Beatus Rhenanus.[92] Gaguin was an approving witness to Giovanni Pico della Mirandola's famous visit to learn the *modus Parisiensis* in 1485, and his correspondence even extended to Ficino.[93] The influx of grammarians and rhetoric teachers during the 1480s earned their living either by lecturing informally and in the preparatory grammar schools known as *pedagogies*, or supported by rich patrons: in this group we find Charles and Jean Fernand, François Tisard, Guillaume Tardif, Fausto Andrelini, Paolo Emilio, and later Girolamo Aleandro and Janus Lascaris.[94] The patrons who supported Lefèvre, Budé, and their friends came not from nobility but from rich mercantile families such as the Briçonnets—they too depended on Paris's new humanism as well as university degrees to legitimize their place in a nascent *noblesse de robe*.[95]

The second context on which my account depends is the new printing presses. Lefèvre's circle produced their university books within a uniquely experimental phase in print culture. This culture had changed in the decades since 1469, when

[89] The rise of the university-based public intellectual in the late fourteenth century counters the early narrative of Jacques Le Goff, *Les intellectuels au Moyen Âge* (1957; 2nd edn, Paris: Éditions du Seuil, 1985). See Jacques Verger, *Men of Learning in Europe at the End of the Middle Ages*, trans. Steven Rendall (Notre Dame, IN: University of Notre Dame Press, 2000); Rita Copeland, *Pedagogy, Intellectuals, and Dissent in the Later Middle Ages: Lollardy and Ideas of Learning* (Cambridge: Cambridge University Press, 2001); Daniel Hobbins, 'The Schoolman as Public Intellectual: Jean Gerson and the Late Medieval Tract', *American Historical Review* 108, no. 5 (2003).

[90] e.g. James K. Farge, 'Erasmus, the University of Paris, and the Profession of Theology', *Erasmus of Rotterdam Society Yearbook* 19 (1999): 18–46.

[91] David A. Lines, 'Humanism and the Italian Universities', in *Humanism and Creativity in the Renaissance: Essays in Honor of Ronald G. Witt*, ed. Christopher S. Celenza and Kenneth Gouwens (Leiden: Brill, 2006), 327–46.

[92] Maria Kalatzi, *Hermonymos: A Study in Scribal, Literary and Teaching Activities in the Fifteenth and Early Sixteenth Centuries* (Athens, 2009), 66–85.

[93] Louis Thuasne (ed.), *Roberti Gaguini Epistole et orationes*, 2 vols (Paris: É. Bouillon, 1903), 1:70, and ep. 21.

[94] A useful overview remains Renaudet, 114–59.

[95] Eugene F. Rice, Jr., 'The Patrons of French Humanism, 1490–1520', in *Renaissance Studies in Honor of Hans Baron*, ed. Anthony Molho and John A. Tedeschi (Dekalb, IL: Northern Illinois University Press, 1971), 687–702, as well as the biographical notices in PE.

one of the Collège de la Sorbonne's professors, Johannes Heynlin, invited three Germans to set up a press in the college precincts. The endeavour was directed and financed by one of the leading professors of theology, then serving as the college's librarian, Guillaume Fichet. Heynlin and Fichet proceeded to publish editions that reflected the new learning of Italy, such as the humanist Gasparino Barzizza's *Epistolae* and *Orthographia*, and works by Sallust and Livy, their own letters, Fichet's *Rhetorica*, and the *Orations* of Bessarion.[96] Once historians took this as a sign of Heynlin and Fichet's humanism—until examination of their teaching and book borrowings showed that their regular reading was full of standard thirteenth-century authorities such as Aquinas, Bonaventure, and so on.[97] In other words, the presses served aspirations, not actual university teaching. The first Paris press was installed within college rooms; yet by publishing Cicero and Virgil, it first served the needs of grammar schools rather than university classrooms.

The early print market suggests a slow entry of university staples into print. Manuscript book production soared in the early fifteenth century, partly because of newly abundant paper.[98] With print, the trend continued at Paris, which became the second-largest producer of printed books in Europe by 1500, overtaking Venice.[99] Yet few early printed books were the university staples of logic, natural philosophy, Aristotle's ethics, the large handbooks on theology and law, or major medieval commentators on Aristotle. The university market was already saturated. In fact, when the printing press arrived in Paris in 1469, it met a mature, tightly regulated book trade, in which general booksellers (*librarii*) competed to become part of a sworn group of stationers (*stationarii*) with privileges to sell, rent, and copy books for scholars and students.[100] This system, it seems, kept its monopoly on university books for some decades; as Severin Corsten showed, the universities were late adopters of print.[101] Although leading printers also could be *librarii iurati*, they only printed books for broader liturgical or literary uses—classroom needs were served by manuscript production.[102] This division

[96] Margaret Meserve, 'Patronage and Propaganda at the First Paris Press: Guillaume Fichet and the First Edition of Bessarion's Orations against the Turks', *Papers of the Bibliographical Society of America* 97, no. 4 (2003): 521–88.

[97] Henri-Jean Martin et al., *La Naissance du livre moderne: Mise en page et mise en texte du livre français (XIVe–XVIIe siècles)* (Paris: Éditions du Cercle de la Librairie, 2000), 100–16.

[98] On the rise of paper, see H. Bresc and I. Heullant-Donat, 'Pour une réévaluation de la "révolution du papier" dans l'Occident médiéval', *Scriptorium* 61, no. 2 (2007): 354–83; see also Daniel Hobbins, *Authorship and Publicity Before Print: Jean Gerson and the Transformation of Late Medieval Learning* (Philadelphia, PA: University of Pennsylvania Press, 2009), notes at 8–9, for the growth in fifteenth-century manuscript publication.

[99] The ISTC (up to 1501) records 3,788 editions for Venice and 3,270 for Paris. In the decade beginning in 1501, Parisian production surpassed that of Venice.

[100] Mary A. Rouse and Richard H. Rouse, 'The Book Trade at the University of Paris, ca. 1250–ca. 1350', in *Authentic Witnesses: Approaches to Medieval Texts and Manuscripts* (Notre Dame, IN: University of Notre Dame Press, 1991), 259–338.

[101] Severin Corsten, 'Universities and Early Printing', in *Bibliography and the Study of 15th-Century Civilization* (London: British Library, 1987), 83–123.

[102] One example is Pasquier Bonhomme, whose *Les chroniques de St Denis* (26 January 1476) is the first book printed in French. One of the four 'grand libraire jurés', he nevertheless restricted his production to non-university books. See Philippe Renouard, *Répertoire des imprimeurs parisiens*, ed. Jeanne Veyrin-Forrer and Brigitte Moreau (Paris: M.J. Minard, 1965), 43–4.

broke down in the 1490s. Johann Higman (d.1500) and Wolfgang Hopyl (d.1522) were Lefèvre's first printers, but only Hopyl seems to have been an official, 'sworn' university bookseller.[103] Lefèvre's main printer Henri Estienne took over Hopyl's press in 1503 and so became an important figure among university printers.[104] Yet Josse Bade came to Paris in 1498, and he seems not to have needed the qualification of being a sworn university bookman in order to publish university books, perhaps because the university's hold on the book trade was no longer total.

Therefore the 1490s, when Lefèvre, Clichtove, Bovelles and their students entered print culture, were a period of intense experimentation with books, their formats, typography, and uses. The classic story of the coming of the book assumes a chronological shift from manuscript to print, but recent historiography eschews an account in which one medium simply eclipses another. Rather, the two technologies interpenetrated. Early printers went to great, inefficient lengths to imitate manuscripts precisely; manuscript publication continued in the age of print, with new and specialized functions; and printed books served as a prompt to further handwriting, whether in margins or commonplace books. Historians have recognized the decades centred on 1500 as the period when some of the typography most closely linked to the printed book began to normalize: title pages, folio numbering, paragraphs indentation, tables of contents and other indexes.[105] In the fifteenth century, in sum, printers did not immediately see the university as a market for classroom manuals—excepting new classics from antiquity, mainly studied in preparatory pedagogies.

Print production turns the question of authorship into one of handcraft. As the long historiography on great scholar-printers like the Estiennes attests, thought-craft and handcraft intermingled in the print shop. This context will inform the perspective of this book; student correctors of Lefèvre's works often worked directly alongside the pressmen as *recognitores in officina*.[106] Students such as Beatus Rhenanus were in a real sense apprentices in the craft of making books. I shall argue that it was students in the 1490s who spotted the market and resolved to supply it.

[103] Renouard, *Répertoire*, 204, 206–7. Higman and Hopyl were among the first to print new texts relating to the standard arts curriculum, such as logic and natural philosophy. In 1489, the first logic textbook by a living author to be printed at Paris was Thomas Bricot's abbreviation of the *Sumule logicales*, which became popular at Paris in the following two decades.

[104] Renouard, *Répertoire*, 140–1. See Elizabeth Armstrong, *Robert Estienne, Royal Printer: An Historical Study of the Elder Stephanus* (Cambridge: Cambridge University Press, 1954); Fred Schreiber, *The Estiennes* (New York: E.K. Schreiber, 1982); Fred Schreiber and Jeanne Veyrin-Forrer, *Simon de Colines: An Annotated Catalogue of 230 Examples of His Press, 1520–1546* (London: Oak Knoll Press, 1995).

[105] The artificial line between incunabula (1500 and earlier) and early printed books (1501 and later) is obliterated by Paul Saenger and Michael Heinlin, 'Incunable Description and Its Implication for the Analysis of Fifteenth-Century Reading Habits', in *Printing the Written Word: The Social History of Books, circa 1450–1520*, ed. Sandra L. Hindman (Ithaca, NY: Cornell University Press, 1991), 225–58.

[106] Armstrong Tyler, 'Lefèvre and Estienne', 23.

OVERVIEW

In order to catch a community at work in a moment of criss-crossing transitions in university learning, print culture, and mathematics, this book is organized as a series of concentric rings, moving from broader milieux inward. It begins with the fifteenth-century world in which the community's central figure, Lefèvre d'Étaples, developed the ideals that attracted a circle of students. Lefèvre underwent a 'turn' to mathematics, which I identify as part of medieval narratives of conversion within fifteenth-century religious reform. As a canonical founder of the French Renaissance, Lefèvre usually figures as an epigone of Italian humanism. He did value humanist methods, but I suggest that his larger intellectual aims were instead set by northern reform movements in which mathematics held a special place.

The book then moves to the university, focusing on what Lefèvre's students experienced in his classroom in this moment of transition from manuscript to print. Their project was partly constrained by Lefèvre's aims, but also by the aims and practices of university life. Chapter 3 considers the encyclopaedia set out by the university arts course as a copious ideal, which students navigated through the practices of disputation and lecture. This was a demanding task to manage; I argue that it became even more difficult with the abundance of new print and with the humanist mandate to access the ancient authors directly. Lefèvre's textbooks can be seen as both cause and response to this overabundance. For the first time, a student could encounter the entire arts curriculum through the printed books that his masters had published. Chapter 3 opens a unique window onto this experience: the archive of Beatus Rhenanus, a student central to the circle. Through his student notebook and annotated textbooks, I trace the importance mathematics acquired in his classroom, to the point where he framed mathematical analogies as a new 'art of arts', a method for mastering all knowledge. Mathematics thus emerges in response to the need to contract, compare, and expand the various disciplines of the arts curriculum, as students attempted to meet the university's demands.

The books therefore were made to answer university demands. These books are intriguing objects, with an unusual typography of marginal numbers, tables, diagrams, letters, and indexes, which range across genres of epitome, introduction, commentary, and paraphrase. They represent an experimental phase in the emerging conventions of the printed page. In chapter 4 I argue that these distinctive features reflect the many hands who helped assemble them, as students urged the publication of notes, editing them, augmenting them, and correcting them in Paris's new print houses. The rich visual apparatus also answered the need to condense and order the copia of knowledge described previously. Mathematics in particular offered a methodogical example of collaborative authorship and how to condense vast quantities of knowledge. We find an extreme example of collaborative authorship in the mathematical books of Lefèvre's circle, in part because mathematical narrative was assumed to distinguish between proposition (considered authoritative) and proof (considered commentary, and therefore optional). As learning tools designed to compress the cycle of arts, these books in principle aimed only to lead to other authorities, leaving a question: how could they be sites for new knowledge?

The final chapters centre on the long-term implications of these books for mathematical culture. Chapter 5 suggests that the textbooks for music and cosmography eroded old disciplinary barriers. The typographic regimes of these books marked the high value the Fabrists awarded the senses in learning. To acquire mathematical expertise, a student learned from worked examples, requiring sight, but also hearing and touch, to discuss the case or manipulate a compass. Lefèvre carefully used the *mise-en-page* of his new books to train students in such skills. In Lefèvre's cosmography, these priorities result in a new kind of book—his commentary on Sacrobosco's *Sphere* invents the kind of cosmographical manual that would define the Renaissance, with a visual apparatus that could give readers the techniques needed to use ephemerides and maps. In his effort to formalize Boethius' classic treatise on music, Lefèvre taught readers to pay particular attention to the sensory experience of measuring vibrating strings. This allowed him to present a new visual tool to resolve the classic problem of ancient music: how to divide the tone. With his visual, geometrical method in music, Lefèvre also crossed conceptual boundaries between parts of mathematics, bringing a long-forbidden class of irrational ratios into music and putting pressure on the ancient distinction between discrete and continuous mathematical objects. In their material and visual structures, Lefèvre's introductions remapped the early modern mathematical disciplines.

Chapter 6 shows how mathematical intuitions filtered into Lefèvre's natural philosophy. The mathematical culture described in previous chapters was one that used sensory techniques to idealize the natural world along mathematical lines. This directly conflicted with the Aristotelian assumption that qualities, not quantities, capture what is interesting about natural change. Lefèvre therefore takes an unusual place: he was famous as a champion of Aristotle, but also deeply invested in mathematical learning. I consider how, in a dialogue introducing Aristotle's *Physics*, Lefèvre presented alternatives to what he considered false Aristotelianism. The dialogue offered ideals of friendship and conversation to replace pernicious university practices. In so doing, the dialogue also rehabilitated a medieval approach to natural change that relied on mathematical intuitions. Lefèvre, therefore, presented an alternative to Aristotle: the possibility that mathematical principles possess causal power in the world. His account relied on a constructionist view of human knowledge, in which humans know through the creative activity of measuring the world. In this perspective, all culture is ultimately mathematical.

Like all communities, the one that made these books eventually gave way to the exigencies of time and place. The epilogue points both to its enduring influence on mathematical culture, and to the ways later proponents obscured the university origins of the mathematical culture they took for granted.

2

A Mathematical Turn

In February of 1495, Lefèvre recounted his turn to mathematics in his first printed mathematical work, a commentary on an introduction to astronomy, the *Sphere* of Sacrobosco.[1] In the dedicatory letter to a leading light of the Paris *parlement*, Charles de Bourrée, Lefèvre envisioned restoring the University of Paris with the right study of mathematics. He traced the vision to Paris's swelling community of humanist scholars. The Greek émigré George Hermonymus had 'set aflame' Lefèvre's mind with tales of the value of mathematics in learning. Hermonymus praised Paris, but argued that it lacked only one thing, namely mathematics. 'If we believe Plato in the *Republic*, book 7, [mathematics] is of the greatest importance not only for the republic of letters but also for the civil republic—and Plato thinks those with the best natures especially should be taught in it.'[2] Lefèvre added an example of such excellent wits: 'Who has a more developed *ingenium* than our philosophers?... George of Trebizond indeed seems to deserve the greatest favour in literary affairs, because he turned his *ingenium* to drawing mathematical learning out of the shadows.'[3]

In this academic conversion story—Trebizond literally 'converted' his *ingenium* to mathematical studies—Lefèvre seems at first to reinforce the standard story of the rise of mathematical culture in the Renaissance. A classic story looks to new translations of Archimedes and Diophantus associated with Greek scholars like Hermonymus, the practical promises of mathematics for patrons like Bourrée, and Platonist replacements for Aristotelian school philosophy. But appearances deceive, and Lefèvre's motivations reflect wider attitudes to reform in the fifteenth century.

With the almost innocent, second-hand reference to Trebizond, Lefèvre lets us into a world of learned cosmopolitan rumour. Around 1470, Hermonymus had joined the entourage of Cardinal Bessarion, patron of other Greek scholars such as Theodore of Gaza and George of Trebizond, as well as of talented Western humanists such as Niccolò Perotti, Lorenzo Valla, and Johannes Regiomontanus.[4] But by

[1] Two years earlier he had published a treatise on number theory, the *Elementa arithmetica* of Jordanus, likely in manuscript. See the section 'Collective Authorship' in Chapter 4, and PE, 17.

[2] *Textus de sphera*, a1v (PE, 27): 'Mathemata, inquit, que (si Platoni septimo de republica credimus) non modo reipublice litterarie, sed et civili momentum habent maximum, et in his (ut sentit Plato) precipue erudiendi sunt qui naturis sunt optimis.'

[3] *Textus de sphera*, a1v (PE, 27): 'Sed qui (inquit) nostris philosophantibus mitiore sunt ingenio? Adduxit et Georgium Trapezuntium, qui vel maxime de re litteraria benigne meritus videtur, quod eius ingenium ad mathematicas disciplinas e tenebris eruendas converterit.'

[4] Maria Kalatzi, *Hermonymos: A Study in Scribal, Literary and Teaching Activities in the Fifteenth and Early Sixteenth Centuries* (Athens, 2009), 39–44.

the time Bessarion died in 1472 and Hermonymus had left Rome for England and France in 1475, the avant-garde circle had fractured—with George of Trebizond at the epicentre of its collapse. In part, the problem was his 'morose' nature, as Lorenzo Valla said, for Trebizond quarrelled with half of Bessarion's circle, descending to blows with Poggio Bracciolini.[5] In part, the problem was his commitment to Aristotle; although the Platonism of Pletho stirred Bessarion, to Trebizond it seemed pantheism, the root of the major Christian heresies and even of Islam.[6] When Trebizond wrote a tract against the paganism of the Platonists, his *Comparatio Philosophorum Platonis et Aristotelis* (1458), Bessarion armed for battle. As John Monfasani has shown, Bessarion enlisted his considerable clientele to help him write the extended *In Calumniatorem Platonis* (1469), aimed directly at Trebizond. And he exerted his influence elsewhere to have Trebizond discredited. When Trebizond newly translated and commented on Ptolemy's *Almagest*, Pope Nicholas V had it examined by a scholar who had recently translated some works of Archimedes for him, Jacobus Cremonensis, who excoriated Trebizond's commentary.[7] And when Trebizond attempted to bring his commentary on the *Almagest* to the Hungarian court of Mathias Corvinus, he faced the wrath of another member of Bessarion's circle, the German mathematician Johannes Regiomontanus, who wrote a *Defensio Theonis* against Trebizond, never published.[8]

Lefèvre knew much of this drama—perhaps he heard the tales when he toured Italy in 1492, 1500, and again in 1507—and he nevertheless held up Trebizond as an exemplar of what mathematics might do for letters. Sometime after 1503 he informed the visiting Cambridge theologian Robert Ridley that *In Calumniatorem* was written by a team animated by Bessarion's hatred of Trebizond; that they had fallen out over political reasons; and that Trebizond's astronomical writings had been taken to Hungary.[9] Certainly he knew about the latter. Lefèvre published an edition of Trebizond's *Dialectica* in 1508, which included a short, partisan life of Trebizond, drawing attention to Trebizond's translations of natural philosophy and Ptolemy's *Almagest*. For the *Almagest* 'he also wrote a commentary that was taken up together with the translation by Johannes Regiomontanus in those books in which he defends Theon of Alexandria, drawing up so much jealousy against

[5] 'Trapezuntium et morosum et mihi nescio quam aequum consulere nolo'; cit. John Monfasani, *George of Trebizond: A Biography and a Study of His Rhetoric and Logic* (Leiden: Brill, 1976), 104.

[6] Monfasani, *George of Trebizond* , 158–62.

[7] Cremonensis, like Trebizond, had studied Latin with Vittorino da Feltre. On his role in the translation of Archimedes, see Paul Lawrence Rose, 'Humanist Culture and Renaissance Mathematics: The Italian Libraries of the Quattrocento', *Studies in the Renaissance* 20 (1973): 82–7. Monfasani describes Cremonensis' review of Trebizond in *George of Trebizond*, 105–8. As Monfasani points out, it was the commentary at issue, not Trebizond's translation.

[8] Monfasani, *George of Trebizond*, 195–6. Bessarion had originally recommended that Trebizond rely on the ancient Alexandrian commentator Theon, but Trebizond ignored the advice and attacked Theon. Ironically, Regiomontanus argued that Trebizond had actually plagiarized Theon. The *Defensio Theonis* has been edited by Richard L. Kremer and Michael Shank (http://regio.dartmouth.edu/, updated 29 March 2004).

[9] Ridley recounted what Lefèvre told him in his copy of Bessarion, *In Calumniatorem Platonis* (Venice: Aldus Manutius, 1503), Yale University, Beinecke Library, shelf mark Gfp 66 +y 503b, a1v. The note is edited and translated by Monfasani, 'A Tale of Two Books', 8.

himself that certain people believed he was poisoned at Rome by Trebizond's sons.'[10] Lefèvre's version of the story presented Trebizond as a victim, unjustly deprived of credit for having sparked some of the most worthwhile translations of Bessarion's circle, for Trebizond 'had imitators, beyond the one mentioned [i.e. Regiomontanus], Theodore of Gaza and Giovanni [Andrea Bussi] bishop of Aleria, each as troublesome as the next'.[11] The reasons for this defence of Trebizond are not far to find—to admirers of a humanist Aristotle such as Lefèvre, Trebizond was a suitable champion. Lefèvre's student, the Alsatian humanist Beatus Rhenanus, made this point in an edition of Trebizond's *Dialectica*, observing that Trebizond, 'moved by the impiety of the Platonists, also wrote an excellent work *Comparatio philosophorum*, in which he showed with powerful arguments that the Aristotelian sort of philosophizing is far more suitable to Christian piety than the Platonist sort. For this reason, Cardinal Bessarion railed at him in four books.'[12] Whatever injustices Trebizond may have suffered, in Paris he was renowned as a brilliant defender of Aristotle whose learning had also served the pinnacle of mathematical erudition, Ptolemy's *Almagest*. Therefore Lefèvre found him a suitable antecedent for his own *ressourcement* of university learning.

If Lefèvre was not the French voice of Italian Platonists, what then were his sources? This chapter situates Lefèvre's own turn to mathematics between Italian influences and northern reform movements, before surveying the educational ideals within which his students encountered mathematics.

MATHEMATICS AT PARIS

The fortunes of mathematics at the University of Paris were at a low ebb in the fifteenth century. This contrasts with the fourteenth century, when fashion at Paris swung in favour of mathematics, picking up trends from Oxford particularly associated with figures at Merton College such as Thomas Bradwardine and Richard Swineshead. To young men striving for the ultimate prestige of a doctorate in theology from Paris, while they lectured in the arts schools, this English mathematics

[10] Jacques Lefèvre d'Étaples (ed.), *Georgii Trapezontii dialectica* (Paris: Henri Estienne, 1508), e4r–v: '[Scripsit Trapezontius] et Ptolemaei ingens opus, quod Peri megales pragmateias inscribitur, in quam et commentaria scripta reliquit, quae reprehenduntur unà cum versione à Ioanne de Monte Regio, iis libris, quibus Theon Alexandrinus defenditur, qua re tantam sibi invidiam comparasse scribitur, ut à Trapezontii filiis Romae veneno extinctus quibusdam credatur.'

[11] Lefèvre, *Trapezontii dialectica*, e4r: 'Aemulos habuit, praeter eum quem diximus, Theodorum Gazam ac Ioannem Episcopum Aleriensem, utrunque iuxta infestum'. This account is largely true, but Lefèvre could not have obtained it from sources printed at the time. The short *Vita Trapezontii* goes on to point out that, in the preface to his translation of Aristotle's *De animalibus*, Gaza claimed to have had no help—but that his stylistic additions show that he must have actually had Trebizond open before him.

[12] Beatus Rhenanus to Ioannes Kierherus, May 1509, in Trebizond, *Dialectica* (Strassburg: 1509), a1v: 'Edidit insuper de comparatione philosophorum opus insigne, quorundam Platonicorum impietate motus; quo opere Aristotelicum philosophandi modum quam Platonis longe est Christianae pietati conformiorem, rationibus minime frigidis ostendit. Quare Bessarion Nicenus cardinalis hunc quatuor libris insectatus est.' Note that this was a Strassburg printing of Lefèvre's own Paris edition of Trebizond's *Dialectica*.

of proportions seemed like the latest and best conceptual tool, and one that might be deployed not only in natural philosophy but most interestingly in theology.[13] By 1535, this form of speculative, abstract theology had eroded the traditional identity of Parisian doctors of theology as masters of the sacred page above all else. Even those theologians who did publish on the Bible were nearly all secular clerics rather than priests or members of religious orders.[14] The brightest and best arts masters of fourteenth-century Paris, such as Jean Buridan, Nicole Oresme, and Henry of Hesse (Langenstein), thrived on this prestige moment for mathematical analysis.

The king noticed. In the 1370s, Charles V called on Oresme as a high-profile consultant on astrological concerns at court.[15] Indeed, Charles showed interest in university mathematics as an instrument of rule by supporting the college founded in 1371 by the royal physician, Gervais Chrétien. There Charles endowed two royal lectureships explicitly intended to teach 'the mathematical sciences, that is to say books of the liberal arts of the *quadrivium* which are considered licit by the holy canons or in no way reprobated by the University of Paris'.[16] By limiting teaching to 'licit' mathematics, these statutes carefully steer clear of more illicit forms of judicial astrology—a growing concern in late fourteenth-century Paris. Both Oresme and Langenstein campaigned against such astrology, but their chief argument was that astrology was not mathematically precise enough to distinguish various heavenly causes. Despite their suspicions of physicians and other astrologers, they agreed with them that mathematics could in principle describe heavenly causes.[17] They were cautious about astrology precisely out of a high regard for mathematics.

Other leading intellectuals were less sceptical. Pierre d'Ailly, who was Chancellor of the University from 1389 to 1395, wrote extensively about the papal Schism, ecclesiastical reform, astrology, and cosmology, while advancing as bishop, cardinal, and royal chaplain. Already in his early years as a teaching master at Paris, d'Ailly

[13] John E. Murdoch, '*Mathesis in philosophiam scholasticam introducta*: The Rise and Development of the Application of Mathematics in Fourteenth Century Philosophy and Theology', in *Arts libéraux et philosophie au Moyen Âge: Actes du quatrième congrès international de philosophie médiévale* (Paris: Vrin, 1969), 215–54; John E. Murdoch, 'From Social into Intellectual Factors: An Aspect of the Unitary Character of Late Medieval Learning', in *The Cultural Context of Medieval Learning*, ed. John E. Murdoch and Edith D. Sylla (Dortrecht: D. Reidel, 1975), 271–348.

[14] William J. Courtenay, 'The Bible in the Fourteenth Century: Some Observations', *Church History: Studies in Christianity and Culture* 54, no. 2 (1985): 176–87.

[15] Joan Cadden, 'Charles V, Nicole Oresme, and Christine de Pizan: Unities and Uses of Knowledge in Fourteenth-Century France', in *Texts and Contexts in Ancient and Medieval Science: Studies on the Occasion of John E. Murdoch's Seventieth Birthday*, ed. Edith Dudley Sylla and Michael M. McVaugh (Leiden: Brill, 1997), 108–244.

[16] Paris, Arch. Nat., MS M 164, ed. Longuemare (1916), 274–5: 'tenebuntur illi duo magistri legere de scientiis mathematicis, videlicet de libros de quadrivio artium liberalium licitos per sacros canones vel per Universitatem Parisiensem nullatenus reprobatos'; cit. Jean-Patrice Boudet, 'A "College of Astrology and Medicine"? Charles V, Gervais Chrétien, and the Scientific Manuscripts of Maître Gervais's College', *Studies in History and Philosophy of Science* 41, no. 2 (2010): 102.

[17] G.W. Coopland, *Nicole Oresme and the Astrologers: A Study of His* Livre de Divinacions (Liverpool: Liverpool University Press, 1952); Stefano Caroti, 'La critica contra l'astrologia di Nicole Oresme e la sua influenza nel Medioevo e nel Rinascimento', *Atti della Accademia Nazionale dei Lincei*, Memorie. Classe di scienze morali, storiche e filologiche, 23 (1979): 543–648.

wrote on standard textbooks such as Aristotle's *Meteorology* and Sacrobosco's *Sphere* in which he set out the legitimate uses of astrology as the 'practical' part of astronomy. In later years, he identified the great conjunctions in the heavens as the markers of history's great events, letting one calculate the date of the Antichrist's advent. Laura Smoller has shown how d'Ailly grew only more preoccupied with these topics as the Great Schism dragged on. D'Ailly interpreted the jagged tears in the Church as signs of the final stages of history and the coming of the Antichrist. Using mathematics to read the heavens, he believed, one could hear a divine call to ecclesiastical, social, and personal reform.[18]

Despite royal patronage, mathematics came under criticism both as analytical fashion and as astrological impiety during the Europe-wide turmoil of civil and ecclesiastical politics around 1400. The next chancellor of the University, d'Ailly's student Jean Gerson, saw mathematics and its applications chiefly as a distraction. Both astrology and mathematical analysis originated in profane curiosity, a pride in learning—this preoccupation had distracted clerics from the care of the souls and love of the Church. Gerson directly affected the university. In a famous sermon published as *Contra curiositatem studentium*, Gerson linked such profane obsessions with the analytical mode of philosophizing that had become dominant at the University of Paris. A sign of curiosity, for example, is to mix disciplines, that is, for arts masters to pontificate on theology and for theologians to do pure logic.[19] With such arguments, Gerson took aim at the central accomplishment of fourteenth-century mathematics in Paris: its interdisciplinary uses in other domains. Gerson launched a campaign against such 'sophistry' and 'subtleties' and during his long chancellorship did much—helped by the exodus of the German–English nations—to replace the previous intellectual fashion. Instead of the dangerous fringes of *secta nominalium* (followers of Ockham, etc.) and the *secta formalizantium* (followers of John Duns Scotus), Gerson proposed a safer alternative, the 'well-trod way' (*tritum iter*) of *peripatetici*, followers of Aristotle himself.

In the shadow of the Great Schism, university scholars in fifteenth-century Paris were no longer inclined to test theology with the mathematico-logical innovations of Bradwardine and Buridan. By praising the arch-Aristotelian Trebizond, Lefèvre echoed Gerson's defence against distracting novelties through the study of Aristotle himself.

THE COURT AND ITALY

Two pressures raised the cultural profile of mathematics in the fifteenth century: its practical utility for growing courts and bureaucracies, and the new philosophical

[18] Laura Ackerman Smoller, *History, Prophecy, and the Stars: The Christian Astrology of Pierre D'Ailly, 1350–1420* (Princeton, NJ: Princeton University Press, 1994).

[19] Gerson, 'Contra curiositatem studentium', in Jean Gerson, *Œuvres complètes*, ed. Palémon Glorieux (Paris: Desclée, 1960), 2:239.

sources of certain humanists. Both pressures marked the prefaces in which Lefèvre sought to defend the *utilitas* of mathematics.[20]

Mathematics had become useful because astrology was voguish in fifteenth-century European courts. Gerson repeatedly wrote short tracts aimed directly at the rising popularity of astrologers and magicians at court and in the city.[21] Yet the trend grew for political and medical reasons. French kings learned that magic and astrology could be tools of political and social control—the trial of Jeanne d'Arc attempted to manage her through accusations of divination and collusion with demons.[22] The great Renaissance libraries begin with Charles V's own, and throughout the fifteenth century rulers competed to outstrip their book-buying peers. Such libraries, including that of Matthias Corvinus, included large holdings in alchemy, magic, and astrology—and mathematical instruments.[23]

Astrology and its tools rose in prominence as university-trained physicians became more visible in society. In Lefèvre's day, astrology was a part of university medicine that was standard but contested. The line between useful prognosis and illegitimate divination was often blurred, as we find in the defence of Simon de Phares, tried for illicit astrology in 1495. His *Le Recueil des plus celebres astrologues* offered the king a genealogy of astrology at Paris, from antiquity to his own day. He carefully specified his own studies in astronomy and astrology in the arts faculty at Paris between 1462 and 1465, with Sacrobosco's *Sphere* and Alcibiatis' *Introductorius*.[24] At several points Simon inferred from instruments, astrolabes,

[20] The development of *utilitas* as a Renaissance defence of mathematics is discussed in Chapter 7; see also Angela Axworthy, *Le Mathématicien renaissant et son savoir: le statut des mathématiques selon Oronce Fine* (Paris: Classiques Garnier, 2016), 125–89; Katherine Neal, 'The Rhetoric of Utility: Avoiding Occult Associations for Mathematics Through Profitability and Pleasure', *History of Science* 37 (1999): 151–78.

[21] *De erroribus circa artem magicam* (2 March 1402?), Trilogium astrologie theologizate (1419), *Super doctrinam Raymundi Lulle*, *De observatione dierum quantum ad opera* (1425); *Contra superstitionem sculpturae leonis* (1428). See the chronology Glorieux gives in Gerson, *Œuvres complètes*, 10:73–6).

[22] There are earlier examples, such as the court of Frederick II, but by the late fourteenth century court culture emphasized divination and astrology: Jan R. Veenstra, *Magic and Divination at the Courts of Burgundy and France: Text and Context of Laurens Pignon's 'Contre Les Devineurs' (1411)* (Leiden: Brill, 1997), 21–8. The trend was true outside of France too: Michael D. Bailey, *Battling Demons: Witchcraft, Heresy, and Reform in the Late Middle Ages* (University Park, PA: Pennsylvania State University Press, 2003); Monica Azzolini, *The Duke and the Stars: Astrology and Politics in Renaissance Milan* (Cambridge, MA: Harvard University Press, 2013); Darin Hayton, *The Crown and the Cosmos: Astrology and the Politics of Maximilian I* (Pittsburgh, PA: University of Pittsburgh Press, 2015).

[23] François Avril and Jean Lafaurie (eds), *La Librairie de Charles V* (Paris: Bibliothèque nationale de France, 1968); Marcus Tanner, *The Raven King: Matthias Corvinus and the Fate of His Lost Library* (New Haven, CT: Yale University Press, 2009).

[24] Jean-Patrice Boudet (ed.), *Le Recueil des plus celebres astrologues de Simon de Phares* (Paris: Honoré Champion, 1999), Book VI, nos 19 and 33a (Gerson), 29a (d'Ailly), 104 (Jacques Juin), 105 (Louis de Beaumont de la Forêt). For Louis de Beaumont de la Forêt, Simon de Phares also specifies that he had a Hebrew grammar made, perhaps stressing a detail that his accusers might have found smacking of illicit knowledge. Note that in certain cases, as Boudet shows in his rich notes, Phares reads certain astrologers into existences; in other cases, such as Jacques Desparts (Book VI, no. 69), physicians did not quite hold the rosy appreciation for astrology that he ascribes to them. See the account of Desparts's scepticism about astrology in Danielle Jacquart, 'Theory, Everyday Practice, and Three Fifteenth-Century Physicians', *Osiris*, 2nd series, 6 (1990): 148–50.

and equatoria to the practices of astrology associated with those instruments; for example, he mentioned the early fifteenth-century Paris master of arts Jean Fusoris, whose astrolabes still exist.[25] Simon finished his genealogy with university-trained astrologers he himself knew: the famous teacher Jacques Juin had demonstrated his skill in the use of an astrolabe; even the late bishop of Paris, Louis de Beaumont de la Forêt, had two astrolabes made, one for his rooms and another for travel; and, finally, Guillaume de Carpentras, 'belonging to the German Nation' of the University, had made various spheres for the king of Sicily, the duke of Milan and 'and, finally, he made another one for you, your Majesty'.[26] Simon's defence of astrology depended on a history of links between university experts and courtly uses.

Practical utility is one reason for mathematics that Lefèvre presented to his early patrons, the brothers Jean and Germain de Ganay. Like many fifteenth-century patrons of new humanist letters, they were not members of the old nobility, but from one of the several officer families which had grown wealthy in banking and then powerful in law and royal administration. Germain eventually became Bishop of Orleans, while Jean took high office in the *parlement* of Paris. It was the Ganay brothers who supported Lefèvre's earliest interest in things mathematical. Lefèvre shared with Germain a fascination with the practical power of mathematics in astrology. In 1491, as king's councillor to the *parlement* and a canon of Notre Dame, Germain borrowed an astrolabe from the library of the College de la Sorbonne.[27] Lefèvre made Germain his main interlocutor in the dialogues he wrote during the mid-1490s titled *De magia naturali*, which focused on Pythagorean and Cabbalist approaches to manipulating the natural world. In the very first lines, he framed natural magic as the 'operative' part of philosophy. 'Those called magicians among the Chaldeans, are philosophers to the Greeks. But this seems to be the difference: philosophers favour contemplation and speculation, and focus less on testing the secret effects of philosophy. In contrast, magicians perceive the wonders of nature, so, according to the good law of the Chaldeans of long ago, magic seems to have been nothing but a practical kind of natural philosophy and an operative discipline of action.'[28] The following books were devoted to discussing with Germain the various effects that the magic of number could work in astronomical, medical, and alchemical domains.[29] Although Lefèvre later took

[25] Fusoris was exiled from Paris for his instruments—not because they were astrological, but because he followed a visiting ambassador back to England for repayment, an indelicacy too close to treason in the middle of the Hundred Years War. Emmanuel Poulle, *Un constructeur d'instruments astronomiques au XVe siècle: Jean Fusoris* (Paris: Champion, 1963), 2–4.

[26] Boudet (ed.), *Le Recueil*, no. 128 (603–4): 'et finablement en fist une autre pour vous, Sire.'

[27] Jeanne Viellard, 'Instruments d'astronomie conservés à la Bibliothèque du Collège de Sorbonne au XVe et XVIe siècles', *Bibliothèque de l'École des Chartres* 131 (1973): 591.

[28] BnF lat. 7454, 4r: 'Apud Chaldeos magi dicti sunt, sint qui apud grecos philosophi. Hoc tamen discrimen esse videtur: quod philosophi magis contemplationi speculationisque addicti, minus ad philosophie secretos effectus probandos sese committunt. Magi vero contra nature miracula sentiunt, ita ut bono iure chaldeorum olim magia nichil nisi quedam naturalis philosophie practica et operisque operativa disciplina fuisse videtur.'

[29] On this work see Jean-Marc Mandosio, 'Le *De magia naturali* de Jacques Lefèvre d'Étaples: Magie, alchimie et cabale', in *Les Muses secrètes: kabbale, alchimie et littérature à la Renaissance*, ed.

his distance from magic, natural or otherwise, Germain continued to support some of the famous Renaissance learned magi: Cornelius Agrippa, Trithemius, and the Orphic hymns of Janus Lascaris.[30]

Mathematics was also useful for administrative reasons. Before printing his commentary on Sacrobosco, Lefèvre published several works in manuscript, dedicated to the brothers by February 1494. Lefèvre presented Jean with his newly reworked version of Jordanus' number theory. Jean, president of the Paris *parlement*, had 'often reminded me, rightly, how many uses the understanding of arithmetic prepares and offers the disciplines, and how much of it lies hidden in darkness, so that its study can be avoided in no discipline'.[31] Lefèvre listed several ways mathematics might benefit public life: arithmetic and geometry were so important to the ancient Roman emperors that they had set up surveyors 'who held the technique of measurement for public use, which the wisdom of numbers completes'.[32] To the public man, Lefèvre held out the public uses of mathematics. In a second letter written in 1494 to Germain, Lefèvre repeated the theme. 'Most judicious Germain, I dedicated the arithmetical work to your brother, who measures out French justice, because, as the Aristotelians show, all justice requires one to distribute in geometrical proportion or by arithmetical commutation, and arithmetic is just like a certain mirror and rule of justice.'[33]

Besides practical utility, new intellectual currents flowing through Italy also stressed the propaedeutic utility of mathematics. Thanks to Bessarion's patronage, a larger range of texts were newly available—Regiomontanus set up a printing press in Nuremberg specially aimed at publishing the complete range of available mathematics since antiquity.[34] As humanists began to examine Greek manuscripts, they rethought their philosophical allegiances. Platonism, for instance, was not necessarily centred on number, but it could be. Plato became most mathematical in the hands of the Venetian professor of Latin letters Giorgio Valla. His massive *De expetendis et fugiendis rebus* (1501) surreptitiously translated large sections of

Rosanna Camos Gorris (Geneva: Droz, 2013), 37–79. I thank the author for allowing me to see this article before publication. See also Letizia Pierozzi and Jean-Marc Mandosio, 'L'interprétation alchimique de deux travaux d'Hercule dans le "De magia naturali" de Lefèvre d'Étaples', *Chrysopoeia* 5 (1992–6): 190–264.

30 Sebastiano Gentile, 'Giano Lascaris, Germain de Ganay e la "prisca theologia" in Francia', *Rinascimento*, 2nd ser., 26 (1986): 51–76, at 71–3.

31 Lefèvre, *Elementa*, a1v: 'Paucis me digne ammonuisti, clarissime vir, quot cognita commoditates adducat disciplinarum parens arithmetica, quotque ignota relinquat tenebras, ut quem nullius discipline fugiat studium.'

32 Lefèvre, *Elementa*, a1v: 'qui podismorum rationem ad publicam utilitatem tenerent quam numerorum perficit sagacitas.'

33 Jacques Lefèvre d'Étaples, *Introductio in metaphysicorum libros Aristotelis*, ed. Josse Clichtove (Paris: J. Higman, 1494), a1v: 'Arithmeticum opus consultissime Germane tuo fratri dicatum est gallicanum iusticiam moderanti quod ut peripatetic probant, omnis iusticia geometrica ratione dispensando utitur aut arithmetica commutando, estque arithmetica tanquam quoddam iusticie speculum ac regula.'

34 Menso Folkerts, 'Regiomontanus' Role in the Transmission and Transformation of Greek Mathematics', in *Tradition, Transmission, Transformation: Proceedings of Two Conferences on Pre-Modern Science Held at the University of Oklahoma*, ed. F. Ragep, Sally Ragep, and Steven John Livesey (Leiden: Brill, 1996), 83–113.

Greek Platonists such as Proclus' *Commentary on the First Book of Euclid*, and unfolded as one long argument for mathematics as the foundation for the whole arts cycle. But by 1501, Lefèvre had already come to similar conclusions, through different sources.

Lefèvre travelled three times to Italy, beginning in the winter of 1491–2. Later he would claim that Giovanni Pico della Mirandola and Ermolao Barbaro had drawn him to Italy, and Symphorien Champier cited Lefèvre's conversations with Pico, Barbaro, and Ficino as the mythic point where Italy's literary glory passed on to France.[35] All three Italians were at the height of their careers. Barbaro was in Rome, a patrician exiled from Venice, and famous for his new translations of the fifth-century philosopher Themistius, who had paraphrased Aristotle in mildly Platonist language (1481).[36] In Florence, Marsilio Ficino had devoted his life to translations and commentaries on Plato, but also had written widely influential works on the role of astrology and natural magic in care of the self. The youthful Pico, also in Florence under the protection of Lorenzo de Medici, had combined Cabbala and biblical exegesis in the *Heptaplus* (1489) and the *De Ente et Uno* (1491). All three men would die before Lefèvre returned to Italy in 1500. From the older Italian generation, Lefèvre did adopt some commonplaces on the value of eloquence in philosophy, and his efforts to present a simplified set of Aristotelian textbooks, corrected against the latest Greek texts, are deeply marked by the desire for direct access to the ancients.[37] Barbaro's Themistian paraphrases gave Lefèvre a new classical precedent for textbooks, and immediately on arriving back in Paris in 1492 Lefèvre published a collection of paraphrases for all of Aristotle's natural philosophy.

War catalysed French use of Italy in the fall of 1494, when Charles VIII entered Italy, claiming ancient dynastic right to Naples, and sparking a series of wars between France and Italy that would last long into the next century.[38] In February 1495, in the commentary on Sacrobosco, Lefèvre cited the ongoing campaign. He noted that the king had sent back relics to be deposited in the great shrine of France, the Cathedral of St Denis just north of Paris. Lefèvre had been there:

> All the best and brightest of the University of Paris were there in reverence: the Rector, theologians, lawyers, physicians, and the philosopher leaders of the various Nations, with the most distinguished of their students. Also there were the important men of city and state, churchmen and civilians. The crowd of people that flowed in from

[35] Jacques Lefèvre d'Étaples, *Decem librorum Moralium Aristotelis tres conuersiones: prima Argyropili Byzantii; secunda Leonardi Aretini; tertia vero Antiqua per Capita et numeros conciliate: communi familiarique commentario ad Argyropilum ad lectio* (Paris: Higman and Hopyl, 1497), f8r: 'cuius viri [Pico] videndi illectus amore et Hermolai Barbari me glorior italiam petiisse'. Symphorien Champier, *Duellum epistolare: Galie & Etalie antiquitates summatim complectens* (Venice, 1519), a4r.

[36] Barbaro's *Castigationes Plinianae* was not printed until 1492 (Rome: Eucharius Silber).

[37] For example, Lefèvre presented his edition of Aristotle's logical *Organon* (1503) with Barbaro's line about drinking only from the purest *fontes*, both on the title page and in prefatory letters. On this phrase see Luca Bianchi, 'Continuity and Change in the Aristotelian Tradition', in *The Cambridge Companion to Renaissance Philosophy*, ed. James Hankins (Cambridge: Cambridge University Press, 2007), 49–71, at 57 and 69 n. 19.

[38] David Potter, *Renaissance France at War: Armies, Culture and Society* (Woodbridge: Boydell Press, 2008), 27–30.

everywhere was so great that it was almost impossible to find a place. And I came humbly through the pressing crowd, to pay reverence. I add this because I can only rejoice that such things should happen in our time—such things happen in only the rarest of times.[39]

Even for the mild-natured Lefèvre, success in Italy stoked a sense that France's time had come. Charles VIII also brought back Fra Giovanni Giocondo, the Franciscan architect who rebuilt a bridge over the Seine and designed several other royal projects, and who lectured at length on Vitruvius—Lefèvre and his friend Guillaume Budé, who also began his Greek studies with George Hermonymus, attended the lectures.[40]

The campaign of 1494 also brought French patrons to Italy. Jean de Ganay accompanied Charles VIII, negotiated with Pope Alexander VI to help the French through Rome to Naples, and presided with the pope over Charles's coronation on 20 May 1495.[41] In the wake of the French conquest, Italian intellectuals joined Parisian scholars clamouring for Ganay patronage. The ageing philosopher of the Medicis, Marsilio Ficino, wrote to Germain before the French armies arrived in Italy, offering to send manuscript copies of his recently printed translations of Plato and Dionysius the Areopagite, as well as his commentary on the *Parmenides*, *Timaeus* and the *Sophisti*. In October of 1494 he reported that the books would be sent in a few days—the *Parmenides* commentary was already copied, and the *Timaeus* was on its way. The next letter offered the wars as an excuse for the late delivery of the books, since 'that disruption of this age intercepted both your letters to me and often mine to you'. Ficino noted he had met Germain's brother Jean with the French king, and he would keep at work on the copies 'as hard as I can between these clashes'.[42] In short order, Ficino also wrote a note to Jean, pleading for peace among philosophers and closing with avowals of devotion: 'I beg you, my noblest Jean, to ardently love your loving Marsilio.'[43] The Ganays became one of Ficino's links to Paris, where his translation of Athenagoras was printed in 1498 along with a dedicatory letter to Germain.[44] Ficino's successor in the Florentine

[39] Lefèvre, *Textus de sphera*, c10v: 'Alme parisiensis academie electissimi quique rector, theologi, iurisperiti, medici, nationum capita philosophi cum suorum studiorum insignibus reverenter affuerunt. Magnifici quoque status et urbis et civitatis, tum ecclesiastici tum civiles. Tantusque omni ex parte affluxit populous ut vix locus capere sufficeret, et nos inter turbam pressi humiliter ad oscula venimus. Hec adiecimus quod talia nostris seculis contigisse non gaudere non possumus, que vel rarissimis obtingere solent temporibus.'

[40] Budé later drew on these lectures in his antiquarian projects: Vladimir Juřen, 'Fra Giovanni Giocondo et le début des études vitruviennes en France', *Rinascimento*, 2nd ser., 14 (1974): 101–15. Lefèvre mentioned Giocondo's lectures on Vitruvius, 'superiore anno' in 1501: Jacques Lefèvre d'Étaples, *Libri logicorum ad archteypos recogniti cum novis ad litteram commentariis ad felices primum Parhisiorum et communiter aliorum studiorum successus in lucem prodeant ferantque litteris opem* (Paris: Wolfgang Hopyl and Henri Estienne, 1503). (This work was published in two stages, the first in 1501.)

[41] For Jean, the campaign was a key step to the highest post in France, Chancellor, which he gained in 1508. On the Ganay brothers, see Rice, 'The Patrons of French Humanism', 691–2; Ernest de Ganay, *Un Chancelier de France sous Louis XII: Jehan de Ganay* (Paris, 1932); and references in PE, 17.

[42] 'tamen inter hos strepitus in Platonicis commentationibus assidue quoad possum'. The letters can be found in Marsilio Ficino, *Opera omnia* (Paris, 1541), 981, 984, 987.

[43] Ficino, *Opera*, 988: 'te obsecro splendidissime mi Ioannes, ut Marsilium ames ardenter amantem'.

[44] Paul Oskar Kristeller, *Studies in Renaissance Thought and Letters* (Rome: Edizioni di storia e letteratura, 1956), 51–4.

Accademia, Francesco da Diacceto, even wrote a letter to Germain later titled 'Apology for Plato against the Philosophers of Paris'.[45]

Beyond a love of eloquence and unfiltered classical texts, Lefèvre found in his new Italian friends some sources for intellectual reform, albeit on his own terms. Lefèvre may have had a hand in the Paris edition of Ficino's *De vita libri tres* (1493?); he certainly added his own commentary to the Paris edition of the Hermetic works that had been translated by Ficino, whom he 'revered like a father' (*tamquam patrem veneratur*). Lefèvre praised the fabled Egyptian prophet, priest, and king Hermes for having shown a *praeparatio evangelica* fully consonant with the teachings of Moses, as Ficino introduced Hermes. Just as importantly, Lefèvre argued, Hermes 'argues that God is intelligible in a rational way to man alone, and God himself should be hunted out from the centre, limits, place, and forms of things'.[46] Such study was the basis for the contemplative life. In his second edition of the work, published in 1505, Lefèvre added further scholia and warnings about straying from Christian orthodoxy, but retained the idea that Hermes could lead one to personal reform. Where the Hermetic author spoke of a *regenerationis mysterium*, Lefèvre exhorted one to 'read this in a Christian manner... remember to thank God the Reformer with all your mental might, completely burning with love for him, having become wholly spirit and reborn by the spirit'.[47]

Like Ficino, Lefèvre believed that the ancient wisdom of Hermes and similar authorities could be the source of a purer philosophy, a theme explored at length by Gentile, D. P. Walker, and Charles Schmitt, among others.[48] But the motivation for such *ressourcement*, it must be stressed, was to recover the most powerful tools to care for Christian souls. It was for this purpose that Ficino turned to Iamblichus and other theurgic Platonists such as Theon of Smyrna in his own popular *De vita libri tres*. For the physicomagus Ficino, care for the soul is not a metaphor; the soul is a matter of both body and mind, and can be controlled and oriented through a regime of medicine, diet, prayer, and intellectual exercises. Michael Allen has explained the remarkable power of mathematics in Ficino's psychology, culminating in a Pythagorean-inspired account of the soul as, quite literally, a plane figure which mediates points and bodies. In sum, manipulation of the soul and even the

[45] In another letter to Germain, Francesco da Diacceto tried to reconcile Aristotle and Plato on space. Kristeller, *Studies in Renaissance Thought*, 314, 316.

[46] Jacques Lefèvre d'Étaples (ed.), *Mercurij Trismegisti Liber de potestate et sapientia dei per Marsilium Ficinum traductus ad Cosmum Medicem*, trans. Marsilio Ficino (Paris: Hopyl, 1494), e4r: 'Secundus disserit deum rationali modo soli homini intelligiblem esse, et ex centro, polo, loco et rerum formis, venatur ipsum deum.'

[47] Jacques Lefèvre d'Étaples (ed.), *Pimander. Mercurii Trismegisti liber de sapientia et potestate dei. Asclepius. Eiusdem Mercurii liber de voluntate divina. Item Crater Hermetis A Lazarelo Semptempedano*, trans. Marsilio Ficino (Paris: Henri Estienne, 1505), 36r: 'Deo reformatori toto mentis vigore gratias agere memento, totus eius amore flagra, totus spiritus factus et a spiritu regenitus.'

[48] Lefèvre's promotion of the *prisca theologia* is well known: D. P. Walker, 'The Prisca Theologia in France', *Journal of the Warburg and Courtauld Institutes* 17, no. 3/4 (1954): 204–59; Charles B. Schmitt, 'Perennial Philosophy: From Agostino Steuco to Leibniz', *Journal of the History of Ideas* 27, no. 4 (1966): 505–32; Brian P. Copenhaver (ed.), *Hermetica: The Greek Corpus Hermeticum and the Latin Asclepius in a New English Translation, with Notes and Introduction* (Cambridge: Cambridge University Press, 1992). More bibliography can be found in Dmitri Levitin, *Ancient Wisdom in the Age of the New Science: Histories of Philosophy in England, c.1640–1700* (Cambridge: Cambridge University Press, 2015).

world is a mathematical task. While not all enthusiasts of the ancient wisdom came to this conclusion, their work is replete with numerology—Pico's effort to use especially Cabbala to reinterpret Genesis 1–2 in the *Heptaplus*, for example, repeatedly returns to Pythagorean statements about the Monad and the Dyad in order to unwrap the levels of meaning in the biblical text.[49]

Lefèvre was haunted by the hope for a universal language or code to unlock the hidden springs of nature, to make the crucial shift from theory to practice. In mathematical prefaces, he alluded to the use of numerology to operate on the deep structures of the world and to unveil hidden meanings of scriptures—as late as 1516, Lefèvre still held out *litterae mathematicae* as the key to philosophy and theology: 'this style of philosophizing was the most ancient of all, even before Pythagoras, Plato, and Aristotle, so that it should be seen as all the more worthwhile on account of its antiquity'.[50] This reflects his early preoccupations during what I have called his mathematical turn, when he explicitly reworked Ficino's *De vita libri tres* and Pico's Cabbala in his own fashion, in the Pythagorean magic of the *De magia naturali*.[51] Around the same time that he was thus exploring how numeric harmonies might govern the soul, he also wrote his treatise *Elementa musicalia*, prefacing it with a concatenation of authors from Orpheus to Pythagoras who demonstrated how music has the power to excite and calm. As for Ficino, this is no metaphor; Lefèvre was convinced that such proportions (or analogies) held a mysterious—but ultimately rational—control over mind and body. These themes therefore had practical application as part of moral philosophy. As an example of how this might work Lefèvre used Pico in his commentary on Aristotle's *Ethics*, on the subject of recreation and gentle comportment. When overwhelmed by study, Pico 'used to sing by himself with a lute, and once he had so modulated his mind he dried his tears, greatly consoled'.[52] Lefèvre then supplied one of the hymns which Pico had sung to refresh and rebalance his soul.

When Lefèvre turned away from Ficinian magic and Pican Cabbala around 1504, it was not for sceptical reasons, but precisely because magic was so powerful—and, if done with the wrong daemons, dangerously idolatrous.[53] In any event, a certain

[49] Crofton Black, *Pico's Heptaplus and Biblical Hermeneutics* (Leiden: Brill, 2006).

[50] Jacques Lefèvre d'Étaples, *Euclidis Megarensis mathematici clarissimi Elementorum geometricorum libri xv, Campani Galli transalpini in eosdem commentariorum libri xv, Theonis Alexandrini Bartholamaeo Zamberto Veneto interprete, in tredecim priores, commentariorum libri xiii, Hypsiclis Alexandrini in duos posteriores, eodem Bartholomaeo Zamberto Veneto interprete, commentariorum libri ii* (Paris: Henri Estienne, 1517), a1v: 'Et hic philosophandi modus vetustissimus fuit, ante etiam Pythagoram, Platonem, et Aristotelem, ut vel ex antiquitate cognoscatur augustior.'

[51] Brian P. Copenhaver, 'Lefèvre d'Étaples, Symphorien Champier, and the Secret Names of God', *Journal of the Warburg and Courtauld Institutes* 40 (1977): 189–211; Mandosio, 'Le *De magia naturali*'; Pierozzi and Mandosio, 'L'interprétation alchimique'.

[52] Lefèvre, *Decem librorum Moralium Aristotelis*, f7v-8r, at f7v: 'studio defessus hunc hymnum et similes, solitarius ad citharam canere solebat, et mente pariter emodulatus magno in solacis lachrymas liquabat'. When Lefèvre met Pico in the early 1490s, Pico give him a manuscript of Bessarion's translation of Aristotle's *Metaphysics*, which Lefèvre published in 1515 (PE, 354–8).

[53] It can be no accident Lefèvre denounced the vanity of magic in an edition of patristic works which included the Pseudo-Clementine *Recognitiones Petri*, which presented itself as a disputation between Saint Peter and the the wicked Simon Magus. See Jacques Lefèvre d'Étaples (ed.), *Pro piorum recreatione et in hoc opere contenta, etc.* (Paris: Guy Marchant for Jean Petit, 1504), A1v (PE, 117–20).

wariness about magical numbers did not negate its cognitive value. Writing to Jean de Ganay just before the invasion, Lefèvre insisted that practical benefits only came *after* one mastered the 'wisdom of numbers', or arithmetic. Indeed, 'once, the whole ancient theology depended on numbers as a certain kind of step to divine matters—even now numbers in sacred letters [i.e. the Bible] keep their mystery.' Such contemplative approaches had profound practical implications, Lefèvre hinted: 'take away numbers and the study of numbers, and laws will lay incomplete, justice will stay blind, no rules will be found to measure things out, there will be no entry to heavenly contemplation, the mysteries of sacred letters will stay hidden—indeed, the whole of philosophy, the understanding of human and divine things, will be constricted'.[54] Without arithmetic, catastrophe! Lefèvre drew together commonplaces from the Platonic tradition to fend off such catastrophe. Pythagoras had said no one could know anything without numbers; Plato had placed over his own school the saying that 'no one who lacks mathematics may enter here'.[55] Indeed, Plato had praised the value of numbers throughout the *Timaeus*, 'concerning the nature of things', as in books 8 and 9 of the *Republic*. Even there, Lefèvre pointed out, work remained to be done, since even the ancient Platonic mathematician and commentator Theon of Smyrna had explained only a little. Numbers needed attention. And the recovery should begin in Paris. 'Thus you were clearly right to find that our great crowd of philosophizers at this University of Paris, fond of good letters, has abandoned the path so very necessary for first rising to divine matters and then descending to human affairs.'[56]

NORTHERN CONVERSIONS

Italian humanism offered some sources for a turn to mathematics. Giorgio Valla and humanists of his kind stressed that mathematics was best for sharpening wits, an essential first step to philosophical or theological learning. But Lefèvre's turn

This suggests Lefèvre's worries about magic had to do with its impiety, not a sudden conversion to modern naturalism, for Cabbalistic themes endure in Lefèvre's commentary on the Psalms (1509). See Eugene F. Rice, 'The *De Magia Naturali* of Jacques Lefèvre d'Étaples', in *Philosophy and Humanism: Renaissance Essays in Honor of Paul Oskar Kristeller*, ed. Edward P. Mahoney (Leiden: Brill, 1976), 19–29; Copenhaver, 'Lefèvre, Champier, and the Secret Names', 196–201; Jan R. Veenstra, 'Jacques Lefèvre d'Étaples: Humanism and Hermeticism in the De Magia Naturali', in *Christian Humanism: Essays in Honour of Arjo Vanderjagt*, ed. Arie Johan Vanderjagt, Alasdair A. MacDonald, and Z. R. W. M. von Martels (Leiden: Brill, 2009), 353–62.

[54] Lefèvre, *Elementa*, a1v (PE, 18): 'et prisca Theologia numeris olim ut quibusdam ad divina gradibus tota innitebantur. Tolle igitur numerous numerorumque disciplinam, leges imperfecis, iusticia ceca relinquitur, nulla modulationum reperietur regula, nullus celestium contemplationum aditus, sacrarum litterarum delitebunt mysteria—immo et universa philosophia quam pariter humanorum divinorumque cognitio describitur'.

[55] In fact, Lefèvre seems to have drawn this from memory, saying 'Nemo huc mathematice expers introeat'. Most sources cite Plato as saying 'Nemo huc *geometriae* expers ingrediatur', so Lefèvre here modifies his source.

[56] Lefèvre, *Elementa*, a1v (PE, 18): 'Quapropter patet non ab re delibas hanc numerosam huius almi parisiensi studii philosophantium turbam et bonarum litterarum cupidam, tam necessaria semita tum ad divina assurgendi tum descendendi ad humana esse destitutam.'

was no simple transplantation of Italian humanism into Paris soil. Rather, Lefèvre selectively grafted shoots onto the deeply rooted stock of the hoary old university, nourished by Northern reform movements.[57] Medieval reform overlapped with conversion, a 'turning towards' a more serious form of religious life.[58] For many medieval Christians this inner life-turn entailed entering a monastic form of life, but in the fifteenth century a conversion could involve other options too.[59] For Lefèvre, the university became such an option.

The University of Paris had long been the target of reform movements seeking to keep universities accountable to society's needs. In the shadow of the Great Schism that had divided Europe between three popes, Gerson urged arts masters to go back to basics, following the example of thirteenth-century masters such as Albert the Great as taught by the arts master Jean de Maisonneuve (Nova Domus) and his students.[60] Even among the University's political allies it was an ongoing concern that academic curiosity could hurt the common weal. In 1452 the papal visitor Cardinal d'Estouteville updated the University's statutes to deal with the growing population of students, recovering after a low ebb in the late Hundred Years War (1430s), emphasizing the responsibility of students to maintain modest lifestyles and serve their dioceses.[61] The university often turned to the crown to keep the Paris *parlement* at arms length, which also made it vulnerable to the king's own version of reform: in 1471 King Louis XI expelled the foreign students who would not swear an oath of fealty to him; in 1472 he outlawed the teaching of nominalism—by all accounts, ineffectual, but nevertheless a symbolic assertion of royal prerogative.[62] Meanwhile, the university was also threatened by reform

[57] Studies of Lefèvre's circle have often focused on these movements: besides Augustin Renaudet, *Préréforme et humanisme à Paris pendant les premières guerres d'Italie, 1494–1517* (1916; 2nd edn, Paris: Édouard Champion, 1953), see especially Yelena [Mazour-]Matusevich, 'Jean Gerson, Nicholas of Cusa and Lefèvre d'Étaples: The Continuity of Ideas', in *Nicholas of Cusa and His Age: Intellect and Spirituality: Essays Dedicated to the Memory of F. Edward Cranz*, ed. Thomas P. McTiche and Charles Trinkaus (Leiden: Brill, 2002), 237–63; Yelena Mazour-Matusevich, *Le siècle d'or de la mystique française: étude de la littérature spirituelle de Jean Gerson (1363–1429) à Jacques Lefèvre d'Etaples (1450?–1537)* (Paris: Arche, 2004).

[58] Gerhart B. Ladner, *The Idea of Reform: Its Impact on Christian Thought and Action in the Age of the Fathers* (New York: Harper Torchbooks, 1959). For more recent scholarship, see Spencer E. Young, 'Faith, Favour, and Fervour: Emotions and Conversion among the Early Dominicans', *Journal of Religious History* 39, no. 4 (2015): 468–83, at 472.

[59] John Van Engen, 'Multiple Options: The World of the Fifteenth-Century Church', *Church History* 77, no. 2 (2008): 257–84; John Van Engen, *Sisters and Brothers of the Common Life: The Devotio Moderna and the World of the Later Middle Ages* (Philadelphia, PA: University of Pennsylvania Press, 2008), 14–19.

[60] Gerson is presented as the central figure in the rise of Albertism at Paris by Zenon Kaluza, *Les querelles doctrinales à Paris: Nominalistes et realistes aux confins du XIVe et du XVe siècles* (Bergamo: Lierluigi Lubrina, 1988). Whether or not Gerson was entirely responsible, it is clear that around 1,400 academics more and more identified with schools or *viae*, often placing themselves in historical traditions reaching back to a thirteenth-century figure: Maarten J. F. M. Hoenen, 'Via Antiqua and Via Moderna in the Fifteenth Century: Doctrinal, Institutional, and Church Political Factors in the *Wegestreit*', in *The Medieval Heritage in Early Modern Metaphysics and Modal Theory, 1400–1700*, ed. Russell L. Friedman and Lauge O. Nielsen (Dordrecht: Kluwer, 2003), 9–36.

[61] On this pattern, see Jacques Verger, 'Landmarks for a History of the University of Paris at the Time of Jean Standonck', *History of Universities* 22, no. 2 (2007): 1–13.

[62] Verger, 'Landmarks', 6.

from within. Most college statutes set out hours for prayer, silence, and fraternal accountability, much like monastic communities; d'Estouteville's reforms newly emphasized these responsibilities. As principal of the Collège de Montaigu from 1483, Jan Standonck rebuilt the college's ruined buildings and instated monastic discipline to combat moral decay.[63] By 1500 these reforms had helped it outstrip the College de la Sorbonne as the leading institution of theological study.

Lefèvre himself nearly entered monastic orders in 1491, the same year he visited Italy. The spark was a book by the thirteenth-century Catalan mystic and philosopher of combinatorics Ramon Lull. When a friend from the Pyrenees held out Lull's *Liber de contemplatione*, Lefèvre was 'seized by a desire to read the book'. Indeed, it 'brought me great consolation', to the point that 'it nearly brought me to the point of leaving behind the world to seek God in solitude'. In a dizzy enthusiasm, Lefèvre told all his friends in Paris. He confessed: 'what I had conceived bothered them a great deal'.[64] Lefèvre fell ill from insomnia, and physicians warned him from entering an overly ascetic life.

In the footsteps of Lull, Lefèvre came to see university reform as an alternative to cloistered conversion. Not entirely certain of Lefèvre's new-found passion and concerned for his fragile health, friends dissuaded him from taking vows. 'For this reason', he wrote, 'I turned back to my earlier arts, in order to assist the prayers of holy men. And since I could not do this in solitude, I freely give help by publishing books which form souls in piety.'[65] Lull himself had provided a model of such a conversion or turn towards deeper piety in his own *Vita*, later published by Charles de Bovelles. Lull had turned away from his life as a successful merchant and, after illumination in prayer, decided to travel to Paris, where he learned grammar and began to write innumerable works.[66] Like Lull, Lefèvre would pursue a life of conversion through the *artes* and the *emissioni librorum*, the liberal disciplines and the publishing of books.

Lefèvre and his circle rehabilitated Lull at Paris, printing several important collections of his works.[67] They tracked manuscripts of Lull's works through the

[63] See the analysis of Montaigu's statutes of 1499, 1503, and 1509 in Paul J. J. M. Bakker, 'The Statutes of the Collège du Montaigu: Prelude to a Future Edition', *History of Universities* 22, no. 2 (2007): 76–111.

[64] Ramon Lull, *Contenta. Primum volumen Contemplationum Remundi duos libros continens. Libellus Blaquerne de amico et amato*, ed. Jacques Lefèvre d'Étaples (Paris: Guy Marchant for Jean Petit, 1505), a1v (PE 141–2): 'plurimam mihi attulit consolationem... et paene ad hoc pertraxit, ut demisso mundo Deum in solitudine quaererem... verum propositum quod conceperam (ut accidere solet) interturbaverunt quam plurima.'

[65] Lull, *Contenta*, a1v (PE ep. 45, 141–2): 'Quapropter ad priores artes revolutus ad emerendas sanctorum virorum preces, cum id in solitudine non possim, libenter emissioni librorum (qui ad pietatem formant animos) operam do.'

[66] Charles de Bovelles, *Commentarius in primordiale evangelium divi Joannis; Vita Remundi eremitae; Philosophicae et historicae aliquot epistolae* (Paris: Josse Bade, 1514), 35r–v. Bovelles completed this *Vita Remundi* at Amiens on 27 June 1511. A discussion of possible sources is given by Joseph M. Victor, 'The Revival of Lullism at Paris, 1499–1516', *Renaissance Quarterly* 28, no. 4 (1975): 521–2.

[67] Eugene F. Rice Jr., 'Jacques Lefevre d'Etaples and the Medieval Christian Mystics', in *Florilegium Historiale: Essays Presented to Wallace K. Ferguson*, ed. J. G. Rowe and W. H. Stockdale (Toronto: University of Toronto Press, 1971), 90–124; Victor, 'The Revival of Lullism'. Note, however, that they never published any version of the Lullian *Ars*, presumably because it was already available in print.

libraries of the College de la Sorbonne, the monasteries just outside Paris, of the abbey of St Victor and the Carthusian monastery of Vauvert—Beatus Rhenanus listed works he found in two of five volumes chained in the library of the Sorbonne, which included an earlier *Vita Remundi*.[68] In 1505 Lefèvre and some students, including Beatus, published the first volume of Lull's enormous *Contemplationes*, which he had found in Padua in 1500. Lefèvre intended the book to replace a volume lost by the Carthusians at Vauvert. Yet even though Lull had written works such as the *Nova geometria* (1299), it was Lull's practical value in soul-craft that caught Lefèvre's attention. As he wrote in 1499, 'there are two things which most properly form our lives: the understanding of universals (which the moral disciplines prepare for) and the method of doing.'[69] He presented Lull's art as one of doing, useful for ordering one's actions in piety. Lull, whom he saw as a Pauline fool—'let no one hold you back because that man was an unlearned layman'—provided a model for setting one's soul in order.[70] Lull's work thus represented a tension riven through Lefèvre's work from beginning to end: does intellectual reform begin with heart and habits, or with the intellect?[71]

To a Parisian intellectual in search of models for setting souls in order, the early Victorines proved attractive sources. Their cloister and rich library stood just outside the Port de Sainte Victoire, the nearest gate to the Collège du Cardinal Lemoine. Twelfth-century priors of this community of Augustinian canons, Hugh and Richard of St Victor, had played a leading role in the early *studia* in the nascent university at Paris. They remained popular among mystical and monastic thinkers into the fifteenth century.[72] In 1506 Clichtove published a set of Hugh's pedagogical works, which included advice to novices and a section of his famous *Didascalicon* which described how the linguistic and mathematical arts help not only read the scriptures but also the book of nature.[73] In general, the Victorines offered an unimpeachably Parisian and devout genealogy—and the Victorines, Clichtove pointed

[68] Lefèvre mentions these as repositories of Lull's writings in Jacques Lefèvre d'Étaples (ed.), *Proverbia Raemundi. Philosophia amoris eiusdem* (Paris: Josse Bade, 1516), a2r. Beatus Rhenanus listed the works of Lull, *cathenis ligata*, in the Sorbonne in his copy of Nicolaus Cusanus, *Opuscula theologica et mathematica* (Strassburg: Martin Flasch, 1488), BHS K 951, flyleaf.

[69] Jacques Lefèvre d'Étaples (ed.), *Hic continentur libri Remundi pij eremite* (Paris: Petit, 1499), a1v (PE 76): 'Cum duo sint quae vitam nostrum rectissime instituunt, universalium scilicet cognition (quam morales disciplinae pariunt) et operandi modus.'

[70] Lefèvre, *Hic continentur*, a1r (PE 77): 'neque vos quicquam deterreat quod vir ille idiota fuerit et illiteratus'.

[71] A related tension concerned education itself, and might be put this way: intellectual reform requires a well-ordered mind, and is not guaranteed by many books; yet books are the readiest road to wisdom.

[72] Giles Constable, 'The Popularity of Twelfth-Century Spiritual Writers in the Late Middle Ages', in *Renaissance: Studies in Honor of Hans Baron*, ed. Anthony Molho and John A. Tedeschi (Florence: G.C. Sansoni, 1970), 3–28; Giles Constable, 'Twelfth-Century Spirituality and the Late Middle Ages', in *Medieval and Renaissance Studies* 5, ed. O. B. Hardison Jr. (Chapel Hill, NC: University of North Carolina Press, 1971), 27–60.

[73] Josse Clichtove (ed.), *Opera [Hugonis de Sancto Victore]: De institutione novitiorum. De operibus trium dierum. De arra anime. De laude charitatis. De modo orandi. Duplex exposito orationis dominice. De quinque septenis. De septem donis Spiritus Sancti* (Paris: Henri Estienne, 1506), 24r–v (on the book of nature). The following year Clichtove also published a work by Hughes de Fouilloy, *De claustro animae* (Paris: Henri Estienne, 1507), which he ascribed to Hugh of St Victor. See PE 173–4. Josse

out, could support an Aristotelian view of education: 'In the *Ethics* Aristotle rightly judges that words are no little inducement to spur men—and especially youths born with quick wits—to virtue. For those things which, when they enter the mind, swiftly excite the lazy boy and drive him to action.'[74] Lefèvre found the work of the Victorines especially amenable to his mathematical interests, for in 1510 he published the great mystical treatise of Richard of St Victor *De Trinitate* together with his own long commentary, in which he reorganized Richard's arguments according to the structure of Euclid's *Elements*, in common notions, axioms, theorems, and proofs.[75] He argued for this axiomatic approach in the preface, where he explained that philosophy proceeds according to the three levels of the soul: imagination, reason, and intellect. Imagination deals only with the senses; at the other end, intellect goes beyond sense and even reason to pure ideas—here lie mathematical notions, like God, real yet devoid of the physical. The Victorines presented a way to perfect the soul by ordering the arts.

By 1508, Lefèvre had even found resources for arithmetical reading of scriptures in a contemporary of the Victorines, Odo of Morimond (d.*c.*1200).[76] Beatus Rhenanus wrote to Michael Hummelberg on 15 May 1508 responding to recent news from their old teacher, just retired from Lemoine. 'But I heard that you had most pleasingly corrected the *Analycen numerarium* which draws pious minds up to the higher understanding of sacred things, which (as you write) Lefèvre then rendered intelligible with commentary. This will be an especially useful work. For you know how many mysteries lie hidden in numbers. NUMBER is a wonderful thing! God, the maker of this worldly mass, had number as exemplar of his thought (as Boethius says) and decided to arrange everything by it, and to discover a harmony in numbers arranged in order in every constructive plan. There are many things about the value of numbers in Cusanus.'[77] Beatus and Hummelberg refer to a project on Odo's twelfth-century exegetical manual on numerological meanings in

Clichtove (ed.), *Hugonis de Sancto Victore Allegoriarum in utrunque testamentum libri decem* (Paris: Henri Estienne, 1517).

[74] Clichtove, preface to *Opera [Hugonis]*, a1v: 'In Ethicis censet Aristoteles et recte quidem, amplissime praesul [Jacques d'Amboise, bishop of Clermont], verba provocandis ad virtutem hominibus praesertim liberis et qui dexteriore nati sunt ingenio hauc parum conducere, quod eau bi animum subierint torpentem ocius excitant ad agendumque permovent.'

[75] Jacques Lefèvre d'Étaples (ed.), Richard of St Victor, *De superdivina Trinitate theologicum opus hexade librorum distinctum. Commentarius artificio analytico* (Paris: Henri Estienne, 1510).

[76] Rice indicated that Lefèvre had read Odo's works around the time he wrote the *De magia naturali*: Rice, 'The *De Magia Naturali*', 26. However, I have found no mention of Odo before this letter of Beatus Rhenanus.

[77] James Hirstein (ed.), *Epistulae Beati Rhenani: La Correspondance latine et grecque de Beatus Rhenanus de Sélestat. Edition critique raisonnée avec traduction et commentaire, Volume 1 (1506–1517)* (Turnhout: Brepols, 2013), ep. 4, 24–32: 'Analyticen vero numerariam ad altiorem sacrorum intelligentiam pias mentes evehentem tua recognition elimari lubentissime audivi, quam (ut scribis) suis mox scholiis Faber reddet clariorem. Egregium id opus erit. Non enim ignores, quanta in numeris lateant mysteria. Admiranda res est NUMERUS. Numerum huius mundanae molis conditor Deus (ut divus Boethius ait) primum suae habuit ratiocinationis exemplar et ad hunc cuncta constituit, quaecunque fabricante ratione per numeros assignati ordinis invenere concordiam. De numeri dignitate plurima apud Cusanum.' See also PE, 382–3 and cf. Alfred Hartmann (ed.), *Die Amerbachkorrespondenz* (Basel: Verlag der Universitätsbibliothek, 1942), 1:372.

the Bible, which never was printed. In Lefèvre's circle, interest in numbers did not diverge from philological or exegetical projects, but explicitly complemented them.

In his near-conversion on reading Lull's *Contemplationes* in 1491, Lefèvre had found advice among Paris university reformers steeped in the schooling of a lay movement in the Rhineland: the *devotio moderna*.[78] The movement itself originated in critique of university life, when its founder Geert Grote of Deventer (b.1340), a successful master at Paris, set himself a series of resolutions in 1374. He would seek no more benefices, cast no horoscopes, burn his 'magical books', and not pursue lucrative arts such as medicine, astrology, or law.[79] Instead, he would only read Scripture, and the Church Fathers, and preach. Master Geert's last decade of life reflected a conversion from the university lectures to devotional reading. This wholesale rejection of the university (though not of learning) softened as later generations of the *devotio moderna* pursued these ideals as teachers in grammar schools and even within the University of Paris. When friends dissuaded Lefèvre from the monastery, he requested advice from the Parisian figures who remoulded university life around monastic disciplines: Jean Mombaer, Jean Raulin, Philippe Bourgoing, and Jan Standonck. All of these figures were, like Lefèvre, members of the Picard nation and linked in one way or another with the *devotio moderna*, which ran schools and directed lay men and women who hungered for a deeper spiritual life outside the cloister. Raulin and Standonck had both been first educated in the *devotio moderna*'s schools in Deventer, and now headed the most visible movements of college reform in Paris. Raulin (d.1501) was the master of the Collège de Navarre from 1481 to 1497, when he entered a monastery himself; Jan Standonck (d.1504) entered the Collège de Montaigu in 1476, becoming master in 1483.

Lefèvre helped print two works that relate the movement's occupation with relating inner soul craft to outer life form: Jean Mombaer's *Rosetum Exercitiorum Spiritualium* (Josse Bade, 1510) and Jan Ruusbroec's *De ornatu spiritualium nuptalium* (Henri Estienne, 1512). Both works provided tools for directing the soul as a way to link the contemplative and active lives. Ruusbroec helped the hungry soul to meet Christ within, in a progression from active to contemplative exercises constructed around the command in the Parable of the Ten Virgins: 'look, the Bridegroom comes: go out to meet him' (Matthew 25:6). Lefèvre stressed the book's usefulness, 'not because books make those who contemplate especially perfect, but because they prepare and encourage those setting out on this path', of seeking spiritual perfection.[80] Mombaer offered preachers and priors alike an encyclopaedia for setting out a reading plan, for conducting the offices and a schedule of preaching, and exercises for properly observing the eucharist. But Mombaer began with very concrete advice for differentiating between different

[78] John Van Engen, *Sisters and Brothers*.

[79] John Van Engen (ed.), *Devotio Moderna: Basic Writings* (New York: Paulist Press, 1988), 63–75.

[80] Jacques Lefèvre d'Étaples (ed.), *Devoti et venerabilis patris Ioannis Rusberi presbyteri, canonici observantiae beati Augustini, de ornatu spiritualium nuptiarum libri tres* (Paris: Estienne, 1512), a1v (PE, 277): 'non quod libri contemplativos facant maxime profectiores, sed praeparat et incitant incipientes; deus autem caetera complet'.

talents: 'The different qualities of those who do the exercises are like the various dispositions of minds.'[81] His examples included Gregory the Great, Anaxagoras, Carneades, Archimedes, and—among those writing *in proximis diebus*—Dennis the Carthusian and Ruusbroec himself. By reflecting on how their dispositions also shaped their eating habits, Mombaer underlined the physicality of these intellectual and spiritual temperaments. In fact, Mombaer's concern for the physical aspects of contemplative work permeates the *Rosetum*, for he draws on the monastic memory tradition to describe the visual learning, and teaches memory techniques such as a 'Chiropsalterium', to master the Psalms by mapping them onto the hand.

Gerson became an authority in fifteenth-century efforts to link inner and outer conversion, both within and without the university. For example, the *Rosetum* quoted his line that learning in the religious life is different from the universities, for 'the school of our religion is not the school of theological or philosophical speculation, but instead of simple Christian learning and devotion'. For this reason, he wrote, learning is 'not for illuminating the intellect but for inflaming the affects'.[82] Lefèvre presented Gerson as a combination of pastoral love and intellect, 'a devout man of weighty authority, cloaking a religious mind under a secular habit'.[83] Lefèvre wrote this in his preface to Ruusbroec, where he tried to mollify readers who knew that Gerson had once censored Ruusbroec for presenting mystical union with God in insufficiently qualified terms of identity.[84] Brushing aside differences, Lefèvre argued that Gerson simply had not read Ruusbroec's Latin. If he had, he would have had a different opinion. For 'grammarians who read this will judge its author to be especially elegant for that age, rhetoricians will find him copious, philosophers will find him skilled in the secrets of nature, astrologers will find him one who knows the times, physicians will find he knows sickness and health, and theologians will find him knowledgeable in divine affairs.'[85] Lefèvre turned Ruusbroec into an emblem of how devout learning should bring together practical and speculative knowledge, in precisely the manner he saw Gerson act as university chancellor.

There remains incongruity in Lefèvre's description of Gerson. To be sure, they shared central priorities. Both hoped to rid the university of useless speculation; both advocated personal reform in the university as a means to social reform in general; both revered Dionysius the Areopagite as Saint Paul's student in philosophy

[81] Jean Mombaer, *Rosetum exercitiorum spiritualium* (Paris: Josse Bade for Jean Petit and Jean Scabelarius, 1510), a1r.

[82] Mombaer, *Rosetum*, a3r: 'schola nostre religionis non est schola theologice vel philosophice speculationis, sed christiane simplicitatis discipline et devotionis; non venimus ad illuminandum intellectum sed ad inflammandum affectu.'

[83] Lefèvre (ed.), *De ornatu*, A2r (PE, 277): 'Gerso, vir devotus et auctoritate gravi, sub saeculari habitu mentem religiosam gerens'.

[84] On Gerson's critique, see Bernard McGinn, *The Varieties of Vernacular Mysticism (1350–1550)* (New York: Crossroad, 2012), 86–95.

[85] Lefèvre (ed.), *De ornatu*, A2r (PE, 277): 'Et grammatici qui hunc legunt iudicabunt auctorem [Ruusbroec] pro illa tempestate apprime elegantem, rhetores copiosum, philosophi secreta naturae callentem, astrologi cognitorem temporum, medici morborum ac sanitatum, theologi rerum divinarum.'

and the first bishop of Paris, making the French patron saint the starting point in their own programmes. And both strove to help youths order the various levels of their souls. But Lefèvre and Gerson focused on different parts of the soul. At the end of his sermon *Against Vain Curiosity* and at the beginning of his bestselling *Mystical Theology*, Gerson argued that change begins in a posture of repentance rooted in the affects. Since the Fall into sin blunted man's cognitive powers, the affective mode (*experientalis*) is a purer way to knowledge of God in this life—and by following the highest affective power of the soul, *synderesis*, experience can give content to the intellectual powers. Thus the affective powers enable the intellective powers to achieve a *cognitio dei experimentalis*.[86] Put imprecisely, Gerson believed the heart (*affectus*) has the power to reform the head (*intellectus*).

In contrast to Gerson (and like Lull), Lefèvre focused on the operations of the intellect. In 1501 he divided ways of doing philosophy into two: rational and intellectual, conforming to the soul's powers of reason and intellect. The intellect here emerges as a power[87] that can see clearly what reason sees in confused and shadowy ways; rational philosophy processes the many inputs of the senses, while intellectual reasoning transcends the oppositions of the many in the simple unity of truth; 'therefore rational philosophy seems better understood to man, even though it is far lower and submissive to intellectual philosophy'.[88] In 1510, Lefèvre expanded his twofold division to fit three powers of the soul, matching them to modes of theology: imagination, reason, and intellect. While imagination dwells on the sensory experience where all understanding begins, to remain stuck in imagination would be idolatry. 'Therefore God', he wrote, 'accepts these two modes of living in the contemplative life within us; but he damns and repudiates the third.'[89] At the next level are the logical tools of rational inquiry, which, if misused, can tempt one back to errors, the 'inane little fantasies' and 'ravings unworthy of thick barbarians' of imagination—here Lefèvre explicitly warns against logical games in the manner of the fourteenth-century followers of Bradwardine and Richard Swineshead.[90] The goal of contemplative reasoning therefore is instead the simple, intellectual mode of understanding, found in the 'humble style' of Richard of St Victor's *De trinitate*.[91]

Lefèvre time and again offered mathematics as a tool for sharpening the intellect in service of this intellectual philosophy. Here lay a deep historical irony, for even

[86] This sketch is based on Steven E. Ozment, *Homo Spiritualis: A Comparative Study of the Anthropology of Johannes Tauler, Jean Gerson and Martin Luther* (Leiden: Brill, 1969), 59–83.

[87] Lefèvre describes Aristotle's psychology in terms of its powers or various *vires*, not rather than the 'parts' of Albertist faculty psychology. See his account of *De anima* in Lefèvre, *Totius Aristotelis philosophiae naturalis paraphrases*.

[88] Lefèvre, preface to Charles de Bovelles, *In artem oppositorum introductio* (Paris: Wolfgang Hopyl, 1501), a1v (PE, 95): 'Non ab re igitur videatur rationalis philosophia homini cognatior, tametsi longe sit intelletuali submissior atque deiectior.'

[89] Lefèvre (ed.), *De superdivina Trinitate*, 2r (PE 225). On these tropes, see further the section 'Friendship and Physics' in Chapter 6.

[90] Lefèvre (ed.), *De superdivina Trinitate*, 2v (PE, 226–7): 'inanes phantasiolas'; 'indigna crassae barbariei deliramenta'.

[91] Lefèvre (ed.), *De superdivina Trinitate*, 2v (PE, 226): 'quandoquidem egregiam et Deo signam sub humilitate stili complectitur'.

though Gerson had played a central role in the demise of fourteenth-century mathematics as a logical tool, Gerson supported a party of Albertists who proposed another form of mathematical speculation. Gerson supported the *tritum iter* of the Albertists such as the Flemish master Jean de Maisonneuve, who during the next decades dominated the philosophy of the arts faculty at Paris.[92] When his students migrated especially to the young university of Cologne in the next few decades, they took with them a distinctive account of the soul, in which the highest forms of understanding occur without images (*phantasmata*), transcending the unreliable imagination or fantasy.[93] In efforts to enunciate this position (which echoed Platonic sources such as Proclus and Dionysius), such Albertists displayed a pronounced interest in mathematics. Heymeric de Campo, a Flemish master of the Albertist *bursa laurentiana* at the University of Cologne, listed one hundred and one kinds of theological reasoning: prominent on his list are 'Pythagorean', 'Geometrical', and 'Lullian' forms of theology.[94] Throughout these works, Heymeric found mathematics useful precisely because it offered tools for helping the soul to figure God. Geometrical theology, in particular, was a useful resource because geometry yielded abstract images such as triangles and circles—such objects gave the mind models for intellectually 'conjecturing' the most abstract and non-material objects of all, the attributes of God. Heymeric cited as sources the twelfth-century philosopher Alan of Lille as well as Ramon Lull. None of this would have pleased Gerson.

Lefèvre's most prized exemplar of all was one of Heymeric's pupils at Cologne, Nicholas of Cusa.[95] Like Gerson, Cusa found his chief inspiration in Dionysius the Areopagite; but, unlike Gerson, he was convinced that devotional reform could begin with the intellect rather than the affects.[96] Cusanus did have some connections to Paris, probably reading Ramon Lull's works in the library of the Collège

[92] Gilles Gérard Meersseman, *Geschichte des Albertismus*, vol. 1: *Die Pariser Anfänge des Kölner Albertismus* (Rome: R. Haloua, 1933); Zenon Kaluza, 'Les débuts de l'albertisme tardif (Paris et Cologne)', in *Albertus Magnus und der Albertismus: Deutsch philosphische Kultur des Mittelalters*, ed. Maarten J. F. M. Hoenen and Alain de Libera (Leiden: Brill, 1995), 207–95.

[93] Katharine Park, 'Albert's Influence on Late Medieval Psychology', in *Albertus Magnus and the Natural Sciences*, ed. James A. Weisheipl (Toronto: Pontifical Institute of Mediaeval Studies, 1980), 522–35.

[94] Ruedi Imbach, 'Das *Centheologicon* des Heymericus de Campo und die darin enthalten Cusanus-Reminiszenzen', *Traditio* 39 (1983): 466–77. Hemeric's account of the *Theologia geometrica* is transcribed on 475.

[95] On Cusanus as a chief inspiration in the circle, see Emmanuel Faye, 'Nicolas de Cues et Charles de Bovelles dans le manuscrit "Exigua pluvia" de Beatus Rhenanus', *Archives d'histoire doctrinale et littéraire du Moyen Âge* 65 (1998), 415–50; Kent Emery, 'Mysticism and the Coincidence of Opposites in Sixteenth- and Seventeenth-Century France', *Journal for the History of Ideas* 45, no. 1 (1984): 3–23; Reinhold Weier, *Das Thema vom verborgenen Gott von Nikolaus von Kues zu Martin Luther* (Münster: Aschendorff, 1967), 23–60; Stephan Meier-Oeser, *Die Präsenz des Vergessenen: Zur Rezeption der Philosophie des Nicolaus Cusanus vom 15. bis zum 18. Jahrhundert* (Münster: Aschendorff, 1989), 36–61.

[96] Ludwig Baur, *Nicolaus Cusanus und Pseudo-Dionysius im Lichte der Zitate und Randbemerkungen des Cusanus. Cusanus-Texte. III. Marginalien. 1* (Heidelberg, 1940). The difference between Gerson and Cusanus was made clear in a debate between Wenck and Cusanus, surveyed in Bernard McGinn, *The Harvest of Mysticism in Medieval Germany* (New York: Crossroad, 2006), chapter 10.

de la Sorbonne around 1428. He also had studied with Heymeric, who had spent formative years with Maisonneuve in Paris.[97]

Lefèvre came to see Nicholas of Cusa as the rightful heir to the Christian philosophy of Dionysius, precisely because he modelled religious reform by means of the intellect. Already in 1499, when he commented on Dionysius' complete works, Lefèvre was familiar with the *De docta ignorantia* of Cusanus. In a section on the *Divine Names*, Dionysius presents his distinctive contribution, a 'negative theology' which is based on the observation that although one can name God, when examined closely every name falls short of completely describing God. In his gloss, Lefèvre offered Cusanus as the chief exponent of this negative theology.[98] In fact, when Lefèvre presented the 'intellectual philosophy' in 1501, he made Cusanus its sole modern expositor. Aristotle was the chief authority in 'rational' philosophy; but in intellectual reasoning, where opposites coincide, one should rely on the Presocratics Parmenides, Anaxagoras, Heraclitus, and Pythagoras—and the Christian witness from Saint Paul to Cusanus: 'Aristotle is the life of studies, but Pythagoras is the death of studies, a death superior to life; the latter properly teaching without speech, while the former teaches by speaking; but silence is act and speech is privation. In Paul and Dionysius there is much silence, and so also in Cusanus and the ὁμοουσίῳ of Victorinus. In Aristotle, however, there is little silence but rather a multitude of words; for silence speaks and words are silent, and even if they did speak, silence would simply be silent.'[99] Cusanus was important because his account of learned ignorance developed this negative theology. In his magisterial edition of Cusanus's works, Lefèvre once again mapped out his threefold account of thought: 'The first teaches in silence; the second teaches with modest words; the third clangs in many voices.... For the first, the immensity of light is darkness, whose ignorance is greater than knowledge; for the second, the finite light is light, whose knowledge is to be possessed in particular; for the third, the darkness seems to be light, and the opinion belonging to this part is a knowledge much worse than ignorance. Where to go from this? As you know... the theology of Cusa entirely

[97] Evidence for both Heymeric's and Cusanus's time in Paris are given in Rudolf Haubst, 'Der junge Cusanus war im Jahre 1428 zu Handschriften-Studien in Paris', in *Mitteilungen und Forschungsbeiträge der Cusanus-Gesellschaft*, ed. Rudolf Haubst (Mainz: Matthias-Grünewald-Verlag, 1980), 198–205; Eusebius Colomer, 'Zu Dem Aufsatz von Rudolf Haubst "Der Junge Cusanus war im Jahre 1428 zu Handschriften-Studien in Paris", in *Mitteilungen und Forschungsbeiträge der Cusanus-Gesellschaft*, ed. Rudolf Haubst (Mainz: Matthias-Grünewald-Verlag, 1982), 57–70. See further Maarten J. F. M. Hoenen, 'Academics and Intellectual Life in the Low Countries: The University Career of Heymeric de Campo (†1460)', *Recherches de Théologie Ancienne et Médiévale* 61 (1994): 173–209.

[98] Jacques Lefèvre d'Étaples, *Theologia vivificans, cibus solidus. Dionysii Celestis hierarchia. Ecclesiastica hierarchia. Diuina nomina. Mystica theologia. Undecim epistole. Ignatii Undecim epistole. Polycarpi epistola una* (Paris: J. Higman and W. Hopyl, 1499), 48v–51r.

[99] Bovelles, *In artem oppositorum introductio*, a1v (PE, 95): 'Aristoteles studiorum vita est, Pythagoras autem studiorum mors, vita superior; hinc rite docuit hic tacendo, ille vero loquendo, sed silentium actus est et vox privatio.... In Paulo et Dionysio multum silentium, deinde in Cusa et Victorini ὁμοουσίῳ, in Aristotele autem silentii perparum, vocum multum; nam silentium dicit et tacent voces, siquidem voces dicerent, simpliciter taceret.'

pertains to that first intellectual theology. By none other are we helped more to the holy inner recesses of Dionysius.'[100]

And it was mathematics that unfolded this paradoxical coincidence of speech and silence in the intellect. Lefèvre downplayed Cusanus' humanist style as 'an eloquence not so much legal as Christian' and downplayed the cardinal's accomplishments as an erudite canon lawyer in order to dilate on his mathematical interests: 'no one', Lefèvre wrote, 'penetrated deeper than he did into the mathematical disciplines'.[101] Besides his various works on geometrical questions, Lefèvre cited *De docta ignorantia* and *De conjecturis* for their myriad uses of mathematical examples to present contemplative truths—pointing out that Cusanus was not entirely unique, since Luca Pacioli had elaborated similar reflections on the attributes of God in *De divina proportione*. It was this mathematical side of Cusanus that Lefèvre isolated as the origin of Cusanus' intellectual philosophy.

In short, Lefèvre and Gerson shared a general vision of how to reform the university within the larger fifteenth-century debates over the *cura animarum*. Lefèvre even echoed Gerson's polemics against mathematico-logical speculation, especially the fourteenth-century exercises of *sophismata* of the *calculatores* in the tradition of the Merton School.[102] But while Gerson repeatedly urged a reform through penitential orientation of the affects, Lefèvre pursued a different means for university reform, focusing on the ordering of the intellect.

UNIVERSITY IDEALS

The Fabrists therefore drew on Italian humanists and monastic reform. But they promoted their positive ideals in the university district of Paris, the Latin Quarter. Their seasons followed university calendars; their days were filled with the rhythms of college life, structured around lessons, communal meals, and times of prayer. The daily life of a college aimed to discipline adolescent men in both senses: to keep them out of trouble, and to tutor them in the cycle of disciplines that made up the arts course. Young men joined a college to be tutored in just enough of Aristotle's logic, natural philosophy, moral philosophy, and perhaps mathematics, in order to gain their BA in university-wide examinations by disputation administered

[100] Jacques Lefèvre d'Étaples (ed.), *Haec accurata recognitio trium voluminum, Operum clariss. P. Nicolai Cusae Cardinalis* (Paris: Josse Bade, 1514), 1:aa2v (PE, 346): 'Prima in silentio docet; secunda in sermonis modestia; tertia in multiloquio perstrepit.... Primae lucis immensitas tenebrae, cuius ignoratio potior est scientia; secundae finitum lumen lumen est, cuius scientia precipua possessio; tertiae tenebrae lumen apparent, estque huius partis opinio, scientia et multo magis illa ignoratione deterior. Quorsum haec? Ut intelligas sapientissime pater, theologiam Cusae ad primam illam intellectualem theologiam totam pertinere, et qua nulla magis iuvamur ad sacra Dionysii Ariopagitae adyta'.

[101] Lefèvre (ed.), *Operum Cusae*, aa2v (PE, 345): 'Mathematicas disciplinas nemo eo profundius penetravit.'

[102] Note that this language against the *calculatores* also had a tradition in Italian humanism; Carlo Dionisotti, 'Ermolao Barbaro e la Fortuna di Suiseth', in *Medioevo e Rinascimento: Studi in onore di Bruno Nardi*, 2 vols (Florence: Sansoni, 1955), 1:219–53. See the sections 'Copia in Practice' in Chapter 3 and 'Friendship and Physics' in Chapter 6.

by the arts faculty. At the Collège du Cardinal Lemoine, they studied the books Lefèvre, Clichtove, Bovelles and their colleagues wrote.

In those books, Lefèvre held out certain ideals for university learning, aims which ordered the practices we will encounter in later chapters.[103] A student might first meet these ideals in classifications of the arts, which Clichtove set out in a booklet, *De divisione artium* (1501).[104] The student would meet the medieval classifications of the arts, lists of the 'servile' mechanical and liberal arts, leading to the speculative arts of natural philosophy, mathematics, and metaphysics. He would also find some *regulae* to help him order the arts, stressing, for example, the greater worth of 'doctrinal' over 'mechanical' arts (Rule 2). He would learn the goals of learning: 'Metaphysics is the bound and ultimate aim of all the sciences, and towards it all the sciences are ordered in a wonderful kind of beauty as if towards the highest pinnacle of philosophy' (Rule 9).[105] The way to attain such lofty goals, our student would find, was to begin 'not with sophistic puzzles' (Rules 12 and 13), but by demonstrations from certain first principles (Rules 15). The student would encounter the explicit aim of university learning as an effort to help him scale the steps of knowledge to their summit, metaphysics or theology.

These grand contemplative aims ordered all other disciplines below theology, but paradoxically also set a greater weight on the scaffolding that helped students make the ascent. An early misstep could be fatal. Our student would learn to pay special attention to first principles, which are a particular concern throughout Lefèvre, Clichtove, and Bovelles's works.[106] The model experience of such principles is mathematical. At the beginning of his geometry handbook, Bovelles framed the experience of mathematical principles as a kind of conversion experience that should accompany all learning:

> For these are indeed the three starting points for grasping a discipline: mastering, turning, and stance. These accommodate a rough mind to the disciplines and introduce it to them. And so first the mind should be mastered; then once mastered it should be turned; and finally—once turned towards the object it now sees—it should be held fast and set firmly in place.[107]

[103] Lefèvre and his colleagues never wrote a systematic account of education as did some Italian grammar school teachers, or Juan Luis Vives later. Cf. the works assembled in Craig W. Kallendorf (ed.), *Humanist Educational Treatises* (Cambridge, MA: Harvard University Press, 2002), and the classic survey of E. Garin, *L'educazione in Europa (1400–1600)* (Bari: Laterza, 1957).

[104] The book's publishing history is given by Jean-Pierre Massaut, *Josse Clichtove, l'humanisme et la réforme du clergé*, 2 vols (Paris: Société d'Edition 'Les Belles Lettres', 1968), 1:33–4. I use a late edition: Josse Clichtove, *Isagoge de artium scientiarumque divisione* (Paris, 1520). See also Heikki Mikkeli, 'The Aristotelian Classification of Knowledge in the Early Sixteenth Century', in *Renaissance Readings of the 'Corpus Aristotelicum'*, ed. Marianne Pade (Copenhagen: Museum Tusculanum Press, 2001), 118–21; Anne-Hélène Klinger-Dollé, *Le* De sensu *de Charles de Bovelles: conception philosophique des sens et figuration de la pensée* (Geneva: Droz, 2016), 112–15.

[105] Clichtove, *Isagoge,* C2v: 'Methaphisica omnium scientiarum meta est et finis ultimus, ad quam ut summum philosophie apicem cetere omnes miro quodam decore ordinantur.'

[106] See especially the section 'Principles' in Chapter 6, and also Richard J. Oosterhoff, '*Idiotae*, Mathematics, and Artisans: The Untutored Mind and the Discovery of Nature in the Fabrist Circle', *Intellectual History Review* 24 (2014): 1–19.

[107] Jacques Lefèvre d'Étaples, Josse Clichtove, and Charles de Bovelles, *Epitome compendiosaque introductio in libros arithmeticos divi Severini Boetii, adiecto familiari [Clichtovei] commentario dilucidata.*

The language of intuiting first principles helps Bovelles to prepare the student for the argument that geometry supplies the easiest and most apt form of knowledge, and therefore the most reliable knowledge. Unless, of course, a student wanted knowledge for the wrong reasons (for money or fame or to appear learned), which would surely lead to error. Mathematical conversion becomes the paradigm for all learning.

Since the experience of mathematical principles is an archetype of learning in general, our student would find mathematics offered as a propaedeutic medicine of the mind.[108] Lefèvre described his introduction to Boethius' *Arithmetica* as such a medicine, 'for just as skilled physicians give healthful potions and strong medicines that lead to better health', such introductions give easy access to philosophical understanding, which is 'just like the perfect health of our mind'.[109] Lefèvre presented the mathematical game of Rithmimachia as a refreshing recreation for exhausted students to recover their mental powers, 'as entirely befits medical advice'.[110] The propaedeutic power of mathematics here responds to Renaissance debates over the active and contemplative life and the role of scholarly otium, explaining why Trebizond might exemplify how mathematics sharpens the 'best natures'.[111] The utility of mathematics is not chiefly in practical expertise, but in regulating the soul as it realizes its larger goals of knowledge.

The university therefore supplied the intellectual scaffolding to help a student climb from particulars to universals—he was to be the *kind of person* capable of such knowledge, with the *habitus* and virtues that required.[112] These were traditional aims, which Lefèvre and Clichtove supported through a careful reading of Aristotelian moral philosophy as found in the *Nicomachean Ethics* and the *Politics*. In his commentary on Lefèvre's schematic and sometimes enigmatic *introductio* to

Praxis numerandi certis quibusdam regulis (auctore Clichtoveo). Introductio in geometriam Caroli Bovilli. Astronomicon Stapulensis (Paris: Wolfgang Hopyl and Henri Estienne, 1503), 49r (PE, 91–2): 'Haec enim tria sunt (praestantissime vir) capessendae disciplinae initia: subiectio, conversio, status, quae rudem quemque animum disciplinis accommodant atque ad eas introducunt, utpote primum subiiciendus est animus, subiectus vero convertendus, conversus denique in eo ad quod conversus est (quodque intuetur) obiecto stabiliendus immutabilique statu confirmandus.'

[108] Sorana Corneanu, *Regimens of the Mind: Boyle, Locke, and the Early Modern Cultura Animi Tradition* (Chicago: University of Chicago Press, 2012). She sees the seventeenth-century English empiricists as heirs to Platonic, Stoic, and other classical accounts of 'care of the self'. While I would agree the fundamental significance of the *cultura animi* tradition, I would suggest that the more important context is the university framework within which such visions of philosophy were implemented. Cf. Guido Giglioni, 'Medicine of the Mind in Early Modern Philosophy', in *The Routledge Handbook of the Stoic Tradition*, ed. John Sellars (London: Routledge, 2015), 189–203.

[109] Lefèvre, *Elementa*, h7v (PE, 34): 'Ut enim periti medici potiones digerentiaque fortibus praemittunt pharmacis quo firmiorem inducant valtudinem... velut quondam perfectam nostrae mentis sanitatem.' Note that Lefèvre seems to refer here to all sorts of introductions, though his argument is primarily for arithmetic.

[110] Lefèvre, *Elementa*, h7v (PE, 37): 'Tale profecto consilium medicum decuit.'

[111] These themes endure at the heart of historiography on humanism. More recently, historians have begun to read Renaissance thinkers in light of Pierre Hadot on spiritual exercises and *personae*, e.g. Alexander Lee, *Petrarch and St. Augustine: Classical Scholarship, Christian Theology and the Origins of the Renaissance in Italy* (Leiden: Brill, 2012). See also references to philosophy as a way of life in Eckhard Kessler, *Die Philosophie der Renaissance: Das 15. Jahrhundert* (Munich: C.H. Beck, 2008), 31–93.

[112] On universals as the aim of knowledge, see Lefèvre, *Introductio in metaphysicorum*, and the sections 'Principles' and 'A Mathematics of Making' in Chapter 6.

Aristotle's *Ethics*, Josse Clichtove explained just how the Aristotelian psychological faculties can be reduced to the Augustinian triad of intellect, memory, and will, 'in which shine forth most brightly the traces of highest divinity in man'.[113] But these faculties require cultivation. Clichtove defined moral philosophy as the discipline responsible for such cultivation of the soul. Moral philosophy depended on the natural powers of the soul and supplied rules to help order one's life to reflect the divine. The moral *ars* supplied the theory for action, he said. After all, one ignorant of the rules of painting cannot craft an image, and one without experience in the art of harmony cannot properly pluck the strings of a lute; 'so it behoves those who are not degenerate in their soul to embrace the study of moral philosophy and to freely attend to its laws and precepts.'[114] Knowledge of God is the end of all knowledge, accessible only in a rightly-ordered soul. One orders their soul by developing intellectual virtues through long habit in the disciplines.

In the first place, care for the soul meant following the best authors. Lefèvre tucked a list of recommended authors deep into his commentary on Aristotle's *Politics*.[115] He began with fifteenth-century grammarians such as Niccolò Perotti, and a smattering of poetry and letters from antiquity that had long been usual student fare, such as Virgil, Prudentius, Cicero, and Pliny the Younger, alongside chaste moderns like Mantuanus and Filelfo. He suggested some history, such as Josephus and various hagiographers. With this grammar school training in place, Lefèvre next sketched the authors of the university course: Aristotle himself, read from 'a good edition' (*ad litteram non precariam*). The quadrivium should be read from Boethius (arithmetic and music), Euclid (geometry), and the *Almagest* of Ptolemy (astronomy). Then a student would be ready to consider the rest of Aristotle: *Acroamaticorum Physicorum, Ethicorum, Politicorum, Economicorum Aristotelis e fonte puri bibantur liquores.* By saying that Aristotle should be 'drunk from a pure source', Lefèvre deploys a humanist commonplace.[116] But by modifying Aristotle's works as *acroamatica*, he alludes to their oral teaching within a close circle of students—college teaching, not public print.[117] The third and final group of texts Lefèvre suggested were those of the *vita contemplativa*, leading into theology. One should begin with Aristotle's *Metaphysics*. This would enable one to 'purify one's

[113] Jacques Lefèvre d'Étaples, *Artificialis introductio per modum Epitomatis in decem libros Ethicorum Aristotelis adiectis elucidata [Clichtovei] commentariis* (Paris: Wolfgang Hopyl and Henri Estienne, 1502), 5r: 'in quibus maxime elucet in homine summe divinitatis vestigium.' For the broader European framework, see Jill Kraye, 'Renaissance Commentaries on the Nichomachean Ethics', in *The Vocabulary of Teaching and Research between Middle Ages and Renaissance*, ed. Olga Weijers (Turnhout: Brepols, 1995), 96–117; David A. Lines, *Aristotle's Ethics in the Italian Renaissance (ca.1300–1650): The Universities and the Problem of Moral Education* (Leiden: Brill, 2002).

[114] Lefèvre, *Artificialis introductio*, 5r–v: 'ut neque quis recte imaginem sine artis pictorie regulis depixerit, aut artis modulatory imperitus recte cytharam non pulsaverit; hoc moralis philosophie stadium amplectantur oportet, eius leges et precept libenter audient, qui degeneres animo non sunt.'

[115] This section is also quoted and discussed by Alexandre Clerval, *De Judoci Clichtovei, Neoportuensis, doctoris theologi parisiensis et carnotensis canonici: vita et operibus (1472–1543)* (Paris: A. Picard, 1894), 55–6. See also Renaudet, *Préréforme*, 485–8; Guy Bedouelle, 'Jacques Lefèvre d'Étaples', in *Prosateurs latins en France au XVIe siècle*, ed. Jacques Chomarat (Paris: Presses de l'université Paris-Sorbonne, 1987), 38–41.

[116] Cf. Ovid, *Epistulae ex ponto* III.v.18.

[117] Cf. Aulus Gellius, *Noctes atticae*, II.2.

mind and trample down vices by exercising the senses, choosing an active life and noble habits' while reading the Church Fathers: Cyprian, Hilary, Origen, Jerome, Augustine, Chrysostom, Athanasius, Nanzianus, Damascene, 'and others'. With such a training—'if one has a more noble mind'—perhaps one might want 'higher contemplations, and so little by little ascend through the books of Nicolas of Cusa and the divine Dionysius'.[118]

But the reading list only offered a skeleton of the disciplines. As a unity of body and intellect, the soul also required social and physical discipline. Along with many other humanists, Lefèvre was deeply concerned about the event that had long been at the centre of university life, the disputation. The only way a student could demonstrate his prowess was by passing a series of disputations. University reformers since the age of Abelard worried that the 'science of doubt' fostered a contentious atmosphere, that youths were encouraged to outsmart their peers rather than seek truth.[119] Some within the university, such as Gerson, worried about specific genres of debate; other humanist authors from Bruni to Vives, years after their own university studies, presented their programme as a reform of disputation. Echoing such critiques, Lefèvre listed the good authors a youth should read as an alternative to the 'pernicious' debates and 'squabbles of contentious men'.[120] He urged students to pursue models of dialectic full of both grace and truth. He found a particularly wonderful example in Trebizond's *Dialectica*, and when he published it in 1508 he offered as justification his own experience in Rome in 1492. On entering a house, he had witnessed two youths debating with such eloquence that he was impelled to ask their teacher how they had learned. After Lefèvre agreed that 'the more persistent minds of youths always should be steeped in the better sort of studies', the teacher referred him to George of Trebizond's little book on dialectic as the secret to their conversation.[121] What struck Lefèvre was the moral quality of the youths: 'I enjoyed the generous liberality of the youths, and have preserved the memory of these youths up to this day as something most dear to me.'[122]

This was the ideal of generous sociability that students encountered at Cardinal Lemoine. In an introduction to the *Nicomachean Ethics*, Clichtove spent extra time

[118] Jacques Lefèvre d'Étaples, *Politicorum libri octo. Commentarii. Economicorum duo. Commentarii. Hecatonomiarum septem. Economiarum publ. unus. Explanationis Leonardi [Bruni] in Oeconomica duo* (Paris: Estienne, 1506), 123v–4r: 'In his autem mente purgata et sensibus exercitatis, actione vite consentanea, et morum honestate calcatis viciis, si mens generosior, elevatiores contemplations affectet, paulatim ex libris Cuse surgat, et divini Dionysii et si qui sunt similes.'

[119] The historiography on medieval disputation is large, but see now Alex J. Novikoff, *The Medieval Culture of Disputation: Pedagogy, Practice, and Performance* (Philadelphia, PA: University of Pennsylvania Press, 2013); Olga Weijers, *A Scholar's Paradise: Teaching and Debating in Medieval Paris* (Turnhout: Brepols, 2015).

[120] On the more literary function of humanist debates over disputation, see Béatrice Perigot, *Dialectique et littérature: les avatars de la dispute entre Moyen Age et Renaissance* (Paris: Champion, 2005); Anita Traninger, *Disputation, Deklamation, Dialog: Medien und Gattungen europäischer Wissensverhandlungen zwischen Scholastik und Humanismus* (Stuttgart: Franz Steiner Verlag, 2012).

[121] Lefèvre, *Trapezontii dialectica*, a1v–a2r (PE, 190): 'et tenaciores adulescentum animi melioribus... semper institutis imbuendi.'

[122] Lefèvre, *Trapezontii dialectica*, a1v–a2r (PE, 191): 'Usus sum, fateor mi Fortunate, generosa adulescentis liberalitate, et memoriam iuvenum in hanc usque diem mihi quidem quam carissimam servavi.'

on the social virtues. Friendship (*amicitia*) earned the most attention, as mentioned in Chapter 1. Another was the virtue of 'even temperedness' (*affabilitas*), the mean between obsequious flattery and contentious anger. Even more important was the virtue of 'tact' (*comitas*), the art of saying the right thing, offering the tasteful joke without falling into scurrilous laughter or rustic coarse humour. A youth with tact was urbane, witty, lively, and good-humoured. Clichtove observed that it is impossible to spend a life in unceasing labour; one needed lively games and recreation, including public games and spectacles—though he hastened to add that 'all these have lapsed into a terrible state now and in place of such games are intemperate games' such as dice and dances.[123] The sort of *comitas* to be practised in college should be displayed in urbane speech, cultivated and aimed to please listeners rather than to hurt them.[124]

This social discipline also implied discipline of the body. Colleges regulated ball-playing within their walls. Students learned that it was bad taste to joke through exaggerated gestures, as 'histrionics and mimes who gesture like play-actors'.[125] Instead, Lefèvre especially held out music as a means for disciplining the soul and so also the body. He made these connections explicit in his textbook on music, in a concatenation of twenty-eight ancient authorities, beginning with Hermes Trismegistus and ending with Boethius, who all commended musical practice as a way to soothe the soul. He repeated well-known examples of the pyscho-therapeutic power of music, such as Pythagoras and Orpheus on the taming of drunken and wild youths, or the effect of music on the elephants of India.[126] While such examples seem purely literary, they gain meaning when compared with his tale of Pico della Mirandola's use of music to restore his own soul when exhausted by studies, mentioned earlier in this chapter.

Aristotle is on Lefèvre's list of musical authorities. Although not given much space in the history of music, Aristotle makes some sense as an authority in light of his *Politics*. There he responded to Plato's *Republic*, which—as Hermonymus said—gave the quadrivium a key role in the education of rulers. Aristotle's *Politics* therefore discussed music at length, a fact that Lefèvre highlighted in his commentary. In the tradition Lefèvre and Aristotle shared, music was the harmonic application of numerical ratios. Because music so powerfully moved listeners, it surely reflected the hidden mathematical structures of nature. Because of this metaphysical commitment, Lefèvre expected that mathematical studies should play an important role in helping students discipline themselves.

The commentary on the *Politics* helps us glimpse why Lefèvre devoted tremendous energies to recovering mathematics. His plea for recovering both mathematics

[123] Lefèvre, *Artificialis introductio*, 25r–v: 'Sed nunc omnia in deterius lapsa sunt et loco talium ludorum exhibentur ludi intemperantie'.

[124] Lefèvre, *Artificialis introductio*, 26r: 'Finis enim comitatis: recreation, relaxation, quiesque quedam est; non igitur intendit comis audientes dolore afficere.'

[125] Lefèvre, *Artificialis introductio*, 26r: 'histriones et mimos qui scenicos gestus agunt damnandos precipit'.

[126] Lefèvre seems to have taken this list from Martianus Capella's list in *De nuptiis philologiae et mercurii* (PE, 31–2).

and natural philosophy takes place within an argument for why contemplative study is necessary to the virtuous man of action. All the disciplines interlock to serve both active and contemplative modes of life:

> One may quite rightly make rational disciplines instruments for both sorts of life. For how can anyone be wise and understand the virtues, if he completely lacks the knowledge of letters given by grammar? How can anyone offer correct and suitable counsel, or persuade one of virtue or the duties of virtues, if he has never thought about the discipline of rhetoric? How will one avoid error in controversy, or put together reasons for what to do, or deal with deliberations, if he is helpless in discourse, which logic ought to give? In pursuing contemplation, these same tools set out and attain a nobler end as well.[127]

The virtuous ruler and his advisers are clearly in view, much as in the prefatory letters discussed earlier in this chapter. It was no coincidence these are notes on the *Politics*. But this is no prefatory letter aimed to catch the eye of a patron. Here, deep in the scholia of an Aristotelian commentary, Lefèvre strives to persuade colleagues and students of the importance of their calling. And we get an almost raw sense of his motivations, as he argues that natural philosophy and mathematics have fallen into misuse:

> However, now—and for many years—it has been the case that the dignity of these instruments and means to wisdom has not been preserved, namely these disciplines of natural philosophy and mathematics. They grow sterile, because they are not directed to their ends, but are occupied with superfluous matters.[128]

The relevant misapplications of natural philosophy and mathematics no doubt include the 'sophistic' logic of previous generations.[129] The solution is to set these disciplines back to their rightful ends—which, in this context of Aristotelian moral philosophy, is the acquisition of intellectual and moral virtues, in service of the common good of the kingdom. These larger, overarching goals are the motivations for the years Lefèvre spent restoring mathematics, and its special place in his vision of university reform.

An indication of Lefèvre's approach to mathematics is found in the longest stretch of commentary Lefèvre offered on the *Politics*. There Aristotle discussed the passage in Plato's *Republic* on the 'fatal number' which somehow governed the change from one political regime to another.[130] Readers of Plato since antiquity

[127] Lefèvre, *Politica*, 112v: 'Rationales tamen disciplinas non absurde quispiam ad utramque vitam fecerit instrumenta. Quomodo enim quisquam prudens et virtutum cognitor, qui litteras quas grammatica donat prorsus ignorabit? Quomodo recte et accommodate consultabit, quomodo ad virtutem et virtutum officia persuadebit, si nichil discipline rhetorice delibaverit? Quomodo in controversiis falsum effugiet, quomodo verum ratiocinatione in agendis collegit, quomodo rationem subducendo verum attinget, nisi illi assit disserendi praesidium, quod a logica vendicetur oportet. Sed hec eadem instrumenta in consequenda contemplatione, nobiliorem sortiuntur, assequuntur que finem.'

[128] Lefèvre, *Politica*, 112v: 'Verum nostra tempestate et ante per plurimos annos, longe aliter evenit ut horum instrumentorum et mediorum ad sapientiam, que sunt naturales et mathematice discipline, servetur dignitas, quod ad suos fines eas non dirigunt, sed circa superflua occuperati, sterilescunt.'

[129] See the first section of this chapter and the section 'Friendship and Physics' in Chapter 6.

[130] Plato, *Republic*, 8.546a1–d3; Aristotle, *Politics*, 5.1316a1b26.

had been fascinated by the number, and later generations of commentators who would repeatedly cite Lefèvre's interpretation.[131] Marsilio Ficino had constructed layers of meaning around the number, embedding its solution within an account of the soul as a geometrical *habitus*, which could be manipulated through computational forms of theurgic magic.[132] Precisely because daemons, skilled in thus manipulating souls, influenced the *ingenia* of individuals, they also governed the change of political systems. Lefèvre took a more sceptical approach. He reiterated Aristotle's arguments against the power of such a number—crucially, Plato's explanation ignores many other obvious causes of political change. But then Lefèvre devoted five large folio pages to finding the number.

Much of this harmonized with Ficino's account. Lefèvre quoted the relevant passage from Plato, in Ficino's translation. He began by teaching his reader the fundamentals of Pythagorean number theory, setting out terms such as sesquitertial and sesquialter proportions (diatesseron 4:3, and diapason 3:2, respectively). This dense primer introduced the basic ontology of numbers from Nicomachus, such as 'perfect' numbers, masculine and feminine numbers (i.e. odd and even numbers), as well as the figurate numbers, those that can be arranged into regular shapes, such as square, spherical, side, and diagonal numbers. These terms in hand, Lefèvre proceeded to an interpretation. He offered the same value for the fatal number of Plato as Ficino had: 1728. Plato had thought such a number should enfold many other numbers within it. This number, he explained, had many of the requisite properties: it was cubed itself ($=12\times12\times12$). Furthermore, its digits encoded a three-fold harmonious proportion: adding the initial one to the number's end, it became the square of 9 ($9\times9=729$); and removing 1 from the last two digits made it the square of 3 ($3\times3=27$). 'Therefore you see how this great number is made from the roots of the 3, the 9, and the 12. It has neither excess nor lack in the 12; it lacks 1, to make the 9; and it is abundant by 1 to make the 3.'[133]

Despite arriving at the same value, Lefèvre's account of the fatal number differed sharply from Ficino's. For Ficino, it was all about the metaphysics of soul hidden in Plato's account, so even as he explained the number 1728, he continually connected its various components to the powers and modes of change within the soul. For Lefèvre, the number is all about the mathematics. The number itself was notorious for its difficulty. Lefèvre recalled the proverb for something extremely difficult and incomprehensible: 'it is more obscure than Plato's number' (*numeris*

[131] e.g. Carl Ernst Christopher Schneider, *De numero Platonis commentationes II* (Wratislavia: Schoene, 1821), included as the third volume of Schneider's edition of the *Republic* (1833).

[132] Ficino discussed these passages in two places: his *argumentum* for Book 8 of the *Republic*; and the commentary he printed as a discrete treatise, *De numero fatali*: Marsilio Ficino, *Opera omnia Platonis* (Venice: Andrea Toresani de Asula, 1491), 225r; Marsilio Ficino, *Commentaria in Platonem* (Florence, 1496), unnumbered, final quire. These are edited and translated in Michael J. B. Allen, *Nuptial Arithmetic: Marsilio Ficino's Commentary on the Fatal Number in Book VIII of Plato's 'Republic'* (Berkeley, CA: University of California Press, 1994); Michael J. B. Allen, 'Ficino, Daemonic Mathematics, and the Spirit', in *Natural Particulars: Nature and the Disciplines in Renaissance Europe*, ed. Anthony Grafton and Nancy G. Siraisi (Cambridge, MA: MIT Press, 1999), 121–37.

[133] Lefèvre, *Politica*, 87v: 'vides igitur quopacto hic magnus numerus confectus est ex radicibus ternaria, novenaria et duodenaria; in duodenaria neque excrescens, neque deficiens; in novenaria, uno deficiens; in ternaria, uno abundans.'

Platonis obscurius). His exegesis of the number focused on unpicking this difficulty. Having succinctly arrived at the number 1728, he explained its mathematical properties without a whisper on its metaphysical powers, and then changed to a new mathematical topic: the qualities of 'diametrical' numbers. In fact, by the end of the few pages of commentary on Plato's number, he had explained nothing of how Plato's number might actually bring about change in political regimes, and instead had reflected on several key series in number theory, including conjectures for long-sought values such as the fifth to tenth perfect numbers. Lefèvre stressed the intellectual value of the exercise, even as he averred that 'it is vain to prophesy from [numbers] and to hope in a futile way to examine divine secrets'.[134] It was not his intention to open up the mind of Plato and the secrets of the universe; rather his goal was to use Aristotle's moment of Platonic weakness to get students to deploy their arithmetical skills in mental exercise—to acquire the habit needed for intellectual virtue.

* * *

By expecting students to exercise their wits rather than explain them, Lefèvre stepped away from Ficino and toward George of Trebizond as the example of mathematics as a fundamental step in erudition. Though attracted to the Ficinian thought that mathematics might let one somehow have direct access to physical causes or human souls, he presented a more prosaic view of mathematics as an intellectual tool for disciplining the soul as it navigates the disciplines—in short, to help students order their studies.

The motif of conversion helps us see the fifteenth-century motivations behind Lefèvre's mathematics. His project was of a piece with his broader programme of printing the texts of the Lullian, Rhenish, and earlier Parisian reforms, but set 'turning the soul' towards God firmly within the university context. The university now presented new constraints and opportunities, as the medieval *cursus* met new print technologies and humanist educational ideals. Like earlier fifteenth-century voices for intellectual reform such as Jean Gerson, Lefèvre hoped to bring about conversion through texts, and so he attended carefully to matters of genre, to the material ways that texts are made and experienced.[135] Unlike earlier editors and authors, Lefèvre and his colleagues reimagined these textbooks during an experimental move from manuscript to print in the university classroom. To see this in motion, we turn to Beatus Rhenanus using Lefèvre's textbooks at the Collège du Cardinal Lemoine.

[134] Lefèvre, *Politica*, 87r: 'verum vaticinari ex illis querere, et futili coniectura divinum scrutari velle secretum, vanum'.

[135] Daniel Hobbins, 'The Schoolman as Public Intellectual: Jean Gerson and the Late Medieval Tract', *American Historical Review* 108, no. 5 (2003): 1308–37.

3

Copia in the Classroom

In 1504, Beatus Rhenanus (1484–1547) copied lectures on natural philosophy into his student notebook, adding a colophon: 'this introduction to the *Physics* of Aristotle was taught (*lecta*) in the Collège du Cardinal Lemoine, at Paris, and written down by me'.[1] He described a series of circles (Figure 3.1). 'The first circle: the concavity of the moon's orb, which contains everything below, which the Philosopher defines in the eight books of *Physics*. The elements are placed in this first circle.'[2] Beatus' manuscript notes do not include this figure—but one has not far to search. The lecturer labelled it with a mnemonic, 'NACAMILUT', which refers to *natura*, *causa*, *motus*, *infinitum*, *locus*, *vacuum*, and *tempus*. These terms and circles are given in a diagram which prefaces another of Beatus' books, a printed textbook on natural philosophy (Figure 3.2). There, in his characteristically neat hand, Beatus also filled the printed page with thick blocks of annotations.

These prose notes, printed textbooks, and marginalia present a view of the Renaissance classroom that pedagogues rarely explained, preferring to list illustrious authors rather than the lowly textbooks and practices that introduced them.[3] Usually they idealized teaching and learning, as does the dialogue we find later in Beatus' book, which is the first edition of the *Totius Aristotelis philosophiae naturalis paraphrases* to include Clichtove's commentaries (1502). I will examine this dialogue more closely in Chapter 6. Here what matters is how teacher and student interact. The dialogue tells of a teacher who places before his students a *figura*, which lists all the characteristics of the sublunary, sensible world—the world Aristotle explains in his eight books on *Physics*. By way of introduction, the teacher questions his students about the figure, eliciting from the boys a set of basic Aristotelian definitions for nature, cause, motion, and so on. By correctly defining these terms, the student in the dialogue constructs a faithful image of nature.[4]

[1] BHS MS 58, 163v: 'Parrhisiis in Cardinali Monacho haec in physicen Aristotelicam introductio lecta est et a me Beato Rhino<wer> litteris mandata.'

[2] BHS MS 58, 147r: 'Primus circulus: concavum orbis lunae quod continet omnia inferiora de quibus philosophus determinat in octo libris phisicorum, in dicto circulo ponuntur primo elementa.'

[3] e.g. Anthony T. Grafton, 'Textbooks and the Disciplines', in *Scholarly Knowledge: Textbooks in Early Modern Europe*, ed. Emidio Campi, Simone De Angelis, and Anja-Silvia Goeing (Geneva: Droz, 2008), 11–14.

[4] Jacques Lefèvre d'Étaples and Josse Clichtove, *Totius philosophiae naturalis paraphrases* (Paris: Henri Estienne, 1502), sig. q1r–v.

With this *figura*—the very one set at the beginning of Lefèvre's own textbook on natural philosophy—the fictive teacher leads the student into natural philosophy by the same mechanism we find in Beatus' notes, by discussing a prefatory diagram that supplies a synopsis of the whole book.[5] The dialogue matches Beatus' experience, evoking the rich practices of visual memorization and spoken dialogue that characterized education well into the Renaissance.[6]

This chapter will end with the way mathematics fits into the practices that structured the *cursus* at the Collège du Cardinal Lemoine. These practices drew on Ficino, Lull, and Cusanus, mentioned in the last chapter, and the new printing presses, discussed in the next. But they emerged in the university classroom. Humanist pedagogy, combined with printing presses, is often seen as a force for simplification; the movement *ad fontes puriores* appeared to wash away obscurantist commentary and scholastic *quaestiones*, leaving students alone with the authors. In fact, I shall argue, these motivations drowned the student in an overabundance of newly printed texts. The encyclopaedia—literally, the circle of learning—was the site for social and material practices for managing this copia. To simplify, students and masters at Lemoine flirted with mathematics, understood as analogy, as a new method for navigating the arts.

UNIVERSITY

In the spring of 1503, the eighteen-year-old Beatus Rhenanus arrived in Paris to study at the Collège du Cardinal Lemoine. What is now called the Left Bank was simply known as 'the University' in the Middle Ages, a term which masked a crowd of allegiances, filling warrens of alleys, walled-in colleges and convents, and churches. The University quarter was dominated by the church of St Geneviève, whose grand steps were the site of university processions, regular sermons and debates, and the conferral of degrees. As many as ten thousand arts students swarmed Paris around 1500;[7] the past century of schism and war had taught rulers that they needed clerics and secretaries, and across Europe universities had expanded rapidly to meet demand. City officials nervously eyed this growing population of potential vagabonds, and by the middle of the fifteenth century faculties pressured 'martinets' or unaffiliated students to find formal lodgings.[8] Students had several affiliations. Each arts student belonged to one of the four Nations, according to origin and language. Each Nation maintained lecture rooms

[5] See Chapter 4. [6] See note 56.

[7] L. W. B. Brockliss, 'Patterns of Attendance at the University of Paris, 1400–1800', *Historical Journal* 21, no. 3 (1978): 503–44, at 511.

[8] In 1463, for example, the faculty of arts required all students to have a fixed residence: Roger Chartier, Dominique Julia, and Marie-Madeleine Compère, *L'education en France du XVIe au XVIIIe siècle* (Paris: Société d'édition d'enseignement supérieur, 1976), 152.

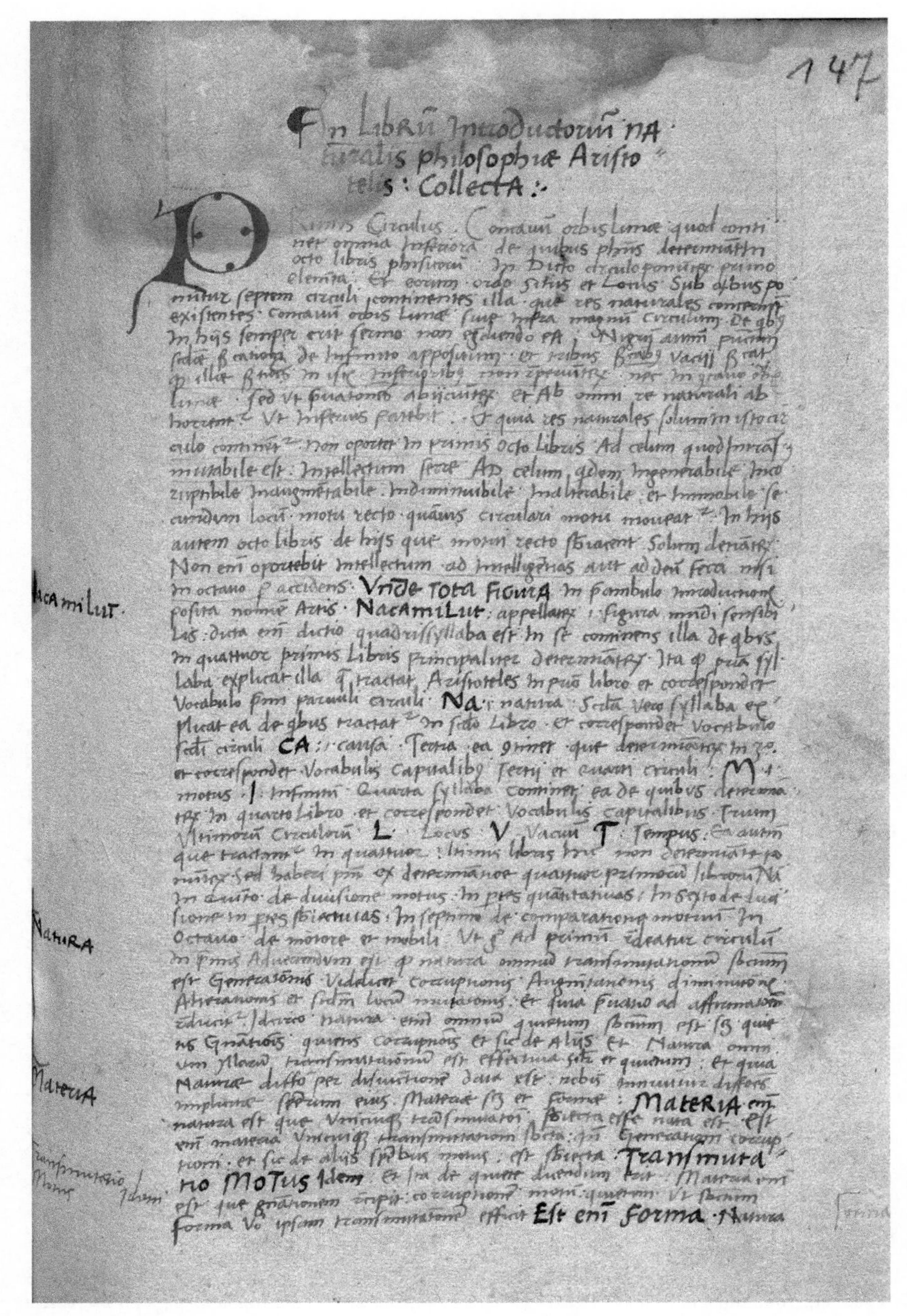

Figure 3.1. Beatus Rhenanus' student notebook: a page from Beatus Rhenanus' *Cahier d'étudiant*, BHS MS 58, fol. 147r, with the mnemonic 'NACAMILUT' in the margin. This acronym labels circles found in Beatus' textbook on natural philosophy (see Figure 3.2).

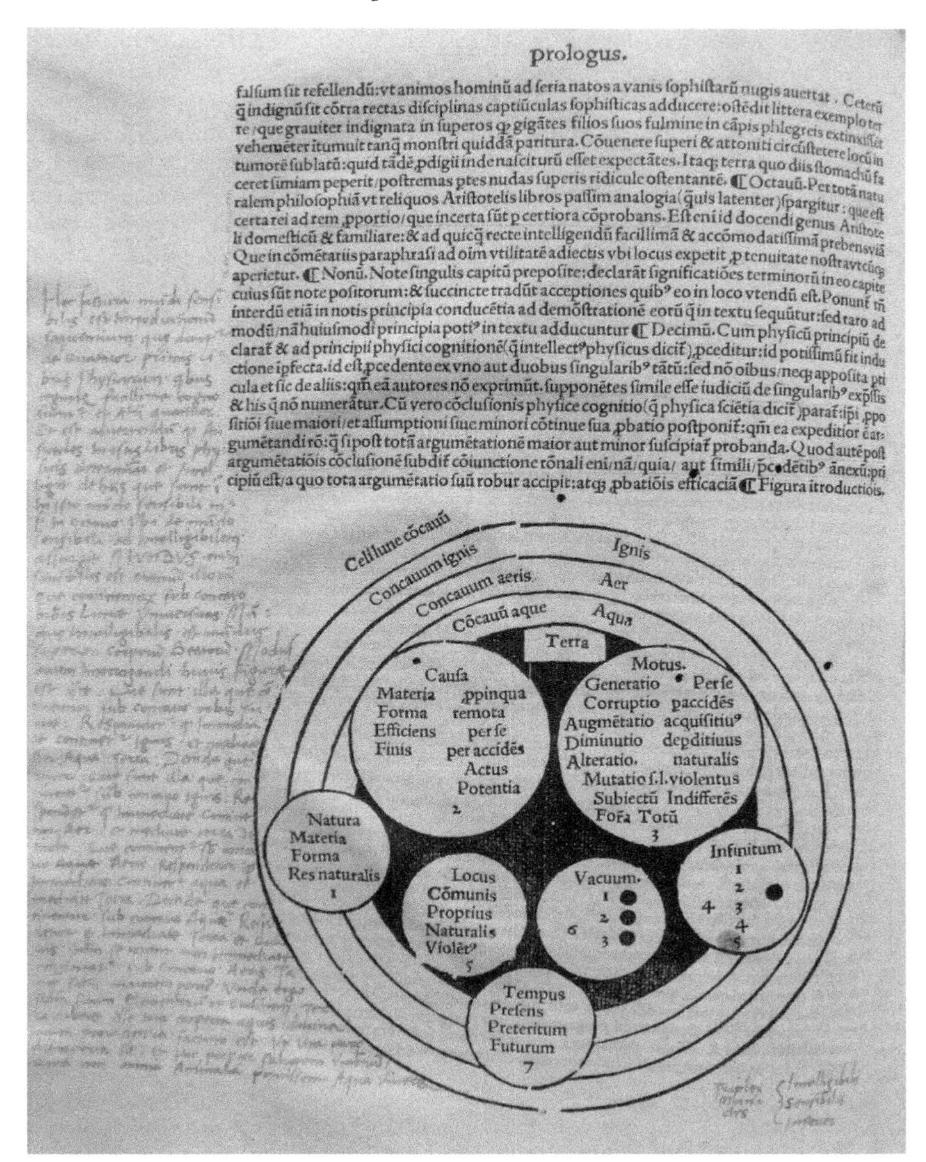

prologus.

falſum ſit refellendũ:vt animos hominũ ad ſeria natos a vanis ſophiſtarũ nugis auertat. Ceterũ q̃ indignũ ſit cõtra rectas diſciplinas captiũculas ſophiſticas adducere:oſtẽdit littera exemplo ter re/que grauiter indignata in ſuperos ꝙ gigãtes filios ſuos fulmine in cãpis phlegreis extinxiſſet vehemẽter itumuit tanq̃ monſtri quiddã paritura. Cõuenere ſuperi & attoniti circũſtetere locũ in tumorẽ ſublatũ:quid tãdẽ ꝓdigii inde naſciturũ eſſet expectãtes. Itaq; terra quo diis ſtomachũ fa ceret ſimiam peperit/poſtremas ptes nudas ſuperis ridicule oſtentantẽ. ¶ Octauũ. Per totã natu ralem philoſophiã vt reliquos Ariſtotelis libros paſſim analogia(q̃uis latenter)ſpargitur:que eſt certa rei ad rem ꝓportio/que incerta ſũt p certiora cõprobans. Eſt enĩ id docendi genus Ariſtote li domeſticũ & familiare:& ad quicq̃ recte intelligendũ facillimã & accõmodatiſſimã prebens viã Que in cõmẽtariis paraphraſi ad oim vtilitatẽ adiectis vbi locus expetit ꝓ tenuitate noſtra vtcũq; aperietur. ¶ Nonũ. Note ſingulis capitũ prepoſite:declarãt ſignificatiões terminorũ in eo capite cuius ſũt note poſitorum:& ſuccincte tradũt acceptiones quib⁹ eo in loco vtendũ eſt. Ponunt tñ interdũ etiã in notis principia conducẽtia ad demõſtrationẽ eorũ q̃ in textu ſequũtur:ſed raro ad modũ/nã huiuſmodi principia poti⁹ in textu adducuntur ¶ Decimũ. Cum phyſicũ principiũ de claratr & ad principii phyſici cognitionẽ(q̃ intellect⁹ phyſicus dicit)ꝓceditur:id potiſſimũ fit indu ctione ipfecta.id eſt ꝓcedente ex vno aut duobus ſingularib⁹ tãtũ:ſed nõ oibus/neq; appoſita pti cula et ſic de aliis:qm̃ eã autores nõ exprimũt. ſupponẽtes ſimile eſſe iudiciũ de ſingularib⁹ expſſis & his q̃ nõ numerãtur. Cũ vero cõcluſionis phyſice cognitio(q̃ phyſica ſciẽtia dicit)parat:ipſi ꝓpo ſitiõi ſiue maiori/et aſſumptioni ſiue minori cõtinue ſua ꝓbatio poſtponit:qm̃ ea expeditior ẽ ar gumẽtandi rõ:q̃ ſi poſt totã argumẽtationẽ maior aut minor ſuſcipiat probanda. Quod autẽ poſt argumẽtatiõis cõcluſionẽ ſubdit cõiunctione rõnali enĩ/nã/quia/ aut ſimili/pcedẽtib⁹ ãnexũ:pri cipiũ eſt/a quo tota argumẽtatio ſuũ robur accipit:atq; ꝓbatiõis efficaciã ¶ Figura itroductiõis.

Figure 3.2. Beatus Rhenanus' textbook on natural philosophy: Lefèvre and Clichtove, *Totius philosophiae naturalis paraphrases, adiecto commentario* (1502), BHS K 1199, 2v.

in the rue de Fouarre, which were less and less used as teaching moved into college rooms.[9] Many students found residences outside colleges, but increasingly they

[9] The reforms of Cardinal Estouteville of 1452 expected arts lectures to be held in the Rue de Fouarre (via Straminis); by 1542, statutes of the faculty of arts recognized colleges as the site of lectures (Chartier, Julia, and Compère, *L'education en France*, 151). Certificates of study from 1512 occasionally say where students studied, e.g. 'in the rue de Fouarre, in colleges, at table (?), and many other

were also expected to belong to a college, which had since the early fourteenth century become the economic, social, and intellectual hubs of university life, regulating daily rhythms with times of communal eating, reading, and prayer.[10] Repeated efforts to enforce such regulations hint that they were often breached; there was much more to do in this part of town. There were ancient, great libraries. Some belonged to the colleges themselves—one of the biggest belonged to the Sorbonne, one of the oldest colleges associated with the theology faculty—but others belonged to the many convents. In particular, the Collège du Cardinal Lemoine was a short step away from the Gate of St Victor, named after the twelfth-century cloister just outside the city walls, which boasted one of the finest libraries in France. University stationers had long held bookshops along the main street of the Latin quarter, Rue St Jacques, and by the 1490s this was the hub of the print trade, with booksellers also keeping stalls on the main bridges to the Ile de Paris, leading across the river to the apartments of the King, the *parlement*, and the houses of rich Parisian businessmen and nobles—a path towards advancement that lured many young clerics even before obtaining a degree.[11]

Lefèvre developed his programme of intellectual reform during an unusually long seventeen years as a regent master at the Collège du Cardinal Lemoine, an arts college linked to the Picard Nation.[12] Due in part to Lefèvre, Lemoine was influential in the sixteenth century as one of the few larger colleges that became *collèges de plein exercice*, in which students could study grammar and the entire arts course.[13] The founding statutes of 1302 stipulate only six regent masters to be paid room and board, besides about fourteen *boursiers* or poor students given lodgings and study. Most regent masters only held their office for a few years, before moving on to richer positions, perhaps in one of the dioceses north of Paris that supported

places' (*in vico Straminis, in collegiis, in mensa, et pluribus aliis locis*). James K. Farge (ed.), *Students and Teachers at the University of Paris: The Generation of 1500. A Critical Edition of Bibliothèque de l'Université de Paris (Sorbonne), Archives, Registres 89 and 90* (Leiden: Brill, 2006), no. 224. Some contemporary university puffery suggests that the Nations used their rooms on the Rue de Fouarre for examinations, and no longer lectures: Robert Goulet, *Compendium on the Magnificence, Dignity, and Excellence of the University of Paris in the Year of Grace 1517*, trans. Robert Belle Bourke (Philadelphia, PA: University of Pennsylvania Press, 1928), 87. After the sixteenth century, with the exception of the *collèges de plein exercice*, colleges lost the central place they had briefly enjoyed in Paris teaching; for the later situation, see L. W. B. Brockliss, *French Higher Education in the Seventeenth and Eighteenth Centuries: A Cultural History* (Oxford: Clarendon Press, 1987).

[10] Foundation statutes usually indicate such rhythms; they were equally the focus of reformers such as Standonck. The requirement of attending daily masses in college is mentioned by Goulet, *Compendium*, 63.

[11] For a sketch of Paris, its university, and the book trade at this time see Simone Roux, *La Rive gauche des escholiers (XVe siècle)* (Paris: Éditions Christian, 1992), and Annie Parent, *Les métiers du livre à Paris au XVIe siècle* (Geneva: Droz, 1974).

[12] Charles Jourdain, 'Le collège du Cardinal Lemoine', *Mémoires de la Société de l'histoire de Paris et de l'Ile-de-France* 3 (1876): 42–81; Nathalie Gorochov, 'Le collège du Cardinal Lemoine au XVIe siècle', *Mémoires de Paris et l'Ile-de-France* 42 (1991): 219–59.

[13] Marie-Madeleine Compère, 'Les collèges de l'université de Paris au XVIe siècle: structures institutionnelles et fonctions éducatives', in *I collegi universitari in Europa tra il XIV e il XVIII secolo. Atti del convegno di studi della commissione internazionale per la storia della Università, Siena-Bologna, 16–19 maggio 1988*, ed. Hilde de Ridder-Symoens and Domenico Maffei (Milan: Guiffre editore, 1991), 101–18.

the college, or in one of the colleges linked to the three higher faculties of law, medicine, and theology. Regent masters might be assisted by outside MAs who had spent two years studying to teach as licentiates in a higher faculty, scraping together an income while they studied elsewhere, as Josse Clichtove did at Lemoine before he became a doctor of theology in 1506. Alongside regent masters and outside masters, Lemoine and other colleges rented rooms to 'pedagogues', grammar teachers who helped younger students prepare for the arts course. As the most senior regent master, Lefèvre wielded great influence in this community.

Beatus Rhenanus' own path to Lemoine reveals how Lefèvre's college became the centre of a network extending far beyond Paris. A butcher's son, he had studied at the municipal school in his hometown of Schlettstadt, in Alsace, where one of Lefèvre's old pupils, Hieronymus Gebwiler (*c.*1473–1545) became headmaster in 1501.[14] Gebwiler seems to have groomed Beatus for Lemoine. In fact, several schoolmates from Schlettstadt accompanied him, including Johannes Sapidus and Michael Hummelberg as well as the two sons of the famous printer Amerbach, Basil and Bruno. All of them would become important figures in Rhineland humanism after their study at Cardinal Lemoine.[15] Meanwhile, Gebwiler was promoted to positions at Strassburg, where he republished several of Lefèvre's textbooks. The college community extended deep into the heart of Rhineland humanism.

Within Paris, Lemoine housed a tight-knit band of scholars. When Beatus arrived in 1503, Lefèvre was at the height of his powers, and both Clichtove and Bovelles were already popular teachers as well. A register from 1512, some years after all of them had moved on from Lemoine, still bears their signatures certifying how many years arts masters had trained: Lefèvre certified twenty-one students, while Josse Clichtove signed for sixteen, and Bovelles for twelve. Even more tellingly, all of the principals of Lemoine still living in 1512 had been Lefèvre's students.[16] Lemoine was by no means tightly sealed from the rest of the university. Older scholarship often presented Lefèvre's Lemoine as a bastion of the humanist avant-garde. The register shows that students might study at one college under one master before moving on to other colleges—particularly Boncour, where Lefèvre had previously been regent, and St Barbe. Most students relied on one college, but some played the field, as Louis Lamengier did. He studied three and a half years with Pierre Tataret, the chief defender of Duns Scotus, as well as with Lefèvre.[17] In the next section, we will turn to the sons of the Basel printer Johann Amerbach, who split their time between the more traditional lectures at the Collège de Lisieux

[14] The Latin school at Sélestat (Schlettstadt) had been founded in 1441, when the city hired as its first headmaster Ludwig Dringenberg, trained by the brothers of the *Devotio moderna* in Deventer. A succinct introduction is found in Robert Walter, *Trois Profils de Beatus Rhenanus: L'homme, Le Savant, Le Chrétien* (Sélestat: Les amis de la Bibliothèque humaniste de Sélestat, 2002), with up to date references in James Hirstein (ed.), *Epistulae Beati Rhenani: La Correspondance latine et grecque de Beatus Rhenanus de Sélestat. Edition critique raisonnée avec traduction et commentaire, Volume 1 (1506–1517)* (Turnhout: Brepols, 2013).

[15] See their biographies, for example, in CE.

[16] Robert Auvray, Nicolas de Grambus, Jean de Moliendo, Jean Pelletier, and Thibauld Petit; listed in Farge (ed.), *Students and Teachers*, s.v.

[17] Farge (ed.), *Students and Teachers*, 397b, c.

in the morning and the innovative lectures at Lemoine in the afternoon. Beatus himself seems to have found Lemoine satisfactory on its own. In his letters, he would name Lefèvre, Clichtove, and Bovelles as his teachers.[18]

The student notes of Beatus are the closest we get to any classroom in Renaissance Paris. Straight upon his arrival, Beatus immediately bought books covering the main disciplines of the university arts curriculum: mathematics, logic, and natural philosophy, followed a year or two later by metaphysics and ethics—he bought them all in editions recently published by his teachers Lefèvre, Clichtove, and Bovelles. In the Bibliothèque humaniste de Sélestat, we can still read over two hundred and fifty of the books that Beatus bought before and during his study at Paris between 1503 and 1507.[19] Beyond statutes and library lists, then, the most complete evidence we have of Paris teaching in this period is Beatus' student library: the books he bought as a student, the annotations he wrote in their margins, and a codex of various lecture notes and short treatises.[20] Such a complete student library is, to my knowledge, unique.[21]

This archive is all the more valuable because Beatus Rhenanus experienced the university classroom at a moment of change in university life—the shift from

[18] Faye has suggested other teachers as well in 'Beatus Rhenanus lecteur et étudiant'.

[19] Listed by Gustav Knod, *Aus der Bibliothek des Beatus Rhenanus: ein Beitrag zur Geschichte der Humanismus* (Leipzig, 1889).

[20] Few studies have tried to decipher annotations in these books, though see studies by Emmanuel Faye and James Hirstein, as well as Christoph C. Baumann, 'Dictata in quinque predicantes voces. Ein Kommentar zur Isagoge des Porphyrius in der Aufzeichnung des Beatus Rhenanus', PhD thesis, Universität Zürich, 2008.

[21] Some medieval student notes exist, and we have collections of student textbooks that share similar sets of annotations, suggesting standard lecture notes, for example from Leipzig in the 1510s, Basel in the 1520s, and Paris in the 1580s: Jürgen Leonhardt, 'Classics as Textbooks: A Study of the Humanist Lectures on Cicero at the University of Leipzig, ca. 1515', in *Scholarly Knowledge: Textbooks in Early Modern Europe*, ed. Emidio Campi, Simone De Angelis, and Anja-Silvia Goeing (Geneva: Droz, 2008), 89–112; Inga Mai Groote, 'Studying Music and Arithmetic with Glarean: Contextualizing the Epitomes and Annotationes among the Sources for Glarean's Teaching', in *Heinrich Glarean's Books: The Intellectual World of a Sixteenth-Century Musical Humanist*, ed. Iain Fenlon and Inga Mai Groote (Cambridge: Cambridge University Press, 2013), 195–222; Anthony T. Grafton and Urs Leu, *Henricus Glareanus's (1488–1563) Chronologia of the Ancient World: A Facsimile Edition of a Heavily Annotated Copy Held in Princeton University Library* (Leiden: Brill, 2013); Anthony T. Grafton, 'Teacher, Text and Pupil in the Renaissance Class-Room: A Case Study from a Parisian College', *History of Universities* 1 (1981): 37–70; Ann Blair, 'Lectures on Ovid's Metamorphoses: The Class Notes of a 16th-Century Paris Schoolboy', *Princeton University Library Chronicle* 50 (1989): 117–44; Jean-Marc Mandosio and Marie-Dominique Couzinet, 'Nouveaux éclairages sur les cours de Ramus et de ses collègues au collège de Presles d'après des notes inédites prises par Nancel', in *Ramus et l'Université* (Paris: Rue d'Ulm, 2004), 11–48. Lecture notes, even where extant, are a forbidding source which few have considered carefully, though see a preliminary fivefold taxonomy outlined in Jean Letrouit, 'La prise de notes de cours sur support imprimé dans les collèges parisiens au XVIe siècle', *Revue de la Bibliothèque National de France* 2 (1999): 47–56. More generally, see Ann Blair, 'The Rise of Note-Taking in Early Modern Europe', *Intellectual History Review* 20, no. 3 (2010): 303–16. For outstanding examples in the history of science, see Darin Hayton, 'Instruments and Demonstrations in the Astrological Curriculum: Evidence from the University of Vienna, 1500–1530', *Studies in History and Philosophy of Science Part C*, 41, no. 2 (2010): 125–34; Darin Hayton, *The Crown and the Cosmos: Astrology and the Politics of Maximilian I* (Pittsburgh, PA: University of Pittsburgh Press, 2015), chapter 3; Owen Gingerich, 'From Copernicus to Kepler: Heliocentrism as Model and as Reality', *Proceedings of the American Philosophical Society* 117, no. 6 (1973): 513–22.

manuscript to print.[22] As argued above and detailed in the next chapter, it was not until the 1490s that masters and printers began to experiment with the new printing presses for the standard arts course. By 1503, a student at Cardinal Lemoine could read the entire arts cursus reading only the printed textbooks of his masters. Beatus Rhenanus' archive allows us to detail the ways the new printed textbooks grew out of, and meshed with, their manuscript context. In Beatus' lecture notes, we also see an early use of printed textbooks in the classroom. The *figura* used by Beatus' master (and the idealized teacher in his textbook) reminds us that either his teacher lectured from the printed book or, perhaps, Beatus himself had a copy open during the lecture.

COPIA IN PRACTICE

When a youth joined the University of Paris around 1500, he faced a tremendous challenge to know a great deal. The medieval arts curriculum was already challenging in scope. It inherited from antiquity the ideal of the complete cycle of learning for the free citizen—the *encyclos paideia*, or encyclopaedia—modified to suit Europe's clerical elite. To gain a degree, a student would have heard at least three years of lectures, culminating in a series of formal disputations. Statutes imposed a demanding discipline on the *artes* through two practices: *lectio* and *disputatio*.

Lectio grew out of the ancient practice of a teacher reading and glossing authoritative texts, in whole or in part. The student proved his attendance after the fact, by swearing an oath and getting his master to provide signed testimony. At Paris, in order to be licensed to teach, a *licentiatus*, the student walked in procession to the church of the ancient Abbey of Sainte-Geneviève, swore before the chancellor that he was twenty-one years of age, unmarried, and had heard at least one ordinary course (or two extraordinary[23]) on Priscian, Porphyry's *Isagoge*, the *Categories* and

[22] This moment comes with grand historical claims. The usual assumption is that print democratized knowledge by making textbooks quickly and widely available, a position both nuanced and promulgated by Elizabeth Eisenstein, *The Printing Press as an Agent of Change: Communications and Cultural Transformations in Early-Modern Europe* (Cambridge: Cambridge University Press, 1979), e.g. 102, 432.

[23] On ordinary and extraordinary lectures see the official statutes of 1215, CUP 1:78–9. These are translated in Lynn Thorndike, *University Records and Life in the Middle Ages* (New York: Columbia University Press, 1944), 27–30. In principle, ordinary lectures addressed first the *Organon* and then the natural philosophy of Aristotle, and these lectures were taught by licensed masters alone. They were held through the morning to 3 p.m. on non-holidays from 10 October to the first Sunday of Lent—discounting the large number of saints' days, ordinary teaching days totalled about seventy-five days of lessons. Even after deducting the high holy days (Christmas, Easter), when all classes were expressly forbidden, this left considerable time for 'extraordinary' lectures. These took place in the afternoons and evenings (after 3 p.m.), the numerous feast-days (discounting Sundays, about forty-five from October to March), September (before ordinary lectures began in October), and Lent. These figures are deduced from Charles Thurot, *De l'organisation de l'enseignement dans l'Université de Paris au Moyen-Âge* (Deis, 1850), *passim*. That many feast-days were in fact spent in study is seen in the statutes at Erfurt where statutes expressly forbad lectures during the high feasts of Christmas and Easter—evidently this was a danger: H. Wissenborn, *Acten der Erfurter Universität*, vol. 2 (Halle: O. Hendel, 1881). A precise total of extraordinary days of lectures is nearly impossible to calculate, but in principle there were at least as many days available for extraordinary lectures as for ordinary ones. Normally it

On Interpretation of Aristotle, and his *Prior* and *Posterior Analytics*. Moreover, he should have heard Boethius' commentaries on Aristotle's *Topics* and [Ps-Aristotle's] *Divisiones*, Priscian on accidents, Donatus' grammar called *Barbarismus*, Aristotle's *Physics*, *On the Heavens*, *On Generation and Corruption*, the *Meteorology*, *On the Soul*, *On Sense and Sensation*, *Memory and Reminiscence*, *On the Length and Shortness of Life*, and 'at least 100 lessons' on mathematics and astronomy.[24] As curricular reforms from 1452 make clear, the 'greater part' of the *Ethics* were also to be read before a bachelor could be licensed to teach.[25]

In practice, *lectio* could take many forms. Each master could interpret these statutes as he saw fit. It is usually unclear whether *lectio* refers to a master's spoken teaching or to the student's own reading under some master's oversight. To deal with the expansive curriculum, medieval masters typically also used digests, shorter florilegia and textbooks, perhaps records of spoken teaching, certainly reworked by students as they prepared for disputations. By the early sixteenth century, we have accounts of instruction by glossing happening outside lecture halls. In these cases, a student might read widely, borrowing glosses and copying notes from other students or from his master's books, prepared either independently or in a sort of tutorial.[26] The abbreviations and summary textbooks necessary to this form of study had long been available. Since the thirteenth century, students used *summulae* such as the standard logical textbook by Peter of Spain. By the fourteenth century, many masters gave up any attempt at compendious, continuous coverage of the text, and limited themselves to questions that interested them such as the *Summule naturalium* of Paul of Venice. Other works excerpted and digested favourite interpreters, such as the tradition of *Summule naturalium* of Pseudo-Albert the Great.[27] This jigsaw of authorities was a great amount to master in a few short years for young students, who often arrived at university at perhaps fourteen years of age, barely competent in Latin.

By May 1503, when Beatus Rhenanus arrived at Cardinal Lemoine, however, there were two new reasons that *lectio* might overwhelm a student. The first was the printing press. Although many presses were first set up in university towns in the 1470s, it took until the 1490s for masters to begin rethinking the structure of textbooks in printed formats.[28] The best known example of a textbook reimagined in print is Gregor Reisch's dialogues on the arts cycle, the *Margarita philosophica*

was on these days that students heard lectures on Aristotle's *Nicomachean Ethics* and mathematical works, which could be taught by either masters or bachelors who had been licensed to teach. For Italian universities, cf. Paul F. Grendler, *The Universities of the Italian Renaissance* (Baltimore, MD: Johns Hopkins University Press, 2002), 143–6.

[24] CUP 2:678: 'Item, quod audivistis centum lectiones de Mathematica ad minus (*Istud per facultatem sic est interpretatum quod sufficit audivisse unum librum totalem mathematice, sicut tractatum de Spera, et alium librum actu audire cum spe audiendi usque ad finem sine fraude*).' Part of the oath of inception is given in César-Égasse du Boulay, *Historia Universitatis Parisiensis* (Paris, 1673), 4:273–4.

[25] The oaths for bachelors and licentiates are given in CUP 2:278.

[26] See the wonderful study of Groote, 'Studying Music and Arithmetic with Glarean'.

[27] Martin Grabmann, *Methoden und Hilfsmittel des Aristotelisstudiums im Mittelalter* (Munich: Verlag der Bayerischen Akademie der Wissenschaften, 1939).

[28] See section on 'Making in Universities and Print Shops' in Chapter 1 and Chapter 4.

(1503), written after his lectures at the University of Fribourg around 1490. It took time to reimagine the classroom around print. Beatus' book purchases show him spanning manuscript and printed textbook cultures during this period. His earliest studies are shown in a manuscript of Virgil's *Eclogues* from the late 1490s.[29] Beatus had copied the text of Virgil in his own hand in wide-spaced lines; between the lines he jammed glosses from his master's lectures, spilling over margins. Since antiquity, this had been a standard way to learn, while also acquiring one's own copy of a text. But around 1500 Beatus began to buy printed books, in truly remarkable numbers. While a century earlier the average student might have owned six or seven volumes, Beatus' purchase of over 250 printed books would shame most undergraduates today.[30]

The second new source of overwhelming *lectio* at Lemoine was the pressure to address the original texts of the ancient *auctores*. In his programmatic reading lists, Lefèvre characteristically focused on authors, rather than commentators—a theme now deeply embedded in the historiography of Renaissance humanism. Fifteenth-century Italian humanists such as Bruni, Vergerio, Guarino, and da Feltro gave lists of classical authors, Christian and pagan, who should be read free from commentary. Historians have placed Lefèvre in this tradition, chiefly focusing on the one passage of his commentary on Aristotle's *Politica* which provides a recommended list of the 'good authors' a youth should read.[31] The names are only a few: Boethius, Ptolemy, and Aristotle, *e fonte puri bibantur liquores*. Such standard university authors are sandwiched between the letters of Cicero, Pliny the Younger, and Francesco Filelfo, first, and the particularly useful Church Fathers, last. Such texts, 'drunk from a pure source', seem to present Lefèvre's humanist credentials as a kindly pedagogue intent on streamlining studies.

But even a glance at the contents of Lefèvre's books suggests that far from simplifying students' lives, the call *ad fontes* actually added to reading. In fact, Lefèvre's explicit ideals of learning disprove Lefèvre's advice by passing over in silence the details of texts and classroom. What books did students hold? What did masters present in lectures? By 1506, when he wrote this advice, Lefèvre had, with students, produced new textbooks for nearly all of the disciplines he mentioned, and he was about to produce editions of many of the church fathers and medieval mystics he recommended. In the case of nearly every school text, Lefèvre's books were much more than mere editions. In earlier handbooks on mathematics, logic, natural philosophy, and ethics he compressed the long books of Boethius and

[29] *Cahier écolier*, BHS MS 50.

[30] On medieval student ownership, see C. H. Talbot, 'The Universities and the Mediaeval Library', in *The English Library Before 1700: Studies in its History*, ed. F. Wormald and C. E. Wright (London: Athlone Press, 1958), 73; Jacques Verger, 'Le livre dans les universités du Midi de la France à la fin du Moyen Âge', in *Pratiques de la culture écrite en France au XVe siècle: actes du colloque international du CNRS, Paris, 16–18 mai 1992, organisé en l'honneur de Gilbert Ouy*, ed. Monique Ornato and Nicole Pons (Louvain-la-Neuve: Fédération internationale des instituts d'études médiévales, 1995), 403–20; cit. Daniel Hobbins, *Authorship and Publicity Before Print: Jean Gerson and the Transformation of Late Medieval Learning* (Philadelphia, PA: University of Pennsylvania Press, 2009), 20.

[31] Lefèvre, *Politica etc.*, 123v–124r, and see my discussion in the section 'Northern Conversions' in Chapter 2.

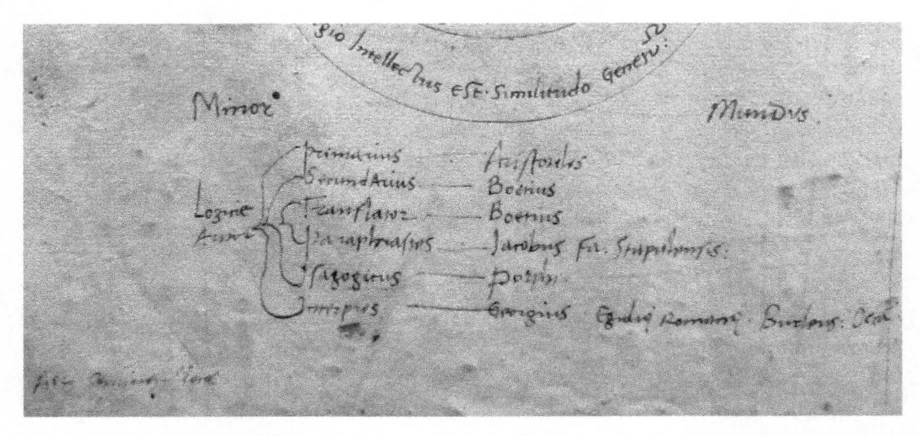

Figure 3.3. Beatus orders the authorities: Lefèvre, *Libri logicorum* (Paris: Hopyl and Estienne, 1503), BHS K 1047, guard page, detail.

Aristotle into short, digestible introductions to the author's works. Some later books on Aristotle's logical *Organon* (1502) and the *Ethica Nicomachea* (1497) also included the Aristotelian texts, often alongside Lefèvre's epitomes, paraphrases, and commentary. In other cases, notably the mathematical works, Lefèvre's short introductions took on a life of their own; from 1503, students at Cardinal Lemoine could study from a compendium of mathematical textbooks that were *not* Boethius, Euclid, or Ptolemy, and sometimes only distantly related to them.[32]

Such introductions increased the workload. With these new books, Beatus and his fellow students were accountable for *all* of Aristotle, not just the customary florilegia and digests of Aristotle, normally limited to questions of analysis that were current or fashionable in disputation. A telling note in Beatus' logic book elaborates the hierarchy of authorities one might find (Figure 3.3). In logic, the original author (*primarius*) is Aristotle, while his authoritative commentator (*secundarius*) is Boethius. Beyond these are several other figures: the translator, the paraphrast—here, Lefèvre himself—and the author of the discipline's introductory manual, the *isagogicus* Porphyry.

Lowest in this hierarchy is the *interpres*, the commentator who interprets the field of study. Here Beatus lists George of Brussels, Giles of Rome, Walter Burley, and William of Ockham. In fact, if we trace Beatus' notes throughout his library, we find him *also* reading authorities and textbooks that were current in other colleges. One example is his copy of Peter of Spain's *Summule*, abridged by Thomas Bricot and commented on by George of Brussels.[33] The book was spartan, with no prefatory matter, and its thick double columns of text are rather distant from Aristotle's own words: the book abbreviates and explains a thirteenth-century

[32] Lefèvre, Clichtove, and Bovelles, *Epitome*.

[33] Thomas Bricot, George of Brussels, and Peter of Spain, *Expositio Magistri Georgii super summulas* (Paris: Félix Baligault, 1495), BHS K 948. Beatus bought the book in 1502, perhaps in preparation for his studies in Paris the following year.

textbook. Yet it had a place in Beatus' ecosystem of authorities. Whatever the talk of purer sources, then, the study of Aristotle at Lemoine involved far more than Aristotle alone.

University disputations were the second central practice of university learning. Disputations were how the university made students accountable for what they learned by *lectio*. Growing out of the polemical school debates of the twelfth century, disputation had been institutionalized during the course of the thirteenth century as the university's main mode of examination.[34] The practice had tremendous cultural implications. Students would test each other, their masters, and their texts in 'ordinary disputations', while in *disputationes de quolibet* (about anything at all) public debate became spectacle.[35] In earlier generations, masters summed up disputations with a final verdict, but by the fifteenth century each student would end their degree in a series of public disputations that culminated in the student's own *determinatio* or summation. The very term *determinatio* became synonymous with graduation.[36]

Disputations marked a student's progress, and so shaped pedagogy. The most marked effect of disputation was on the format of lectures, as commentary took less the form of glosses than a series of *quaestiones*, in which authors lined up evidence *pro* and *contra* to their *conclusiones* and *corrolaria*. Thus *lectio* came to be based on *disputatio* in scholastic commentary. Even critics of these genres depended on the practice. Humanists from Bruni onward hurrumphed over a university culture that they saw as perniciously disputatious, but they never could bring themselves to reject the genre entirely. Leonardo Bruni famously modelled his own critique in an alternative form of disputation in the dialogue *De disputationum exercitationisque studiorum usu*. Although Lefèvre himself reserved his harshest words for 'sophistic' forms of disputation and studiously avoided reducing any discipline to a few favourite *quaestiones*, nevertheless the form deeply marked his paraphrases, which are most often divided into a series of *conclusiones*.[37] As Chapter 6 will discuss, Lefèvre's most innovative introductions he wrote as dialogues modelling dispute driven by friendship. Even students who were not captivated by the thrust and parry of competition needed these structures to thrive.

In this context, Lefèvre's approach, if copious in content, still raised anxieties over rigour. The register of MAs at Paris from 1512 shows that students of Lefèvre,

[34] See Alex J. Novikoff, *The Medieval Culture of Disputation: Pedagogy, Practice, and Performance* (Philadelphia, PA: University of Pennsylvania Press, 2013); Brian Lawn, *The Rise and Decline of the Scholastic 'Quaestio Disputata'* (Leiden: Brill, 1993). In particular, see the many works of Olga Weijers, notably *In Search of the Truth: A History of Disputation Techniques from Antiquity to Early Modern Times* (Turnhout: Brepols, 2013); Olga Weijers, *A Scholar's Paradise: Teaching and Debating in Medieval Paris* (Turnhout: Brepols, 2015).

[35] In particular, see Novikoff, *The Medieval Culture of Disputation*, chapter 5.

[36] In the statutes of Paris, a bachelor who 'determined' (*determinavit*) is one who successfully concluded their final *disputationes* and so merited the bachelor's degree.

[37] Weijers suggests that by the fourteenth and fifteenth centuries, many authors wrote treatises as an extension of such disputations. Olga Weijers, 'The Development of the Disputation Between the Middle Ages and Renaissance', in *Continuities and Disruptions Between the Middle Ages and the Renaissance* (Louvain-la-Neuve: Brepols, 2008), 140. On Bruni's views, see Leonardo Bruni, *De Disputationum exercitationisque studiorum usu* (Basel: Heinrich Petri, 1536).

Clichtove, and Bovelles came from many other colleges.[38] It would seem that Beatus and fellow students at Lemoine studied texts in the new intellectual style at the college while simultaneously preparing for examinations in the traditional *quaestiones* disputed in other colleges—to gain the experience needed to navigate successfully the Arts Faculty's examinations by *disputatio*. Because students at Cardinal Lemoine did not read more traditional textbooks in logic, the Basel printer Johann Amerbach worried that they might not be enough to pass their disputations. Amerbach sent his boys Basil and Bruno to Paris in May 1501.[39] Johann knew and admired Lefèvre,[40] but apparently he intended his sons to study with Scotist masters, placing them in the care of Ludwig Ber at the Collège de Sainte-Barbe, where Tataret was teaching. The boys had other plans, and began attending lectures at the Collège de Lisieux in short order. Within a month of learning their arrangement, the worried father issued a warning:

> I had in mind, when I sent you to Paris, that if the Scotist school still was strong, you would devote yourself to it—but I hear that it no longer is. I warn you: after you have been made logicians, follow the school more commonly practised, paying it vigorous attention, lest it be said after you return that I sent away young asses, but welcomed back a pair of great asses.[41]

In following letters, Johann vented his frustration that the boys had left Ber's care to follow their own interests. Bruno waited until the end of the school year to allay his father's fears. In a long letter, Bruno expressed his hope that the misunderstanding had cleared up, and to prove his industry he outlined his schedule. Holidays were for rhetoric and mathematics, and on regular mornings he took up logic 'in the classroom' along with 'a great many other studies'.[42] In the afternoons he went to the Collège du Cardinal Lemoine to hear the 'cursus Fabri'. Bruno did admit that at first he had reservations that the Fabrist logic, lacking commentary, would have to be supplemented with nominalist teaching to ensure success in disputations, but he had been assured this would not be necessary.

Such examples mark the beginning of a parting of the ways between schoolmen and humanist critics. On the one hand, the Amerbach boys would return to their

[38] Farge (ed.), *Students and Teachers*, s.v.

[39] The Amerbach boys could have been educated in their hometown of Basel, but their father's ambitions for them say something about the ongoing cultural dominance of the Parisian *academia*, even in humanist style. Amerbach himself had been educated at Paris by Johann Heynlin von Stein—a realist or rather, in Amerbach's terms, a Scotist, and one of the founders of the first Paris press. Amerbach's commitment to printing the ancients, especially the Church Fathers, can be traced to his own Paris days, perhaps followed by some years of travelling in Italy. On Amerbach, see CE, 1:47.

[40] Amerbach's letter to Lefèvre praises his work on Dionysius, in Alfred Hartmann (ed.), *Die Amerbachkorrespondenz* (Basel: Verlag der Universitätsbibliothek, 1942), 1:100.

[41] Amerbach to Bruno and Basil, from Basel, 11 June 1501, in Hartmann (ed.), *Die Amerbachkorrespondenz*, 1:120: 'Fui mentis, quando vos misi Parisius [*sic*], si adhuc viguisset via Scoti, ut ad eandem vos dedissetis, sed audio eam aboluisse. Moneo, postquam efficiemini logici, viam communiorem quae practicatur, eidem operam date fortiter insistendo, ne dicatur, postquam reveneritis, quod pullos asinorum Parisius miserim et magnos asinos reacceperim.'

[42] Hartmann (ed.), *Die Amerbachkorrespondenz*, 1:146: 'qui tum festis diebus rhetoricam et arithmeticam practicam, quem vulgo algorismum nominant, tum quottidie aliquid loyces nobis in camera resumit innumerasque alias lectiones resumeret, si omnibus satisfacere possemus (siquidem lectiones publicas collegii audimus).'

father's printing house in Basel, joining the circle of scholars such as Erasmus and Vives who enshrined a view of the older pedagogy as harsh and intellectually deadening.[43] Rabelais had great fun at the expense of those masters the elder Amerbach valued, making the eminent Tataret the author of *De modo cacandi* in his fabulous catalogue of St Victor's library.[44]

On the other hand Amerbach may have had reason to worry, and certainly other masters shared his concerns. Later in the decade, John Mair, the leading terminist at Paris, prefaced his commentary on the main university theology textbook of Peter Lombard with a dialogue. His choice of this literary genre was a pointed reminder that Mair preferred school method because of its rigour, not because his Latin was not up to humanist standards. Theology was neglected, Mair's humanist protagonist complained, because

> the children of the aristocrats and the rich leave logic and theology behind, and after hearing lectures on the *Summulae* [of Peter of Spain] rush off as swiftly as they can to law. It is easy to find a great bounty attending the *Summulae* at the College of Navarre or the College of Burgundy, but because of the penury of those preparing for the licentiate, at the end of the course the regents depart with their purse empty. And the whole thing is a mistake since once the wheat is abandoned they rush to the chaff.[45]

Mair's point, in 1510, was that traditional logical rigour mattered, and new abbreviations of traditional logic were popular because of old-fashioned greed, not superiority. Whether or not his diagnosis was correct, it shows that Lefèvre's compendious method of teaching attracted students and gained imitators, ambitious scholars like Symphorien Champier, who possibly studied with Lefèvre before he left Paris in 1494, and claimed that Lefèvre was his model and 'spiritual father'. In 1494 Champier began medical studies at Montpellier—while seeking court patronage in Lyon. There Champier had no students to worry about, just nobility to impress, and so he carried Lefèvre's mode of abbreviation further in his second publication, *Ianua logice et physice* (1498).[46] This handbook owed a great deal to Lefèvre's introductions, and Champier explicitly titled a scant fifteen-page section 'An exposition of difficult terms and words in the *Introductiones logicales* of the

[43] e.g. Erasmus' colloquy 'Ichthyophagi', and Vives, *Contra pseudodialocticos*; Rita Guerlac, *Juan Luis Vives against the Pseudodialecticians: A Humanist Attack on Medieval Logic: The Attack on the Pseudialecticians and On Dialectic* (Dordrecht: Springer, 1979).

[44] Rabelais, *Pantagruel*, chapter 7. Cf. Rabelais, *Gargantua*, chapter 37, and the *Quart livre*, chapter 21.

[45] Alexander Broadie, 'John Mair's Dialogus de Materia Theologo Tractanda: Introduction, Text and Translation', in *Christian Humanism: Essays in Honour of Arjo Vanderjagt* (Leiden: Brill, 2009), 424: 'Optimatum et locupletum liberi et logica et theosophia relicta ad leges ocyssime post auditas *Summulas* ruunt. Magnam affluentiam ad *Summulas* in Navvarae Collegio vel Burgundiae facile est reperire, sed ob poenuriam licentiandorum in fine cursus cum bursa vacua regentes discedunt, et totus error est quoniam tritico relicto ad paleas curritur.'

[46] Recently a previously-unknown work by Champier has been found which appears to have been earlier than the *Ianua*. It indicates that Champier saw himself, presumably in the early 1490s, as a nominalist. The position he takes in the *Ianua*, following Lefèvre, suggests that Champier was seduced by Lefèvre's teaching quite early on, away from the nominalist path towards a more humanist approach to the philosophical arts. Brian P. Copenhaver and Thomas M. Ward, 'Notes from a Nominalist in a New Incunabulum by Symphorien Champier', in *Essays in Renaissance Thought in Honour of Monfasani*, ed. Alison Frazier and Patrick Nold (Leiden: Brill, 2015), 546–604.

most eminent arts professor Lefèvre'.[47] If this was the next stage in ever-shorter introductions, Amerbach's and Mair's reservations about catering to student tastes may not have entirely missed the point.

New intellectual fashions had to be adapted to the required statutes and rigorous examination by disputation, and presented Beatus and his fellow students with a challenge: how do you learn everything, fast?

EXERCITATIONES INGENII

The solution to the problem of *copia* was, as we all know, more books and more writing—especially writing notes by hand in printed books.[48] It appears that Beatus Rhenanus began his compulsive book-buying in 1501, when he became Gewiler's teaching assistant, perfecting his studies before going to Paris along with several other Sélestat students, including Bruno and Basel Amerbach, Johannes Sapidus, and Michael Hummelberg. All of them would, with Beatus, become key members of the Rhineland *sodalitas* which welcomed Erasmus as their own a decade later.[49] The central texts that Beatus purchased were, of course, textbooks. In Paris, most strikingly, these were the new textbooks that his teachers Lefèvre, Clichtove, and Bovelles had written as the standard of the *cursus Fabri*, though Beatus also bought other standard textbooks to serve as references (to judge by their drastically sparser annotations). But he also bought smaller pamphlets to supplement his studies. These gesture towards a larger material context—many of them are not simply textbooks, but 'how-to' reference books for the ambitious student.

Several of his little books belonged to a venerable tradition. As medievalists know well, students—and indeed the most ambitious medieval philosophers, from Aquinas to Buridan—had long used *florilegia* as a quick way to navigate the usual doctrines that structured a domain of study. Probably in his last year at the grammar school of Sélestat, Beatus bought one by Pseudo-Bede, a *Repertorium or general table of the authority of Aristotle and the Philosophers*.[50] Most such florilegia in his collection, however, have to do with the moral ends of knowledge. For example,

[47] Symphorien Champier, *Ianua logice et physice* (Lyons: Guillaume Balsarin, 1498): 'Expositio terminorum seu vocabulorum difficilium in eminentissimi artium professor Iacobi Fabri in logicem introductiones.' The section closed with a short dialogue between Symphorien and Lefèvre.

[48] The observation is made pointedly in Peter Stallybrass, 'Printing and the Manuscript Revolution', in *Explorations in Communication and History*, ed. Barbie Zelizer (London: Routledge, 2008), 111–18. On the broader implications for note-taking as an effort to both limit and augment knowledge, see also Ann Blair, *Too Much to Know: Managing Scholarly Information before the Modern Age* (New Haven, CT: Yale University Press, 2010); Ann Moss, *Printed Commonplace-Books and the Structuring of Renaissance Thought* (Oxford: Clarendon Press, 1996); Terence Cave, *The Cornucopian Text: Problems of Writing in the French Renaissance* (Oxford: Oxford University Press, 1979).

[49] Rhenanus' *Vita* was written by Johannes Sturm, edited by Hubert Meyer in Jean Sturm, 'Vie de Beatus Rhenanus', trans. Charles Munier, *Annuaire. Les Amis de La Bibliothèque Humanist de Sélestat* 35 (1985): 7–18. On these early years, see Walter, *Trois Profils*. The list of books is in Knod, *Aus der Bibliothek*.

[50] Ps-Bede, *Repertorium sive tabula generalis autoritatem Aristotelis et philosophorum, cum commento per modum alphabeti* (Cologne: Heinrich Quentell, 1495), BHS K 810k.

an alphabetized set of proverbs from Pseudo-Seneca,[51] or even Filipo Beroaldo's *Little Work on Happiness*, a tract which in fact functions as a concatenation of *dicta* from the ancients and Church Fathers.[52] In the context of ideals of education as a *cura animi* in Lefèvre's pedagogical programme (described in the previous chapter), these florilegia supply material for what emerge in Beatus' notes as *exercitationes ingenii*.[53]

While preparing for Paris at the grammar school of Sélestat, Beatus carefully examined Battista Guarino's famous essay on *The Method and Order of Teaching and Learning*, claimed to represent the practices of his father's school in Ferrara. Anthony Grafton and Lisa Jardine argued that the rote learning of Guarino's school actually created docile bureaucrats, and that the claims for the freedom of learning were in fact sales tactics espoused by some of the school's most successful pupils, such as Vittorino da Feltre.[54] If Beatus had his bottom line in view, he never said so, although he did read Guarino for clues on how to conduct a life of learning, perusing with pen the sections on Virgil as a lifelong companion and on Cicero and Plautus as measures of eloquence and wit. These themes are readily associated with the fifteenth-century teachers of the *studia humanitatis*.[55] But Beatus attended most closely to practices related to the art of memory. The art of memory has been widely studied; perhaps less remarked is how Guarino links note-taking practices and the art of memory.[56] Rewriting what one has learned *as if about to teach it* has numerous cognitive benefits: 'For we are more attentive to those things from which we seek praise; for this sort of activity marvellously sharpens the wit, frees the tongue, generates quickness of the pen, spurs a complete attention to the matters in question, strengthens the memory, and finally presents the student with a store-room of explanations and memory resources.'[57]

[51] Ps-Seneca, *Proverbia Senece secundum ordinem Alphabeti* (Strasbourg? 1496?), BHS K 981.

[52] Filipo Beroaldo, *De felicitate opusculum* (Bologna: Plato de Benedictis, 1495), BHS K 810g.

[53] With this phrase I intend to evoke (and refine) recent efforts to frame Pierre Hadot's insights for the early modern period, e.g. Sorana Corneanu, *Regimens of the Mind: Boyle, Locke, and the Early Modern Cultura Animi Tradition* (Chicago: University of Chicago Press, 2012).

[54] Anthony Grafton and Lisa Jardine, *From Humanism to the Humanities: Education and the Liberal Arts in Fifteenth- and Sixteenth-Century Europe* (Cambridge, MA: Harvard University Press, 1986).

[55] For the classic picture, see Eugenio Garin, *L'éducation de L'homme Moderne*, trans. Jacqueline Humbert (1957; Paris: Fayard, 2003).

[56] Frances A. Yates, *The Art of Memory* (Chicago: University of Chicago Press, 1966); Mary Carruthers, *The Book of Memory: A Study of Memory in Medieval Culture* (1990; 2nd edn, Cambridge: Cambridge University Press, 2008); Lina Bolzoni, *The Gallery of Memory: Literary and Iconographic Models in the Age of the Printing Press*, trans. Jeremy Parzen (Toronto: University of Toronto Press, 2001). For bibliography on the *ars memoriae* well into the seventeenth century, see Rhodri Lewis, 'A Kind of Sagacity: Francis Bacon, the Ars Memoriae and the Pursuit of Natural Knowledge', *Intellectual History Review* 19, no. 2 (2009): 155–75. Lefèvre is often spotted on the margins of histories of mnemonic practices, such as Paolo Rossi, *Logic and the Art of Memory: The Quest for a Universal Language*, trans. Stephen Clucas (1983; London: Continuum, 2000), 29, 38.

[57] Underlined by Beatus in Baptista Guarino, *De modo et ordine docendi ac discendi* (Heidelberg: Henr. Knoblochtzer, 1489), BHS K 810e, B3r: 'Attentiores enim ad ea sumus ex quibus laudem venari studemus, hoc exercitationis genus mirifice acuit ingenium, linguam exploit, scribendi promptitudinem gignit, perfectam rerum noticiam inducit, memoriam confirmat, postrem studiosis quasi quadam expositionum cellam promptuariam et memorie subsidium prestat.'

Not only did Beatus underline this passage, but on a blank page he excerpted and reflected at length on Guarino's classical sources for the memory practices of systematic recollection.[58] Guarino had cited the Pythagoreans, who meditated on what had been said and learned every evening, and who devoted one day a month to reviewing everything learned in the previous four weeks. Beatus first identified Guarino's source, then wrote out the passages from *De senectute* where Cicero reported these Pythagorean habits—Beatus called this Cicero's *ars memorativa*. In Beatus' notes, the advice expanded. Beatus also associated this art of memory with the wondrous powers of ancient sages. As Cicero had said, 'I now work as much as I can on orations about augural, priestly, and civil law, and for the sake of my memory, according to the practice of the Pythagoreans, every evening I recall to memory whatever I said, heard, or did during the day. These are the exercises of wit (*exercitationes ingenii*); these are the racecourses of the mind.'[59] These are also, Beatus added, the 'practice of the seers' (*mores presagoreum*) to which Virgil had alluded in various places.[60] Beatus concluded the series of notes with some advice to himself: 'Put these things up in your room [i.e. on the bedroom wall], and you may progress in remembering them, with the help of God. Amen.'[61]

In this context, note-taking practices belong within the larger rhythms of daily discipline, or what Cicero here called the *exercitationes ingenii*. 'No one', as a well-worn line from Sallust put it, 'ever exercised their wit without a body'.[62] Most remarkable about Beatus' perspective on such practices is the seamless interweaving of bookish tools and the bodily concerns of scholars. In fact, right after Guarino in the *Sammelband* we find a handy manuscript collation of notes on Virgil, Plato, and Ovid, relaying their advice on how heavenly influences and bodily habits influence the life of letters—all under the heading 'against decadence and drunkenness' (*contra luxuriam et ebrietatem*).[63]

The importance of the body in *exercitationes ingenii* is perhaps best seen in light of the first book in the *Sammelband*, Ficino's *De vita libri tres*.[64] There Ficino presented the scholar as a medical patient, constantly manipulating the physical influences of the stars through diet and drugs as much as through paper tools. For a youth like Beatus, learning to be a scholar and to manage one's memory was a matter for physical and medical manipulation too, informed by natural philosophy. Another

[58] Blank page inserted after Guarino, *De modo*, BHS K 810e.

[59] Blank page inserted after Guarino, *De modo*, BHS K 810e, quoting from Cicero, *De senectute*, 38: 'nunc quam maxime conficio orationes, ius augurum, pontificium, civile tracto Pythagoreorumque more extende memorie gratia quid quarumque die dixerim, audiverim, egerim, commemoro vesperi. He sunt exercitationes ingenii; hec curricula mentis sunt.'

[60] Virgil mentioned the Sybil in *Aeneid* VI, the practices of a *vates* in *Georgics* IV, and was widely assumed in the Middle Ages to have presaged Jesus Christ in *Ecloques* IV.

[61] Blank page inserted after Guarino, *De modo*, BHS K 810e: 'Supradicta affige cubiculi tuo, et recordare ea proficies auxiliante deo. Amen.'

[62] Sallust, *Catalinae bellum*, 8: 'Ingenium nemo sine corpore exercebat.'

[63] BHS K 810f [=MS 325], Incipit *Publii Vergilii carmen. Contra Luxuriam et ebrietatem.* e.g. 'Ingenium ingenuosque facit ut rapiatur ad artes utque homini culto pectore preset homo, et teste Empedocle dicente amicitia et concordia contineri elementa que procul dubio ea deficiente defluerent.'

[64] Marsilio Ficino, *De triplici vita* (Bologna: Benedetto I Faelli, 1501), BHS K 810a.

companion resource was his copy of Pseudo-Albert the Great's *De secretis mulierum*.[65] No doubt youths opened the pamphlet for titillating details about women's genitalia; yet the core of the book would have interested anyone worried about managing their talents in light of planetary influence, as an account of how celestial bodies affect the development of embryos, including their minds. Another of Beatus' how-to purchases, the *Tractatus utilissimus Artis memorative* of Matteoli Perusino, suggests practical methods for managing memory using cognitive 'rules' as well as medical remedies. The rules reviewed familiar practices from standard authorities, such as Cicero and Quintillian: cultivating attention, delight and wonder in listening, careful subdivision of topics, regular rumination on the subject, etc. But the book would have reminded Beatus that such techniques depended on an order of nature. Reminiscence, for example, should be practised in the same order in which things make an impression on the soul, via the senses. This embodied approach to mental discipline was completed in the second part of the tract, which addressed medical remedies. Here Perusino recommended a dietary regime and various drugs. One avoid foods that make one 'superfluous', such as wine, garlic, and beans; instead one should eat lots of buglossa and ginger, to maintain the wet and cold conditions that keep a memory impressionable. Meanwhile, daily doses of the confection of anacardina in the amount of a chickpea would help the memory, along with small amounts of the 'oil of philosophers'.[66]

Beyond the magical bestseller of Ficino, Beatus skirted the less reputable forms of intellectual magic available, even though such books often promised instant access to the disciplines. Lefèvre's own *De magia naturali* included sections on the names of God, a Cabbalistic theme reprised in his commentary on the Psalms (1509), which Beatus also owned and annotated carefully.[67] This focus on the divine names was a main theme in high humanist magic, following Pico and Ficino. Like those authors, Beatus was surely well aware of the medieval tradition of the *Ars notoria*, which used the names of God and angels to conjure immediate knowledge, the knowledge given to Solomon. Some of the most interesting examples of such texts included visual aids as the focus of theurgic prayer to shortcut the labour of mastering the arts cycle.[68] These practices would have been close to mind when Beatus recalled the 'practices of the seers' in Virgil. But the closest Beatus came to using such shortcuts himself was his purchase some years later of Johannes Foeniseca's *Quadratum sapientiae*, a pamphlet which offered in the space of a few pages all the riches of learning: the seven liberal arts, including the Hebrew alphabet and applications of mathematics, as well as an overview of the 'Mosaic

[65] Ps-Albert the Great, *De secretis mulierum* (Speyer: Johann and Conrad Hist, 1483), BHS K 981a.

[66] Matheolus Perusinus, *Artis memoratiue Matheoli Perusini medicine Doctoris praestantissimi, Tractatus vtilissimus: com nonullis Plinij et Gordani documentis* (Strasbourg: Mart. Schott, 10 Oct.1498), BHS K 981d. Oil of philosophers was made from distilling roasted brick dissolved in oil. *Confectio anacardina* is made from cashew nut extract. Both of these remedies are given by Mesue in his *Antidotes*.

[67] Brian P. Copenhaver, 'Lefèvre d'Étaples, Symphorien Champier, and the Secret Names of God', *Journal of the Warburg and Courtauld Institutes* 40 (1977): 189–211.

[68] Michael Camille, 'Visual Art in Two Manuscripts of the Ars Notaria', in *Conjuring Spirits: Texts and Traditions of Medieval Ritual Magic*, ed. Claire Fanger (Stroud: Sutton, 1998), 110–39.

metaphysics', with all the elements of the created world arranged in a massive, coloured, foldout.[69]

MARGINS AND ENDPAPERS

Beatus' *exercitationes* left his books full of prose annotations, dicta, diagrams, and schemata, as he wrestled the arts cycle into categories, making it ready for use. Early modern students kept notes in two ways: (1) they wrote under the direction of a teacher, glossing as they went through a text; and (2) they independently turned their books into reference devices, underlining or repeating commonplaces and headings in the margins.[70] Beatus did both, though it is sometimes hard to tell the difference between tutored glosses and independent notes. He wrote some of his lecture notes (*lecta*) in a plain paper notebook, which appears to be a fair copy of notes from oral delivery—though likely with a printed referent, as we saw at the beginning of this chapter. Second, he also wrote in his printed textbooks. For example, on his books on natural philosophy, we find large annotations written in the same neat hand he used for his clean notebook on the topic. It is usually impossible to say how much of these could be copied from a fellow student or a tutor's exemplary book, a practice we find a couple of decades later.[71] But some of the marginal glosses surely result from lectures in class. In one case, Beatus wrote out mathematical calculations which include mathematical errors within some of the steps.[72] Oddly, the errors do not affect the outcome; the calculation must have been copied, perhaps from a teacher or an early messy draft. On the same page we find blanks in his sentences, in a list of authorities that would naturally end with 'Regiomontanus'. Beatus must have missed the name while writing at speed, but left himself a space in order to return and fill in the blank—this suggests he was taking dictation *verbatim*.

While many of Beatus' books from these years have the occasional mark or note, only the university textbooks by Lefèvre, Clichtove, and Bovelles are heavily annotated (see Appendix). In those books, roughly a third of the pages are filled with paragraph on paragraph of gloss. Their endpages are covered with notes, paragraphs from other readings, and schemata.

These notes reveal the curriculum as the field of exercise for a whole range of practices for managing knowledge. Historians have considered the epistemic consequences of such practices by looking chiefly to natural history. Collecting, collating, taxonomizing and storing information from texts was integral to the

[69] Joannes Foeniseca, *Quadratum sapientiae: continens in se septem artes liberales veterum. Circulos bibliae iiii. in quibus metaphysica mosaica. Commentaria horum* (Augsburg: Miller and Foeniseca, 1515).

[70] Blair, 'The Rise of Note-Taking'.

[71] This was a mode of teaching employed by Heinrich Glarean in Basel, for example: Inga Mai Groote and Bernhard Köble, 'Glarean the Professor and His Students' Books: Copied Lecture Notes', *Bibliotheque d'Humanisme et Renaissance: Travaux et Documents* 73, no. 1 (2011): 61–91; Grafton and Leu, *Henricus Glareanus's (1488–1563) Chronologia*, 17–19.

[72] Lefèvre, *Textus de sphera* (1500), BHS K 1046c, a1v.

development of early modern natural history, in the accrual of collections that merged texts and objects, such as that by Ulisse Aldrovandi and Gesner, the *Kunstkammern* of merchants and princes, or in the specimen collections of mining companies; we have learned to see Baconian natural history in the context of literary commonplacing.[73] Beatus' notes show how the *cursus* of university disciplines, including mathematics, as the next section shows, could be mobilized not only to categorize knowledge, but to develop habits that would enable one to deploy that knowledge. To set those practices in perspective, we can sample Beatus' approach to these books from the endpapers bound to the front and back of these textbooks.

Endpapers play a special role for these annotations. In the margins, Beatus attends to narrower questions; in endpapers he makes larger connections. Genette and now William Sherman have analysed printed title pages and colophons as thresholds that frame a reader's way into and out of the text.[74] We find manuscript analogues of such thresholds in these endpapers, made by the student's own hand. In endpapers, Beatus considers the place of each discipline and its relation to others; he offers maxims about method and the role of study; he cites paradigmatic authorities.

Consider these in turn, in his books on logic, mathematics, and natural philosophy. First, Beatus wrote notes on endpapers that connect the broader vision of the arts cycle to its role in forming students. Short proverbs incite the student to studious attention. 'In every discipline, one should have truth for teacher.'[75] Fascinating general summaries evoke topics in the Fabrist approach to learning. One presents knowledge as the project of man, the 'little world', making within himself similitudes of the outer cosmos, or 'greater world' (Figure 3.4).

The image emblematizes central aspects of Lefèvre's and Bovelles's view of learning: the analogy of man and world, the hierarchy of the senses, and a constructivist vision of knowledge adapted from Nicholas of Cusa.[76] Above all, such notes present education as fundamentally a project of shaping the soul. Beatus excerpted Themistius' paraphrases on the *Physics*: 'just as medicine is for the body, so is the whole of philosophy for the soul'.[77]

[73] Ann Blair, 'Humanist Methods in Natural Philosophy: The Commonplace Book', *Journal for the History of Ideas* 53, no. 4 (1992): 541–51; Krämer, 'Ein papiernes Archiv für alles jemals Geschriebene: Ulisse Aldrovandis Pandechion epistemonicon und die Naturgeschichte der Renaissance', *NTM Zeitschrift für Geschichte der Wissenschaften, Technik und Medizin* 21, no. 1 (2013): 11–36; Richard Yeo, *Notebooks, English Virtuosi, and Early Modern Science* (Chicago: University of Chicago Press, 2014).

[74] Gérard Genette, *Paratexts: Thresholds of Interpretation*, trans. Jane E. Lewin (Cambridge: Cambridge University Press, 1997); William H. Sherman, 'The Beginning of "The End": Terminal Paratext and the Birth of Print Culture', in *Renaissance Paratexts*, ed. Helen Smith and Louise Wilson (Cambridge: Cambridge University Press, 2011), 65–87.

[75] Lefèvre, *Libri logicorum*, BHS K 1047, guard page: 'In omnibus disciplinis veritas pro doctore est habenda.'

[76] See Chapter 6. The account of knowledge as 'similitudes' of sense, reason, and intellect reflects Bovelles's adaptation of Cusanus, particularly from *De coniecturis*, book 1. See the manuscript tract of Bovelles edited by Emmanuel Faye, 'Nicolas de Cues et Charles de Bovelles dans le manuscrit "Exigua pluvia" de Beatus Rhenanus', *Archives d'histoire doctrinale et littéraire du moyen âge* 65 (1998): 415–50.

[77] Lefèvre and Clichtove, *Totius philosophiae naturalis paraphrases*, BHS K 1199, title page: 'Themistius, primo Physices: Est enim ut corporis medicina, ita animae philosophia perfectio'.

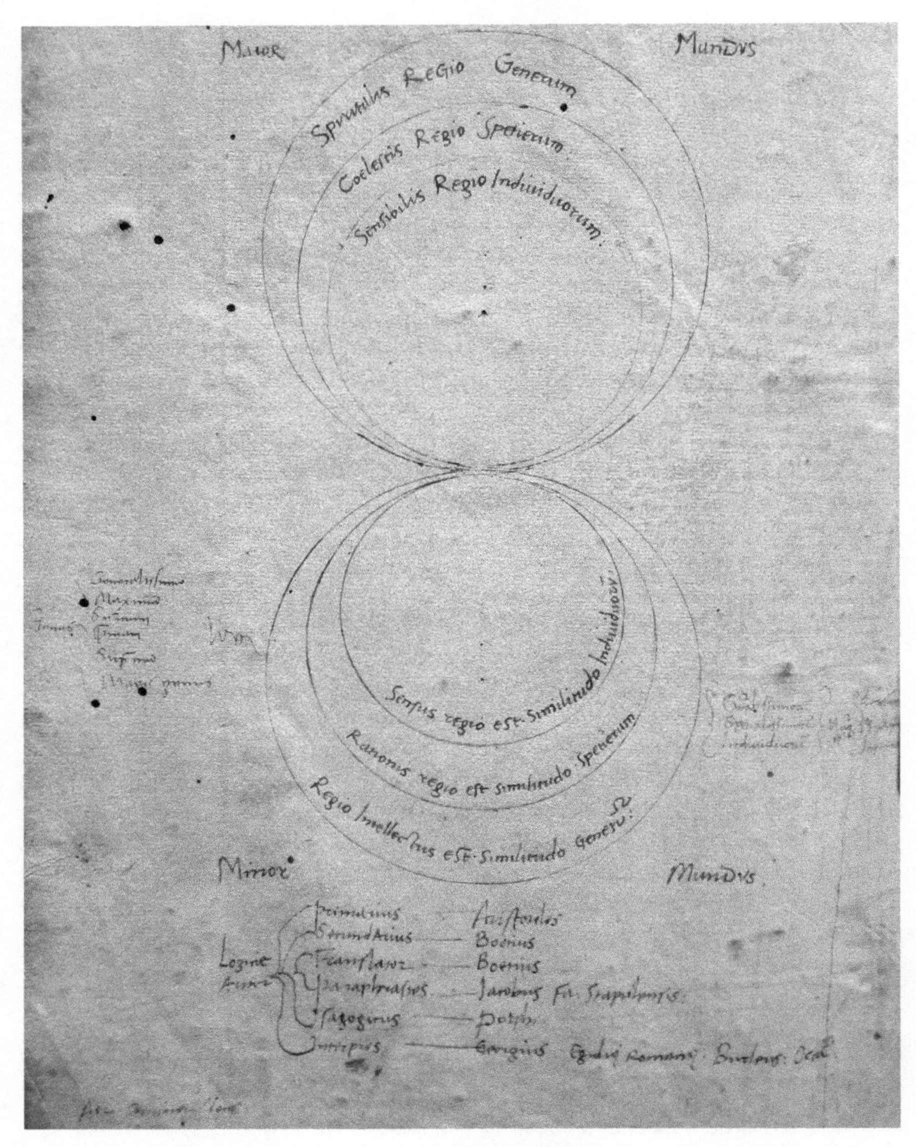

Figure 3.4. The *minor mundus* (man) reflects the *maior mundus* (world): Lefèvre, *Libri logicorum* (1503), BHS K 1047, end page with annotations by Beatus Rhenanus.

Second, on endpages Beatus wrote out visual and textual aids relevant to each discipline. In his natural philosophy textbook, he reports that 'the greatest goal of the present work is (as Themistius agrees) to teach what is universally present in natural things, insofar as they are natural'.[78] Synoptic diagrams are a handy reference

[78] Lefèvre and Clichtove, *Totius philosophiae naturalis paraphrases*, BHS K 1199, title page: 'Presentis operis summa intentio est (Ut themistius paraphrasus astipulatur) ea docere que universaliter insunt in rebus naturalibus, qua naturales sunt.'

for the 'discovery' (*inventio*) of key terms. In an endpage of his mathematical textbooks, Beatus drew a striking image for such invention: 'a figure from Pythagoras himself for finding the *genera* of number'.[79] The figure itself is a series of hammers set handle to head, from small to large, then labelled with the names of various intervals. The hammers evoke the oft retold story of how the inventor of the arithmetical and musical arts discovered that different-sized hammers striking anvils rang out pitches—so the story went—in harmonic intervals of whole integers. Here Beatus deployed the word *inventio* as a pun, since the figure recalls both the story of Pythagoras' supposed discovery and also is an aide-memoire for the specific intervals, from diatesseron and tone to diapason. With such synoptic figures, the endpapers extracted the few bits of information actually relevant to an arts student—a cheat sheet.

A third theme throughout Beatus' notes in these books is the relation of disciplines. These commonplaces are especially found on endpages, where Beatus frequently links his reading in one book with observations or authorities from another. Thus in his logical textbook he reinforces the coherence of knowledge by citing Aristotle's *Physics*, where the Philosopher observes that 'when one thing is inappropriately given, many other things follow, and from a single error many others arise'.[80] In his natural philosophy textbook, Beatus carefully delineated the goals of natural philosophy from those of other disciplines, noting that 'the end of speculative science is to know | the end of practical science is to do'.[81]

Throughout these notes, Beatus' favourite schematic tool is analogy. Analogies perform all the functions just listed: they situate learning in moral context; they help one recall in invention; they compare the disciplines. Typically, analogies visually correlate entities of different kinds, which therefore cannot be properly compared—they in fact belong to different disciplines. A particularly clear example comes from the *Politica* of Aristotle (Figure 3.5).

In fact, Beatus' schema picks out a theme in Lefèvre's commentary. Lefèvre observed that the component 'elements' that make up political science correspond to the parts of other disciplines, such as grammar, logic, geometry, arithmetic, or music.[82] As Beatus' annotation sets out, every discipline divides the world into minimal parts, whether people or letters, which combine into a maximal whole, such as a city or a complete discourse. Here, in his copy of the *Politics*, analogy serves as a way to rehearse the structure of a discipline. The practice of analogizing has an important function: it helps Beatus organize and navigate the disciplines, a function central to mastering university learning.

But such analogizing is hardly casual. This is one of the few times that Beatus carefully titles his schema, explaining its function of explanation 'by analogy also, that is, by proportion and similitude, to grammar' (*per Analogiam quoque id est proportionem*

[79] Sammelband BHS K 1046: 'formula ad inventionem numerorum generu ab ipso pythagoras'.

[80] Lefèvre, *Libri logicorum*, BHS K 1047, endpage: 'Philosophus primo physicorum: Uno inconvenienti dato multa contingunt, et ex uno errore facile multi surgunt.'

[81] Lefèvre and Clichtove, *Totius philosophiae naturalis paraphrases*, BHS K 1199, title page:

'scientiae <	speculativae> finis est <	scire
	practicae	operari'

[82] Lefèvre, *Politica*, 2v.

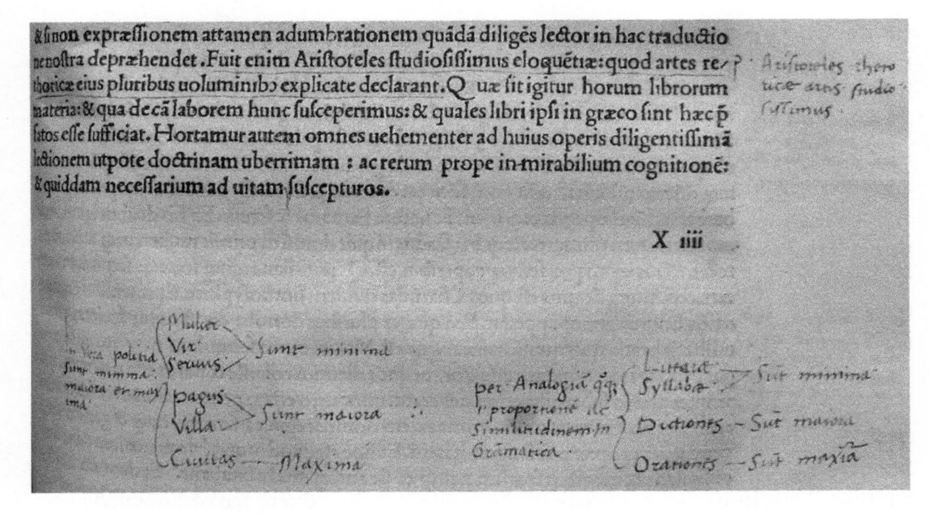

& si non expræssionem attamen adumbrationem quãdã diligẽs lector in hac traductio ne nostra deprætendet. Fuit enim Aristoteles studiosissimus eloquẽtiæ: quod artes re thoricæ eius pluribus uoluminibus explicate declarant. Quæ sit igitur horum librorum materia: & qua de cã laborem hunc susceperimus: & quales libri ipsi in græco sint hæc p latos esse sufficiat. Hortamur autem omnes uehementer ad huius operis diligentissimã lectionem utpote doctrinam uberrimam: ac rerum prope inmirabilium cognitionẽ: & quiddam necessarium ad uitam suscepturos.

X iiii

In the true city there are smallest, larger, and largest parts:	Woman Man Slave	are smallest	By analogy also, that is by proportion and similitude, to grammar:	Letters Syllables	are smallest
	Countryside Farm	are larger		Words	are larger
	City	largest		Discourses	are largest

Figure 3.5. Beatus observes analogies between grammar and city: Aristotle, *Politica*, in *Opera* (Venice, 1496), BHS K1276, 260r.

ac similitudinem in Grammatica). What could it mean that analogy is equivalent to proportion and similitude? There is a clue at the back of Beatus' lecture notes.

ANALOGY AS *ARS ARTIUM*

The archive of Beatus' annotated university books shows us Lefèvre's mathematical project as a reorientation of the medieval university disciplines. In a context of practices for managing the demands of *lectio* and *disputatio*, mathematics at Lemoine answered the need for a universal method between the diverse domains of the arts cycle.

The place of mathematics in Beatus' first year of studies is crucial evidence for a revaluation of mathematics among the Fabrists. As we learned above, students usually focused on logic in their first year of the arts course. In his *Summule* Peter of Spain formulated the orthodox starting point: dialect or logic was the foundational *ars artium*. Logic was the universal science, the 'art of arts', because it supplied analytical tools to every other discipline.[83] In contrast, mathematics was usually an

[83] Most early MSS and printed editions read: 'Dialectica est ars artium et scientia scientarum ad omnium methodorum principia viam habens'. See the apparatus in Peter of Spain, *Tractatus, Called Afterwards Summule Logicales*, ed. L. M. de Rijk (Assen: Van Gorcum, 1972), 1.

afterthought. Recall that, according to the Paris statutes, students were only required to swear that they had read a little geometry and astronomy near the end of the BA course. Some anonymous clerk added to the Faculty's records the tell-tale clarification that students should make this vow 'without lying'—a sure sign that some did lie.[84] Even in Vienna, where Andreas Stiborius and Georg Tanstetter renewed the vibrant astronomical culture of Regiomontanus, they offered only private and extraordinary lectures on mathematical topics.[85] But recall that Parisian masters could interpret the university's statutes as they saw fit: at Cardinal Lemoine mathematics came first, as part of the standard cursus. Beatus acquired his mathematics textbooks immediately on arrival in 1503, and the fact that he annotated this book alongside his logic textbook is itself a significant shift in the valuation of mathematics.

Yet students needed a method for ordering and comparing the disciplines—for mastering the *copia* of the university curriculum. Lefèvre, Bovelles, and Clichtove saw in mathematics the possibility of revisiting the foundations of university education. Accordingly, they set mathematics at the centre of the prescribed curriculum. Some of the most striking evidence for their efforts to give such a role to mathematics is in Beatus' notes. A series of theses at the end of Beatus' *cahier*, now partly obscured by a large water stain, outlines a candidate for a foundational tool that is an alternative to traditional logic: *analogia*. Consider as much of the list as can be deciphered:

8. An analogy is that thing that equals and even surpasses demonstration.
9. Every analogy is a numerical proportion.
10. Nothing is higher than analogy.
11. Aristotle claims that he discovered the secrets of the sciences by analogy.
12. The books that pursue the thread of analogy are the books one should read.
13. Analogy is the art of arts for investigating the sciences.
14. Analogy is suitable for each and every discipline.
15. What is known by analogy is more firmly known than by demonstration.[86]

[84] CUP 2:678. The original requirement reads: 'Item, quod audivistis centum lectiones de Mathematica ad minus.' The added note reads: '*Istud per facultatem sic est interpretatum, quod sufficit audivisse unum librum mathematicae, sicut tractatum de sphera, et alium librum actu audire cum spe audiendi usque ad finem sine fraude.*' The provenance of this note is surmised by Thurot, *De l'organisation*, 51.

[85] Hayton, *Crown and Cosmos*, 70, 72; Hayton, 'Instruments and Demonstrations'. On mathematics in this circle, see Franz Graf-Stuhlhofer, *Humanismus zwischen Hof und Universität: Georg Tanstetter (Collimitius) und sein wissenschaftliches Umfeld im Wien des frühen 16. Jahrhunderts* (Vienna: WUV-Universitüts Verlag, 1996); Helmuth Grössing, *Humanistische Naturwissenschaft: Zur Geschichte der Wiener mathematischen Schulen des 15. und 16. Jahrhunderts* (Baden-Baden: Valentin Koerner, 1983); C. Schöner, *Mathematik und Astronomie an der Universität Ingolstadt im 15. und 16. Jahrhundert* (Berlin: Dunker & Humblot, 1994), esp. chapter 5.

[86] BHS, MS 50, 235r:

[...] analogiam in terra
[...]
4. Analogia [...]
5. Syllogismus a transs [...] [...] iis
6. Habita per analogias [...] ecta sicut per demonstratione
7. Analogia munus

These may only be theses for disputation, not positive teaching. But the list, though partial, is suggestive, and it is clear that these theses explore *analogia* as an alternative master discipline to logic. Analogy is the *ars artium* (nos 13 and 14), and its mode of inquiry is surer than demonstration (nos 8 and 15). Beyond claims about its utility, in this list the clearest definition of analogy is as a numerical proportion or ratio.

I suggest that this list plays on layers of meanings that *analogia* had accrued in medieval usage, ranging from mathematical proportion to figures of speech. The original Greek meaning of ἀναλογία was mathematical, the 'equality of ratios': e.g. one is to two as two is to four (1:2::2:4).[87] Boethius, responsible for the mathematical vocabulary of the Latin West, translated ἀναλογία as *proportionalitas*, and medieval thinkers repeated this as a main meaning of the word. But analogy quickly gained more diffuse meanings. Both Plato[88] and Aristotle[89] used the mathematical sense, but also used the word to make diverse arguments based on qualitative similitudes: 'Dawn is to day as childhood is to man.' This was a second kind of analogy, which some argued was simply a figure of speech. Finally, a third option existed in between, for those who believed analogies pointed to deep realities that unified the terms of comparison. Readers of Pseudo-Dionysius' theology especially viewed analogy as a principle of unity along the hierarchy of being.

8. Analogia res equans atque superans demonstrationem
9. Omnis analogia est in proportione numeri
10. Nichil altius analogia
11. Aristoteles fatetur se per analogiam secreta scientiarum invenisse
12. Qui libri filum Analogie sequuntur legendi sunt
13. Ars artium in investigandis scientiis Analogia
14. Cuilibet discipline Analogia, idonea est
15. Illa que analogia cognoscuntur firmius cognoscuntur quaquam demonstratione

A comparison with H. Bonitz, *Index Aristotelicus* (Berlin: Reimer, 1831) confirms that these are not direct quotations from the works of Aristotle. The term analogia and its cognates, however, are found throughout the Aristotelian corpus, notably at *Posteriora analytica* 1.10 (76a), suggesting its role. A limited set of loci are discussed by Mary Hesse, 'Aristotle's Logic of Analogy', *The Philosophical Quarterly* 15, no. 61 (1965): 328–40.

[87] Euclid, *Elements of Geometry*, V def. 4, cf. VII def. 20. David H. Fowler, 'Ratio in Early Greek Mathematics', *Bulletin of the American Mathematical Society*, New Series, 1, no. 6 (1979): 807–46. Aristotle occasionally uses the term in this sense, e.g. *Poetica* 1457b. This definition is thought to reflect the harmonic theory of Archytas: Andrew Barker (ed.), *Greek Musical Writings*, Vol. II: *Harmonic and Acoustic Theory* (Cambridge: Cambridge University Press, 1990), 46–52. To supplement the overview of this paragraph, see Jonathan James Rolls, 'God and the World: Some Interpretations of the "Transcendental" Analogy of Being in Western Theology From the Thirteenth to the Sixteenth Centuries', PhD dissertation, Warburg Institute, University of London, 1999.

[88] See Plato at *Republica* VII, 534a6. At 531c, Plato may also allude to Archytas, who was his friend. The Platonic tradition had long attended to analogies, and the medieval commentary tradition on the *Timaeus* placed a premium on such tools. David H. Fowler, *The Mathematics of Plato's Academy: A New Reconstruction*, 2nd edn (Oxford: Oxford University Press, 1999), esp. chapter 4; Anna Somfai, 'Calcidius's Commentary to Plato's Timaeus and Its Place in the Commentary Tradition: The Concept of Analogia in Text and Diagrams', in *Philosophy, Science and Exegesis in Greek, Arabic and Latin Commentaries*, ed. P. Adamson, H. Baltussen, and M. W. F. Stone, 2 vols (London: Institute of Classical Studies, 2004), 1:203–330.

[89] For a representative sample, see *Posterior Analytics* 1.5, 74a18–24; *Metaphysics* 5.6, 1016b34–1017a2; *Nicomachean Ethics* 1.6, 1096b29. More than thirty passages in Aristotle deploy the word ἀναλογία in a wide variety of ways: see H. Bonitz, *Index aristotelicus* (Berlin: Reimer, 1831), 47–8.

In this context, *analogia* became a keyword in the great medieval debates over how to speak of God. Lefèvre himself was an influential reader of Pseudo-Dionysius, and this tendency in his thought can be measured in the commentary his student Gérard Roussel wrote on Boethius' *Arithmetica*, which expanded the various 'similitudes' of the order of being as a sign of the universe's underlying mathematical harmonies.[90]

It was possible to distinguish these linguistic and mathematical uses of the terms. Nevertheless, even within one domain, the lines could get perilously snarled. A good example is the main translation of Euclid's *Elements* by Campanus, where book 5 considered ratios. In the edition Lefèvre published in 1517, Campanus expounded on length on ratio 'not only in quantities, but also in weights, powers, and sounds', citing Plato's *Timaeus* and Boethius' *De musica*.[91] Not only does *proportio* extend beyond quantity, then, but he further cites Aristotle's *Categories* on relation (*habitudo*), arguing that ratio is equivalent to relation. Analogy extends to ratio, proportion, relation—the step Beatus made to similitude (*similitudo*) in his diagram on Aristotle's *Politics* is not far.

We can watch Beatus exploring the space between the linguistic and mathematical poles of analogy during his studies with Lefèvre. In his logical textbooks, he marked the linguistic definition that Boethius had made part of the standard school vocabulary.[92] Commentators on Peter of Spain's *Summule* also often mentioned *analogia* in the section on equivocal terms. In the margin of his copy of the *Summule*, Beatus underlined and annotated in large letters a classic example: the term 'man' refers first to a living person, and secondarily to the picture of a man. 'Such an equivocal is usually called, in another way, an analogue.'[93]

But it was in his grammatical studies that Beatus found ways to see analogy as a mathematical term. In the copy of Quintilian's *Institutiones* that Beatus bought in his first year at the Collège du Cardinal Lemoine, he primarily underlined sections dealing with various sorts of verbal analogy. Then, at book 1 chapter 6 Quintilian pointed out that 'reason [*ratio*] is found in analogy and sometimes in etymology'. Beatus observed that *analogia* is 'a Greek term for which a Latin equivalent has been found in *proportio*'. Quintilian also claimed that critical judgement deployed analogy to test 'all subjects of doubt by the application of some standard of

[90] Gérard Roussel, *Divi Severini Boetii Arithmetica duobus discreta libris, adiecto commentario, mysticam numerorum applicationem perstringente, declarata* (Paris: Simon Colines, 1521), e.g. at 29r–31v. In such contexts, I find no connection to the *analogia fidei* as an exegetical principle, raised by Augustine in *De utilitate credendi*, 5. This was taken up by Luther and Calvin, who knew Lefèvre's commentary on Romans (Paris, 1512), but they would have found Lefèvre uninspiring on the main passage cited in later debates, Romans 12:6. There, although Lefèvre recognizes the translation *analogia sive ratio fidei*, he offers no comment.

[91] Lefèvre, *Euclidis Elementa*, 57r: 'Non enim solum in quantitatibus reperitur proportio, sed in ponderibus, potentiis et sonis.' A comparable use of harmony and ratio to move across various disciplinary lines is found in Marsilio Ficino, *Commentary on Plotinus*, vol. 4: *Ennead III*, *Part* 1, ed. and trans. Stephen Gersh (Cambridge, MA: Harvard University Press, 2017).

[92] Boethius described *analogia* in his commentaries on the *Categories*, for instance, in Lefèvre, *Libri logicorum*, BHS K 1047.

[93] Bricot et al., *Expositio super summulas*, BHS K 948, e1r: 'et tale aequivocum alio modo solet vocari analogum'.

comparison about which there can be no question, the proof that is to say of the uncertain by reference to the certain'.[94] Beatus found the same ideas back in his copy of the fifth-century encyclopaedist Martianus Capella. Likely drawing on Quintilian, Capella expended nearly half his account of Grammar (Book 3) on analogy, likewise explaining it as the Greek equivalent of *proportio*. For Capella, analogy was the basis for comparing words in every way: sounds, morphology, and meanings can be measured against each other by analogy.[95] *Analogia* returns again in Capella's discussion of geometry, where he identifies four kinds of proportions, of which analogy is one. Once again, it is this definition Beatus attentively underlined in his own copy.[96]

All this broader reading shows Beatus encountering *analogia* as several things which are nevertheless closely related. As we shall see in a moment, the metaphysical assumptions of Lefèvre's circle appear to draw these meanings closer together, ultimately relating the whole order of being, by analogy, to its divine Creator. Ultimately, it may be futile to expect a well-formed definition of analogy: the term functions as a grab bag of fruitful concepts. But the list of theses crystallizes an attention to analogy that is only hinted at elsewhere. Beatus, or whoever first wrote the list, tries to transform the concept into a larger, if correspondingly vague, interdisciplinary tool. In later chapters this will help explain why the Fabrists repeatedly use the word to do methodological work—it is a candidate for a universal method.

We find analogy as a candidate *ars artium* in the endpages of Beatus' mathematical textbooks, bound together and now shelfmarked as BHS K 1046. The notes fall into the same categories sketched in the previous section. First, there are incentives to study in general: 'Just as a medicine is made stronger with many [ingredients], so a wise man fulfils his nature with all the disciplines.'[97] Second, the value and goals of mathematics in particular. Beatus repeated topoi familiar to every Renaissance student: 'Plato says that he who knows how to count knows all things. Plato also took care to have this epigram carved for the members of his academy: "Let no one lacking mathematics enter here." '[98] Beatus also learned to think of mathematics as having a double use, first as its own proper domain and second as a means to other kinds of speculation: 'The goal of thought in the mathematical disciplines is two-fold: first, that we know and understand. The second and more

[94] *Institutiones* 6.11, 4. Beatus paid special attention to these pages in his copy: Quintilian, *Institutiones cum commento Laurentii Vallensis ac Pomponii* (Venice: Pellegrino Pasquali, 1494), BHS K 1123a.

[95] Martianus Capella, *Martianus Capella and the Seven Liberal Arts*, Vol. II: *The Marriage of Philology and Mercury*, trans. William Harris Stahl and E. L. Burge (New York: Columbia University Press, 1977), 87–104; the book on grammar is at 64–106. Anthony Ossa-Richardson also suggests Varro as a possible source. See Daniel J. Taylor, *Declinatio: A Study of the Linguistic Theory of Marcus Terentius Varro* (Amsterdam: John Benjamins, 1974).

[96] Quintillian, *Institutiones*, BHS K 1123b.

[97] BHS K 1046 (Beatus' *Sammelband* of mathematical works by Lefèvre, Clichtove, and Bovelles, described in Chapter 5), endpages: 'Quemadmodum pharmaca ex multe conficitur potius, ita et sapiens, ex cunctarum disciplinarum consumata natura constat.'

[98] BHS K 1046a, title page: 'Plato eum omnia scire dixit qui numerare sciret. | Plato item pro socibus academiae suae hoc epigramma insculpi curavit: Nemo huc mathematicae expers introeat.' This line seems to come from memory, since Beatus did not simply copy this line from his textbook; there Lefèvre reported Plato's famous epigram differently, giving geometry instead of mathematics.

powerful is that we elevate our minds and intellect beyond ourselves, to philosophize about higher things, and that we invoke certain fine spirits.'[99] This last goal of mathematics, to consider higher things and 'invoke fine spirits' (*elicere subtiles spiritus*), is deeply ambiguous. The word *elicere* could mean 'conjure' or 'call out', suggesting a theurgical invocation of *daemones*; more simply, it could mean 'to draw out', suggesting that it draws on the finer filaments of one's own soul. As mentioned above, Lefèvre believed in the possibility of Ficinian manipulation of *daemones*, but his own interpretation of natural magic, especially after 1500, tends to steer safely clear of such dangerous practices. More likely Beatus was taught that mathematics *could* inform natural magical practices, but that it was best used as a guide to an ascetic, intellectual philosophy which made mathematics the deepest, most general mode of inquiry.

The thick notes Beatus added to the endpages of his mathematics books set a conceptual context for this approach to mathematics as super art. Beatus cites Pythagoras and Boethius (here labelled a Pythagorean), and alludes to the mathematical philosophy of Nicholas of Cusa. In Chapter 5 I will track some of these annotations through the winding paths of the text itself. For now, we may use them to survey the philosophical neighbourhood.

One overriding theme is how mathematics structures the soul's highest operations. 'Just as a straight right angle is the medium and equality between obtuse and acute angles, so in man reason [*ratio*] is a kind of equality and medium between the sensual appetite and the exterior senses which are located to the side [i.e. at the extremes].'[100] Beatus cites Lull as an authority for the view that imagination itself operates in mathematical modes: 'Quantity is the proper object of the imagination itself, according to Ramon Lull.'[101] Describing cognitive structures in mathematical terms was a move made frequently by Bovelles; it implied that reason itself ultimately depends on mathematical procedures.[102]

A second main theme in these endpages is the philosophical theology of Cusanus, often described as Pythagorean. Beatus explains the value of negative theology, central to the tradition Cusanus represented: 'We cannot more aptly philosophize about incomprehensible entities (as divine things are) than by negation.'[103] Cusanus pervades these notes. His favourite maxim, the Hermetic saying that 'God

[99] BHS K 1046, endpages: 'Duplex est finis cognitionis disciplinarum mathematicarum. Unus ut sciamus, cognoscamusque. Secundo et potior ut ex ipsis mentem et intellectum nostrum elevemus ad philosophandum de altioribus, eliciamusque quosdam subtiles spiritus.'

[100] BHS K 1046, endpages: 'Quemadmodum angulus rectus rectilineus, medius est inter obtusum et acutum ipserumque equalitas, sic ratio in hominem equalitas est ac medium quoddam inter sensualem appetitum, et sensus exteriores qui a latere collocantur.' Bovelles makes this point at several places, including *De sensu*.

[101] BHS K 1046, endpage: 'Quantitas est proprium obiectum ipsius phantasiae secundum Raymundum Lulium.' One place that Lull makes this argument is in *De modo naturali intelligendi*: see *Raimundi Lulli, Opera Latina*, ed. F. Stegmüller et al., Vol. VI, CCCM, 33 (Turnhout: Brepols, 1978), 177–223.

[102] Charles de Bovelles provides a similar mathematization of the soul in *Liber de intellectu etc.*, 18r–20r, 23r, 32v–33r, 58r, *inter alia*. Cf. Cusanus, *De docta ignorantia*.

[103] BHS K 1046, endpages: 'Non possumus aptius philosophari de incomprehensibilibus (ut sunt divina) quam per negationem.'

is a sphere of plenitude, whose centre is everywhere and circumference nowhere' is introduced in Beatus' notes as a 'Pythagorean definition of God'.[104] This Pythagorean theology turns on analogies, subtly adapting a classic motif of medieval theology. The Victorines, Bonaventure, Aquinas, and many others found in creation traces of the divine, which allow one to reason about the invisible through visible analogies. The difference is that Beatus focuses on the mathematical forms of analogy: he lines up unity, dyad, and triad which together form an *analogia* to the hierarchy of God, angel, and man.[105]

The point of analogy, whether in mathematics or metaphor, is to compare cases, to reason from one domain to another. Analogy therefore holds out the tantalizing possibility of transgressing disciplinary boundaries, such as that between mathematics and natural philosophy (i.e. physics):

> Those who philosophized intellectually say that the point relates to the line in the same way as the unit to number, so that they say a line is known from a flowing point (and a surface is from a flowing line). A physicist should deny this absolutely.[106]

Beatus proceeds to make the very analogies between natural philosophy and mathematics that 'a physicist should deny'. In Chapter 5 (especially the section 'A Reader') I will closely analyse how this use of analogy underlay Fabrist practices in mixed mathematics, disciplines that burgeoned in the Renaissance. At this point, I want to underline the place of such analogies within an effort to construct a universal philosophy. Beatus observes that such analogies are permitted to those who 'philosophize intellectually'. The 'intellectual philosophy' was Lefèvre's term for the Pythagorean mode of philosophizing that Cusanus had inherited from Boethius, and ultimately from Dionysius the Areopagite.[107] Emmanuel Faye has shown that the annotations of Beatus offer a rich perspective on what Lefèvre and Charles de Bovelles meant by this intellectual philosophy.[108] In particular, I suggest that Beatus' annotations on mathematics set analogy as the master concept of intellectual

[104] BHS K 1046, endpages: 'Diffinitio dei pythagorica: Deus est sphera plenitudines cuius centrum est ubique, et circumferentia nusquam.' A fuller treatment of quotations from Cusanus in Beatus' notes is Faye, 'Nicolas de Cues et Charles de Bovelles'.

[105] BHS K 1046, endpage:

Unitas	- Deus	
Binarius	- Angelus	} analogia
Ternarius	- homo ex actu	
	et potentia conflatus in materia	

[106] BHS K 1046, endpage: 'Qui intellectualiter philosophati sunt, dixerunt punctam eodem se modo habere ad lineam sicut unitas ad numeros, et ex puncti fluxu lineam (ex cuius perfluxu superficiem) percipi dixerunt. Quod quidem physicus constantissime negare debet.

		Motu	- mutatum esse	
In	{	Tempore	- nunc	} per analogiam cognoscaretur quod in aliquod
illorum		Magnitudine	- punctum	diceret necesse est utquod in aliis halucinetur.'

[107] e.g. Lefèvre's prefatory letters to Bovelles's *Introductio in artem oppositorum* (1501) and the *Opera* of Nicholas of Cusa (1514). The main contours of this intellectual philosophy are Reinhold Weier, *Das Thema vom verborgenen Gott von Nikolaus von Kues zu Martin Luther* (Münster: Aschendorff, 1967), 12–60; Emery, 'Mysticism and the Coincidence of Opposites in Sixteenth- and Seventeenth-Century France'.

[108] Faye, 'Beatus Rhenanus lecteur et étudiant'.

philosophy to go beyond Aristotelian method. By elevating analogical processes to a structure of the human mind, analogy becomes coextensive with reason itself. It encompasses the comparison of numbers and lines of Euclid and Pythagoras as well as words and qualities of Vergil and Aristotle; it even turns out to be the only way to speak of God, that is, to do theology. Analogy, modelled on Pythagorean mathematics, reaches towards a universal method, as a way to compare, navigate, and build on the all disciplines of the arts cycle.

* * *

There remain a host of questions of what analogy meant more concretely for the Renaissance history of method, and for the landscape of knowledge, including mathematics and natural philosophy—later chapters take up these questions. Here I wish to make the larger point that these new ways to imagine the relations of the disciplines emerged in the rich context of the Renaissance classroom, within texts set by the medieval cursus.

Such changes happened in the convergence of several practices for managing the *copia* of the classroom. University lectures and disputations were already taxing. Humanist textbooks did not simplify learning, as we have assumed from the educational handbooks by Vergerio and Guarino, whose rhetoric of simplification urged scraping away the excesses of disputatious predecessors. In fact, humanists like Lefèvre required even more of their students than their scholastic colleagues. Furthermore, Lefèvre and Bovelles represented a wisdom that emerged from complete (*perfectio*) mastery of the arts, which would only have intensified student anxieties. To master these disciplines, students like Beatus turned to what practices they could find: *exercitationes ingenii* from memory techniques to drugs to procedures for taking notes.

As we have seen, Lefèvre's circle was motivated by late medieval reform movements in which intellectual growth was closely tied to care of the soul. They echoed the concern of earlier generations that a certain approach to logic dominated the culture of disputation, and therefore foiled university aims. Masters at Cardinal Lemoine looked about for alternative methods to order the various disciplines—Beatus' notes themselves reveal a process of wrestling with analogy as a mathematical mode of bridging and ordering domains. Such notes, of course, had strictly limited circulation. But the culture of Cardinal Lemoine did circulate; it was the source of books read throughout Europe. The making of these books is next.

4

Inventing the Printed Textbook

Between 1490 and 1506, Jacques Lefèvre d'Étaples lived, learned, and worked with his students at the Collège du Cardinal Lemoine in Paris, a ten-minute walk from the print shops of rue St Jacques, where he produced new textbooks for the whole of the arts curriculum. These textbooks also bear the thumbprints of his students and fellow masters at the college, including Bovelles and Josse Clichtove, and the innovative energy of Henri Estienne the Elder's ambitious print house. By the summer of 1503, when the eighteen-year-old Beatus Rhenanus arrived in Paris to study with Lefèvre, a student at Cardinal Lemoine could pass his entire university career reading only the textbooks of Lemoine's masters. On arrival, Beatus immediately bought his major textbooks for the university arts curriculum: mathematics, logic, and natural philosophy, followed a year or two later by metaphysics and ethics—all recently published by Lefèvre, Clichtove, and Bovelles.

It is hard now to appreciate just how novel this experience was. As we saw in the last chapter, Beatus took notes either within, or with close reference to, printed books. Either his teacher lectured on physics directly from the printed book, or Beatus himself had Lefèvre's textbook open during the lecture. This had been impossible only a couple of decades earlier. In fact, Beatus' generation was the first to study the whole arts course in printed books. This point may run counter to our intuitions. After all, the first Paris press had been set up at the Collège de la Sorbonne in 1469. Yet when it came to the central books of the curriculum—such as those on logic—universities were late adopters of print. The earliest books printed at Paris, in 1469, were orthographic manuals and ancient classics such as Cicero, the subject of 'extraordinary' lectures taught by grammarians, rather than the logic, natural philosophy, and moral philosophy taught as the 'ordinary' or term-time curriculum.

For the first time in history, then, Lefèvre's printed books spanned the whole of the university arts course. In 1988 Charles Schmitt pointed out that the textbook came to dominate philosophical study over the course of the sixteenth and seventeenth centuries.[1] In a philosophical world that revolved around masters commentating and students disputing authoritative texts, the move to textbooks was a shift of tectonic proportions—slow, but profound and inexorable. The power of such textbook regimes to reshape entire cultures is well recognized in later cases, such as

[1] Charles B. Schmitt, 'The Rise of the Philosophical Textbook', in *The Cambridge History of Renaissance Philosophy*, ed. Charles B. Schmitt, Quentin Skinner, Eckhard Kessler, and Jill Kraye (Cambridge: Cambridge University Press, 1988), 792–804. See also Patricia Reif, 'The Textbook Tradition in Natural Philosophy, 1600–1650', *Journal of the History of Ideas* 30, no. 1 (1969): 17–32.

Philip Melanchthon's handbooks for Wittenberg, or the Jesuits at the end of the century. The tantalizing hope that all knowledge could be tucked tidily into an ordered compendium culminated in the grand edifices of such pedagogues as Johann Heinrich Alsted or Franco Burgersdijck, as well as the late seventeenth-century universal learning of the polyhistors. In his prescient sketch, Schmitt quite rightly noticed Lefèvre at the origins of this tradition.[2] But Schmitt and recent exploratory studies have not yet traced the defining characteristics of these books: how they were put together; their visual figures and tables; and their composition.[3] Even Ann Blair's history of scholarly reference works lies oblique to the university mainstream.[4] Unless we wade into the ocean of books designed directly for the university, it will be impossible to consider what larger intellectual changes the new printed page could bring about.

Lefèvre's books take us to the sources of these shifts in the manuscript culture of the medieval classroom. So far I have argued that the arts cycle—*especially* with the new humanist call to sources—was a challenge of condensation and memorization for an ambitious student like Beatus; we can sense this challenge and its answer in the 'little' texts Beatus possessed, his student notes, and the ways he and his teachers at Lemoine flirted with analogy as a conceptual tool for navigating the disciplines. In this chapter, I will show how this context shaped these textbooks in two ways. First, motivated by these concrete demands of university learning, students themselves turned to the printing presses. Senior students and junior colleagues collectively authored the experimental textbooks of Lefèvre's circle. Second, this authorial mode had cognitive consequences for the genres and organization of these textbooks, and hence for later debates on method.

COLLECTIVE AUTHORSHIP

Lefèvre's students were the publishing coterie who brought his manuscripts into print. Studies of early fifteenth-century authorship suggest that manuscript publication could happen in two stages and contexts. Around 1400 a bestselling author such as the university chancellor Jean Gerson could first publish something by circulating a work in his intimate community or coterie. But when a work reached a Europe-wide readership, it was because a second community had copied, reproduced, and publicized it.[5] Premodern publicity thus depended on links between

[2] Schmitt, 'Philosophical Textbook', 795.

[3] Eckhard Kessler, 'Introducing Aristotle to the Sixteenth Century: The Lefèvre Enterprise', in *Philosophy in the Sixteenth and Seventeenth Centuries: Conversations with Aristotle*, ed. Constance Blackwell and Sachiko Kusukawa (Aldershot: Ashgate, 1999), 1–21; David A. Lines, 'Lefèvre and French Aristotelianism on the Eve of the Sixteenth Century', in *Der Aristotelismus in der Frühen Neuzeit: Kontinuität oder Wiederangeignung?*, ed. Günter Frank and Andreas Speer (Weisbaden: Harrassowitz Verlag, 2007), 273–90. See note 93.

[4] Ann Blair, *Too Much to Know: Managing Scholarly Information before the Modern Age* (New Haven, CT: Yale University Press, 2010).

[5] Daniel Hobbins, *Authorship and Publicity Before Print: Jean Gerson and the Transformation of Late Medieval Learning* (Philadelphia, PA: University of Pennsylvania Press, 2009), chapters 6 and 7. On the complexities of manuscript publication, see especially 155–6. At 187–93, Hobbins describes

intimate contexts and wider networks. In Lefèvre's network, manuscript and print reflected such inner and outer communities. We see in his textbooks a transitional period in the history of publishing: first, works were circulated within the circle in manuscript and, second, the community republished them in print. Students located in this privileged inner circle thereby had the opportunity to become authors.

Several of Lefèvre's works were published first or only in manuscript. In 1504, Beatus Rhenanus copied a treatise by Lefèvre titled *Compendium analogiarum in De anima* into his student notebook.[6] The term 'compendium' at first glance suggests the text could be mere class notes, since in the same manuscript Beatus assigned that term to some arguments on natural philosophy that he claimed were from lectures.[7] But the manuscript also bears marks of publication: the signature markings copied with the *Compendium analogiarum*; the unusually neat hand in which Beatus wrote; and the presence of both title and colophon. It seems that Beatus copied the tract from a clean exemplar, not from lectures. In other words, the tract circulated at Cardinal Lemoine in manuscript. Its subject matter reinforces the sense of a limited-circulation treatise, for it was no ordinary commentary or school introduction: it claimed to introduce the real—hidden, esoteric—theological meaning of analogies that Aristotle had given in *De anima*. Lefèvre claimed analogies were hidden throughout Aristotle, and hid his own account of those analogies throughout his own published paraphrases and commentaries; this text, copied into the notes of an intimate student, suggests that he reserved such analogies for manuscript tracts shared only within a circle of insiders.

Although very few manuscripts from Lefèvre's circle are now extant, printed editions sometimes hint at their earlier lives as manuscripts.[8] Lefèvre's first mathematical work, after George of Trebizond convinced him to recover the discipline for the glory of Paris, was an edition of Jordanus de Nemore's *Elementa arithmetica*, a thirteenth-century treatise on number theory. Though first printed in 1496, the book must have been published in manuscript before 1494, because early that year Lefèvre reminded Germain de Ganay that he had already dedicated his edition of the *Elementa arithmetica* to Jean de Ganay.[9]

It mattered that these books suited the patrons. Mathematics was especially important for Jean de Ganay as a kind of 'mirror and rule of justice', because he was an important lawyer in the Paris *parlement*.[10] For Germain, a bishop called to priestly service, Lefèvre offered metaphysics, the study of eternal truths, 'which the

'coteries' as the first context of manuscript publication; coteries then became responsible for the broader publication. On this terminology see also Kathryn Kerby-Fulton and Stephen Justice, 'Langlandian Reading Circles and the Civil Service in London and Dublin, 1380–1427', *New Medieval Literatures* 1 (1997): 59–83.

[6] BHS ms 58, fols 206r–216v. [7] See the opening paragraphs of Chapter 2.

[8] The paucity of original exemplars has led most historians to assume that printers usually destroyed or recycled manuscripts after reproducing them.

[9] Jacques Lefèvre d'Étaples, *Introductio in metaphysicorum libros Aristotelis*, ed. Josse Clichtove (Paris: J. Higman, 1494), sig. a1v (PE, 21): 'Arithmeticam opus, consultissime Germane, tuo fratri dicatum est'.

[10] Lefèvre, *Introductio in metaphysica*, sig. a1v (PE, 21): 'estque arithmetica tamquam quoddam iustitiae speculum ac regula.'

Platonists call ideas', first studied as mysteries by the Egyptian priests and Chaldean magi. Neither treatise was an introductory school text; both addressed advanced parts of the university curriculum. Similarly, the copy of the *Compendium analogiarum in De anima* is a good example of how the socially esoteric circulation of manuscripts suited the esoteric nature of certain topics, as Lefèvre uncovered the hidden meanings of analogies in Aristotle's text. Manuscript publication here reflected not merely shared interests, but a certain intimacy.[11]

In fact, an economy of manuscript esoterica helped the community forge select connections over longer distances. As we saw in the first chapter, the Ganays' tastes ran to esoterica, attracting them to Ficino as well as the mathematical arts. In 1491, Germain was one of the few fifteenth-century borrowers of an astrolabe and planetarium from the Sorbonne's library.[12] Certainly, the Ganay brothers encouraged Lefèvre's interests in mathematics and the more dangerous arts—but in manuscript. Ficino's *prisca theologia*, perhaps with Germain's encouragement, was the starting point for Lefèvre's treatise *De magia naturali*, which Jean-Marc Mandosio has shown was only ever intended for manuscript publication.[13] The text exists in four copies, but only one manuscript has all six books; the others have only the first four books. Mandosio observes that the work was elaborated in several stages. After Book I was written, which expounds 'natural magic' proper, Lefèvre added the ensuing books at the request (or so he says) of Germain de Ganay, who is presented in Book II as the master whose secret teachings on the 'Pythagorean philosophy' of numbers are faithfully reported by Lefèvre. In contrast, the last two books in the Olomouc copy are not a dialogue, but a treatise; they were added, Mandosio suggests, to a text that had been published as a dialogue in manuscript in the early 1490s. The manuscript's esoteric themes of secrets of nature, Cabbala, alchemy, and *magia Pythagorica*, go a long way to explaining why this text was not meant for print; recent scholarship has emphasized that manuscript was the main mode of diffusing magical texts in early modern Europe.[14] Clearly this was not an introductory text; the work reflected Germain's place within a close circle of *cognoscenti*.

[11] Brian Richardson, *Manuscript Culture in Renaissance Italy* (Cambridge: Cambridge University Press, 2009), 1–2.

[12] Jeanne Viellard, 'Instruments d'astronomie conservés à la Bibliothèque du Collège de Sorbonne au XVe et XVIe siècles', *Bibliothèque de l'École des Chartres* 131 (1973): 591: 'XII. Germanus de Ganney: 1491. Die ultimo octobris 1491, ex ordinatione et consensu collegii, habuit magister Germanus de Ganney, consiliarius regis, astralabum cupreum cum certis aliis cupreis circulis planetarum in diversis partibus, et fuerunt extracta de parva libraria.... Restituit.'

[13] Jean-Marc Mandosio, 'Le *De magia naturali* de Jacques Lefèvre d'Étaples: Magie, alchimie et cabale', in *Les Muses secrètes: kabbale, alchimie et littérature à la Renaissance*, ed. Rosanna Camos Gorris (Geneva: Droz, 2013), 37–79. This puts to rest the assumption, traceable back to Lynn Thorndike (*A History of Magic and Experimental Science* (New York: Columbia University Press, 1924), 4:515ff) and Renaudet, that this manuscript was never 'published' because Lefèvre feared the theology faculty after they prosecuted the royal astrologer Simon de Phares in 1495. The affair is recounted in Jean-Patrice Boudet (ed.), *Le Recueil des plus celebres astrologues de Simon de Phares* (Paris: Honoré Champion, 1999), vol. 1.

[14] Federico Barbierato, 'Writing, Reading, Writing: Scribal Culture and Magical Texts in Early Modern Venice', *Italian Studies* 66, no. 2 (2011): 263–76.

These manuscripts of metaphysics, mathematics, and magic dedicated to the Ganay brothers were 'published'. That is, they circulated within Lefèvre's coterie of students, patrons, and friends. The school introductions that made Lefèvre famous first emerged in this context of manuscript publication.[15]

In fact, these later widely published books were first intended for an even more restricted readership of students. The very first work that we know Lefèvre to have written, the introduction to Aristotle's *Metaphysics*, was not printed until at least four years after it first circulated. It was first written in 1490, two years before he printed his *Totius Aristotelis philosophiae naturalis paraphrases*, and one year before he travelled to Italy; but it was not printed until 1494, when it was printed with some advanced dialogues on the same topic.[16] Indeed, the fact that one of Lefèvre's introductions was written before his travels to Italy raises the intriguing possibility that more of these trademark tracts (often taken to confirm dependence on Italian humanists) had been first circulated in manuscript, possibly much earlier. This scenario is all the more likely because metaphysics came at the *end* of the arts curriculum—Lefèvre surely taught the other disciplines earlier, and may well have developed their introductions first. But we have no other indications of these manuscript introductions. In any event, the circulation of such manuscript introductions seems to have been limited to Lefèvre's students at the Collège du Cardinal Lemoine, and perhaps his students from other colleges too. The point to acknowledge is that Lefèvre initially hesitated to publish in print.

But it is equally clear that students were more enthusiastic about print's possibilities, and it was their industry that put these books into print. Lefèvre's reticence and his students' industry can be seen in his short introduction to logic. This book presented the very sorts of logical operations (*suppositiones*, *syncategoremata*, etc.) that he simultaneously decried as sophistic barbarisms. In 1496, and again the following year, his *Introductiones logicales* were printed in a starkly brief format: fifty-six octavo pages—a mere fraction of competing tomes.[17] In the preface, Lefèvre urged students to hurry through this sort of logic like scouts taking stock of an enemy army, not delaying long enough to get caught up in sophisms and other logical tricks commonly deployed in disputations.[18] The title page reveals

[15] An affair relevant to Fabrist patterns of collective authorship is found in Jean-Marc Mandosio, 'La fabrication d'un faux: l'*Introduction à la rhétorique* pseudo-lullienne (*In rhetoricen isagoge*, Paris, 1515)', *Bibliothèque d'Humanisme et Renaissance* 78 (2016): 311–31.

[16] This information is added to Lefèvre's 1515 edition of these introductions, published with Bessarion and Bruni's translations of the *Metaphysica* (Paris: Henri Estienne, 1515). The dedicatory letter is edited in PE, ep. 113.

[17] If fifty-six octavo pages does not seem brief, this is dwarved by competitors in both format and number: e.g. the very popular textbook on the same topic by Jean Buridan, *Summulae*, ed. Thomas Bricot (Paris: 1487), in-folio, 344 pages. Even abbreviated expositions of Peter of Spain are much larger: Johannes Dorp (comm.), *Summulae de dialectica Petri Hispani [per Buridani]* (Paris: 1495), in-quarto, 282 pages; Pierre Tataret, *Expositio in summulas Petri Hispani* (Paris), in-quarto, 177 pages; Thomas Bricot, *Textus totius logices* (Basel, 1492), in-octavo, 301 pages; Thomas Bricot (ed.), *Textus abbreviatus totius logices Aristotelis*, with questions by Georgius of Brussels (Paris, 1495), in-quarto, 545 pages.

[18] Jacques Lefèvre d'Étaples, *Introductiones logicales in suppositiones, in predicabilia, in divisiones, in predicamenta, in librum de enunciatione, in primum priorum, in secundum priorum, in libros posteriorum, in locos dialecticos, in fallacias, in obligationes, in insolubilia* (Paris: Guy Marchant, 1496), a1v

that these octavo pages were 'diligently gathered together by Josse Clichtove', and a postscript to the first edition indicates that Lefèvre's students Guillaume Gontier and David Laux did all the editorial work. Lefèvre evinced a palpable distaste for the subject, and he claimed to have written the text to help students at Lemoine avoid the much longer standard introductions to logic by Peter of Spain, Jean Buridan, and Paul of Venice. Perhaps we may speculate on three reasons why Lefèvre was slow to print these short introductions, while students were eager. First, as we have seen, intimate material was often published effectively in manuscript, in local coteries. Second, Lefèvre may not have thought these introductions worthy of larger circulation, especially in light of his commitment to the 'purer sources' of antiquity. These introductions were effectively cheat-sheets, which allowed students at Lemoine to learn all the terminology needed to pass exams without being swallowed by logic or seduced into its unending rigours. Third, like many of his generation, Lefèvre may not have immediately seen the value of print.[19] Lefèvre likely completed his MA in 1476. In a world where most MAs taught for only a few years before moving on to a higher faculty or a better career, Lefèvre was unusual, and by 1492 he already belonged to an older generation: he may not have been convinced the new printing presses were viable or valuable. On the evidence we have, a definitive picture is impossible; but perhaps students took the initiative to print these manuscripts because it was they who best understood the broader usefulness of such grubby helps, and it was they who could imagine what printing might offer.

Of course, it was not new for students to copy or even publish books. Students had long produced books in the classroom, in the same way that Beatus Rhenanus effectively made his own copy of Virgil's *Georgics* from his teacher's dictation in the Latin school at Sélestat.[20] University students have been compared to artisanal apprentices. Like them, they gained and proved competences by making new physical objects: books. From the thirteenth century at least, students had complained when masters lectured too quickly to let them copy texts *verbatim*, cheating them of the chance to make a reliable copy.[21] In the reading halls, some students copied texts, perhaps for extra money, working within the *pecia* system. This was the hub of university book production: *libraires* dealt in old and new

(=PE, 39): 'Satis enim est ea vel in transcursu (velut qui exploratores hostile agmen transcurrunt) attigisse.'

[19] See the many early criticisms of the new technology in its first generation in Elizabeth L. Eisenstein, *Divine Art, Infernal Machine: The Reception of Printing in the West from First Impressions to the Sense of an Ending* (Philadelphia, PA: University of Pennsylvania Press, 2012).

[20] BHS MS 50, 'cahier écolier'. See Chapter 3.

[21] That this was normal for the later Middle Ages has been challenged by Paul Saenger, 'Reading in the Later Middle Ages', in *A History of Reading in the West*, ed. Guglielmo Cavallo and Roger Chartier, trans. Lydia G. Cochrane (Amherst, MA: University of Massachusetts Press, 1999), 120–48. It is even possible that student requests for teachers to speak at a copyable speed rose in number during the Renaissance: Ann Blair, 'Student Manuscripts and the Textbook', in *Scholarly Knowledge: Textbooks in Early Modern Europe*, ed. Emidio Campi, Simone De Angelis, and Anja-Silvia Goeing (Geneva: Librairie Droz, 2008), 39–73; Françoise Waquet, *Parler comme un livre: l'oralité et le savoir, XVIe–XXe siècle* (Paris: Albin Michel, 2003). See also the taxonomy of university practices given in Olga Weijers and Louis Holtz (eds), *L'Enseignement des disciplines à la Faculté des arts (Paris et Oxford, XIIIe–XVe siècles)* (Turnhout: Brepols, 1997).

books, paper and ink, and would rent out *pecia* (corrected and certified sections of manuscripts) to students and professional scribes.[22]

With printing presses, students helped to produce books in new ways. In particular, there was a new division of labour. The newcomers were pressmen, the handworkers who sweated over the presses. The physical labour of a handpress is prodigious, even for later (and somewhat more efficient) presses. Other new crafts were associated with print too, such as punch-cutting, type-founding, and press-building. But print shops also created new jobs for clerics, needing compositors to set type, and careful readers to correct proofs coming off the press. Thus early print shops were a bustling mix of artisans and university students, and it is no surprise that Lefèvre's first printers, Wolfgang Hopyl and Johann Higman, were themselves university graduates.[23] Early printers frequently found their correctors in universities, often among senior students. The heavy labour of pressmen required more dexterity than literacy, and even compositors did not always read with ease. But to check the proofs, to ensure that they reflected the author in the best possible light, to guarantee every letter was in its place, every solecism safely eluded, to compile swelling indexes, and to compose the prefaces and printer advertisements—these tasks required strong Latin, a keen eye, and a thick skin to withstand competing pressure from authors and printers.[24] In the hierarchy of book-making, correctors belonged to the world of work, the print shop, and Anthony Grafton has shown that although they were often as erudite as their authors, they were more often ignored if not unjustly blamed for errors.[25] They were everywhere, but it takes care to see them.[26]

Like most authors, Lefèvre depended on others to publish his work, but the abundant evidence for his correctors is unusual, a rare window into a scholarly community working alongside the pressmen as *recognitores in officina*.[27] Their

[22] For the structural description of this system, see Jean Destrez, *La Pecia dans les manuscrits universitaires du XVII et XIV siècles* (Paris: Éditions J. Vautrain, 1935); Richard H. Rouse and Mary A. Rouse, 'The Book Trade at the University of Paris, ca. 1250–ca. 1350', in *La production du livre universitaire au Moyen Age: exemplar et pecia: actes du symposium tenu au Collegio San Bonaventura de Grottaferrata en mai 1983*, ed. Louis Bataillon, Bertrand G. Guyot, and Richard H. Rouse (Paris: Éditions du Centre national de la recherche scientifique, 1988), 41–114. My account should be qualified with the fact that evidence for this system trails off in the late fourteenth century.

[23] Higman received the bachelor degree in 1478. See Anatole Claudin, *Histoire de l'imprimerie en France au XVe et au XVIe siècle* (Paris, 1900), 1:350.

[24] The tasks of the corrector are the subject of Anthony T. Grafton, *The Culture of Correction in Renaissance Europe* (London: British Library, 2011), 6–32, and *passim*.

[25] Grafton, *Culture of Correction*. See also Brian Richardson, *Print Culture in Renaissance Italy: The Editor and the Vernacular Text, 1470–1600* (Cambridge University Press, 1994). 'Editor' is Richardson's translation of *correttore*.

[26] Cf. Erasmus' tense reliance on his printers and correctors, already described by Percy Stafford Allen, 'Erasmus' Relations with His Printers', *The Library* 13, no. 1 (1913): 297–322. Recent work, however, tends to focus on how Erasmus adopted an authorial personality that both invoked and effaced the assistance of amanuenses, correctors, and printers: Lisa Jardine, *Erasmus, Man of Letters: The Construction of Charisma in Print* (Princeton, NJ: Princeton University Press, 1993), 99–121; Alexandre Vanautgaerden, *Érasme typographe: humanisme et imprimerie au début du XVIe siècle* (Geneva: Droz, 2012).

[27] These students included, besides those discussed in the following paragraphs, Wolfgang Pratensis; Petrus Porta Monsterolensis (PE, 137); François Vatable (PE, 249–50); Michael Pontanus

traces are found in paratexts: prefatory letters and notes, verses, illustrations, tables, marginal notes, paragraph markings, commentaries, indexes, and other elements meant to frame a reader's use of the text.[28] The visual structures of these paratexts are so important that this chapter will later focus strictly on this phenomenon. Here, it suffices to say that Lefèvre's books, published with the Estienne print dynasty, have long been hailed as pioneering examples of typographic and paratextual ambition.[29]

These paratexts are populated by a whole generation of students and younger colleagues at Lemoine. Josse Clichtove was Lefèvre's earliest and perhaps closest collaborator, and his hand is very often indistinguishable from Lefèvre's own. (It must be said that Lefèvre attempts a more eloquent Latin; Clichtove's Latin is usually clearer if less inspiring.) Clichtove corrected Lefèvre's first printed book in 1492, even adding some of his own verses. In these verses (which disappeared in later editions) Clichtove praised the printer Higman, enjoining the reader to thank this German printer who took up the job at his own expense. He added that he himself had, with a certain 'faithful Bohemian', as well as he could, corrected any errors that had been left 'in lead'—that is, the mistakes of compositors.[30] It seems quite likely that Clichtove's Bohemian helper was one Stephanus Martini de Tyn (Bohuslas Tinnensis), from Prague, who was then a student in the medical faculty at Paris.[31] Not long after this, the Frisian humanist Viglius Zuichemus noted in the illustrious shop of Froben in 1534 that a corrector might work with a reader.[32] We might imagine Stephanus reading the manuscript exemplar aloud, while Clichtove collated it with the proofs. Lefèvre depended on experienced students

(PE, 381–2); Jean Solidus de Cracovie (accompanied him in 1500, copied a vulgate edn of Job, see Renaudet, 506); Jean Multuallis de Tournay, Louis Fidelis (all involved in some way with the 1514 *Cusani Opera*); and Jean Pelletier. In this vein, Rice calls the famous printer Robert Estienne Lefèvre's 'last and most brilliant disciple' (PE, 494). Armstrong Tyler, 'Lefèvre and Estienne', 23.

[28] Useful reflections on paratexts are available in Genette, *Paratexts*; Terence Cave (ed.), *Thomas More's Utopia in Early Modern Europe: Paratexts and Contexts* (Manchester: Manchester University Press, 2008); Helen Smith and Louise Wilson (eds), *Renaissance Paratexts* (Cambridge: Cambridge University Press, 2011). Rice's *Prefatory Epistles* is a masterclass in how editing and commenting on such paratexts can bring into view entire communities and their intellectual landscapes.

[29] Besides studies cited in Chapter 1 note 44, see Ruth Mortimer (ed.), *French 16th Century Books*, 2 vols (Cambridge, MA: Belknap Press, 1964); Fred Schreiber and Jeanne Veyrin-Forrer, *Simon de Colines: An Annotated Catalogue of 230 Examples of His Press, 1520–1546* (London: Oak Knoll Press, 1995); Frans A. Janssen, 'The Rise of the Typographical Paragraph', in *Cognition and the Book: Typologies of Formal Organization of Knowledge in the Printed Book of the Early Modern Period*, ed. Karl A. E. Enenkel and Wolfgang Neuber (Leiden: Brill, 2005), 9–32; Frans A. Janssen, *Technique and Design in the History of Printing: 26 Essays* (Houten: Hes & De Graaf, 2004), 39–56, 75–99.

[30] Lefèvre, *Paraphrases* (1492), after colophon: 'Debetis grates Alemano et adusque Johanni | Higman, qui proprii sumptibus egit opus. | Mendam corripui fido comitante Bohemio | (Ut potui) in plumbo si qua relicta fuit.'

[31] Jean-Pierre Massaut, *Josse Clichtove, l'humanisme et la réforme du clergé*, 2 vols (Paris: Société d'Edition 'Les Belles Lettres', 1968), 1:186 n. 46; cf. PE, 15.

[32] Johan Gerritsen, 'Printing at Froben's: An Eye-Witness Account', *Studies in Bibliography* 44 (1991): 149–50. On aural correction, see Anthony Grafton and Glenn W. Most (eds), *Canonical Texts and Scholarly Practices: A Global Comparative Approach* (Cambridge: Cambridge University Press, 2016), 1–3.

like Clichtove to prepare his volumes for publication, and his early readers could trace their presence in the letters, poems, and colophons in these books.

Many other students followed Clichtove's example, announcing their involvement. Guillaume Gontier, already mentioned as Lefèvre's amanuensis during his travels to Italy, corrected Lefèvre's popular *Ars moralis* (1497), and had his name put on the titlepage. Lefèvre's mathematical books seem to have drawn the widest involvement of students. The *Textus de sphaera* (1495) colophon includes no less than four *recognitores diligentissimi* and *matheseos amatores*: Lucca Walter, Jean Grietan, Pierre Griselle, and Gontier. The number of correctors makes it likely that they made a gift of their time—more likely than Higman paying for four.[33] Another colleague at the college, David Laux, also bore the weight of the larger and much more difficult *Elementa arithmetica*, in 1496, 'the year of the Lord, who formed all things in number and harmony'. The colophon emphasized that the printers Johann Higman and Wolfgang Hopyl had published the lavishly illustrated work with its corrected diagrams 'at their own heavy labour and expense', sweating for the sake of Parisian students—adding, not quite as an afterthought, that they had been helped by 'David Laux, the Briton from Edinburgh, who diligently corrected the whole thing from the exemplar'.[34] Given the rhetoric and financial risk associated with such a labour-intensive, expensively illustrated text, the colophon likely reflects the printers' own sentiments. The book was indeed a labour of love—or at least economic nerve.

Such help obviously benefited the author; but it also benefited students, for they became authors in these paratexts. The rich paratextual apparatus of Lefèvre books meant that there were many opportunities for such authorship. It would not have been unusual for a corrector to be responsible for the detailed table indexing all the *argumenta* of Lefèvre's commentary in his *Textus de sphaera*.[35] Lefèvre might have singled out his friend Étienne for thanks in his first print publication, the *Totius Aristotelis philosophiae naturalis paraphrases* of 1492, but Clichtove took the greater opportunity: the short poem thanking Higman and Estienne was Clichtove's first appearance in print. This was the pattern for a quarter-century of collaboration. In 1517 Clichtove still piggybacked on Lefèvre's authorial gravity, when he took the dozen pages of Lefèvre's *Astronomicon* and augmented it with over a hundred pages

[33] Higman may have been low on funds in 1495 when, perhaps coincidentally, he lent a property deed to the Hôtel-Dieu of Paris as surety for a large loan, which was not recovered from his widow until 1508. Philippe Renouard, *Documents sur les imprimeurs, libraires, cartiers, graveurs, fondeurs de lettres, relieurs, doreurs de livres, faiseurs de fermoires, enlumineurs, parcheminiers et papetiers ayant exercè à Paris de 1450 à 1600* (Paris: Champion, 1901), 88; Émile A. Van Moé, 'Documents nouveaux sur les libraires, parcheminiers et imprimeurs en relation avec l'Université de Paris à la fin du XVe siècle', *Humanisme et Renaissance* (1935): 5–25.

[34] Lefèvre, *Elementa*, colophon: 'Has duas Quadrivium partes et artium liberalium precipuas atque duces cum quibusdam amminiculariis adiectis: curarunt una formulis emendatissime mandari ad studiorum utilitatem Joannes Higmanus, et Uolgangus Hopilius suis gravissimis laboribus et impensis Parisii Anno salutis domini: qui omnia in numero atque harmonia formavit 1496 absolutumque reddiderunt eodem anno, die vicesima secunda Iullii suos labores vbicunque valebunt semper studiosis devoventes. Et idem quoque facit David Lauxius Brytannus Edinburgensis: Ubique ex archetypo diligens operis recognitor.'

[35] On correctors as authors of indexes and other apparatus, see Grafton, *Culture of Correction, passim.*

of dense commentary. Similarly, Charles de Bovelles's first active role in print was not his *Introductio in philosophiam oppositorum* (1501), but from the year before. In 1500 Lefèvre invited him to edit the short *Annulus astronomicus* of Bonet de Lattes, the treatise on a miniature astrolabe that Bovelles would later witness in Rome. Beatus Rhenanus first became a published author in 1505 when he corrected Lefèvre's edition of Lull, which included Beatus' own poem on the title page, followed by Lefèvre's edition of the *Politica*. Lefèvre owed his first steps in print to his enterprising students; they in turn owed their first words as published authors to him.

The most striking examples of such paratextual authorship are found in mathematical books. In the next section I shall suggest that mathematics exemplifies a genre particularly open to joint authorship. But at a more basic level, this is partly so because mathematical books relied so much on paratexts—geometrical figures, diagrams, numbers, pictures of instruments, and tables. In Lefèvre's *Textus de sphaera*, four correctors may have been needed for the large number of diagrams, charts and numbers, always in need of yet one more collation with the original. In fact, in the prefatory letter, Lefèvre expressly praised 'my domestic, Jean Grietan' for his skill in abacus and mathematical practice, as well as his learning in the variety of mathematical disciplines, even saying that he had 'written the work'.[36] The word 'domestic' here suggests that Jean Grietan earned his bursary at the Collège du Cardinal Lemoine as household labour, a common arrangement in the Parisian colleges. But instead of carrying water and wood, young Jean was put to work as a secretary and computer, very likely helping to produce and check the numerous tables of longitudes, latitudes, and zodiacal ascensions that illustrated Lefèvre's commentary.

The experimental form of joint authorship brought unintended consequences. These paratexts aimed to present the author more clearly, but themselves became popular as authoritative texts. This contradicts a central historiographical myth of humanism, namely that going *ad fontes* replaced the medieval obfuscation of glosses with the text itself. The pure sources, so the myth goes, replaced the derivative commentary. After all, the most common medieval paratext was the commentary; it glossed, supported, and literally framed the authoritative text. Lefèvre's own language sometimes supports the myth. He echoed Bruni's call to the *fontes* of Aristotle; he presented the curriculum as a series of *auctores*. But as we found in the last chapter, the textual reality of his classroom was rather more complicated. The call to the purified sources burdened students with additional reading. The irony is compounded in these printed paratexts. Many of Lefèvre's own supports served to comment on Aristotle or Boethius' original text, but in turn were seen as authoritative texts in their own right, sometimes even printed without the original text. In 1502, Clichtove supplied the *Paraphrases on Natural Philosophy* with a commentary that was often longer than Lefèvre's own paraphrase. In the following

[36] Lefèvre, *Textus de sphera*, sig. a1v: 'Affuit levamini domesticus noster Ioannes Griettanus, abaci numerandique peritiae et reliquae matheseos non inscite studiosus; scripsit opus et quasi fesso umerum subiectit Atlanti.'

four decades, it was *this* version of Lefèvre's paraphrases that was most frequently reprinted.[37] Lefèvre's mathematical books supply some of the most dramatic examples of the phenomenon. The various later editions of these books carefully preserved the indexes and prefatory tables. Even as Lefèvre meant this genre of textbooks to ease students into the traditional *auctores*, the unintended consequence of print was to make Lefèvre himself into an *auctor*, replacing the very authors he claimed to serve, as we will see.

As paratexts were reprinted as texts, students themselves became authoritative. Consider the career of Lefèvre's arithmetical *Epitome*. In its first instantiation of 1495, Lefèvre had described the work as 'medicine of the mind' that would prepare the student to read Boethius and Jordanus.[38] That is, the *Epitome* was an introductory support to help read the authorities. Indeed, the work itself comprised a synoptic table, lists of definitions, and a set of tables of terms and propositions, indexing the authorities it was intended to introduce.

In 1503, in its second instantiation, the *Epitome*'s role changed. Lefèvre's tables and glosses gained their own layers of paratexts. This edition centred on Lefèvre's definitions—while dropping the original texts of Boethius and Jordanus.[39] Thus Lefèvre's definitions became the enunciations, followed by Clichtove's commentary in smaller type: paratext had become text, demanding its own explication.

Even the prefatory epistles took on a life of their own. The edition of 1503 acquired a new matrix of letters to patrons and friends. Clichtove added a letter to Jean Molinar, another arts master at Cardinal Lemoine, and another to one Philippe Prévost, 'co-militioni in philosophie studio'. Authors used letters throughout the volume to tie themselves to a larger community of patrons, colleagues, and authorities (see Figure 4.1). Such letters could themselves become authoritative texts, worthy of copying and comment. Lefèvre had originally used these letters to suggest an approach to arithmetic; in the 1503 edition, Clichtove added a new letter written as a commentary on Lefèvre's original letter.[40] Then, Clichtove's letter itself—a gloss on a preface to an introduction to an authority—was excerpted and recombined with other letters elsewhere. The Dutch humanist Johannes Caesarius added it to a Deventer redaction of the 1503 edition, along with his own introductory letters and excerpts from Augustine (see Figure 4.1).[41] In another handbook from Cologne, Clichtove's letter commending his own *De praxi numerandi*

[37] For full bibliography, see PE. [38] See Chapter 1.

[39] Lefèvre, Clichtove, and Bovelles, *Epitome*.

[40] The *Epitome Boetii* was republished separately in Basel, Vienna, and Paris in 1500, 1533, 1536, 1541, 1549, and 1553 (twice). The same is true of the *Introductiones in suppositiones*, first published in 1497, which was republished with Clichtove's commentary many times from 1500; editions in the 1540s still retain all the paratextual apparatus. Interestingly, editions of Clichtove's own introduction to logic, which earned the commentary of Johannes Caesarius, did not have the same rich paratextual apparatus as late as 1560: Josse Clichtove and Johannes Caesarius, *Introductio in terminorum cognitionem, in libros Logicorum Aristotelis, authore Iodoco Clichtoveo Neoportuensi, una cum Ioannis Caesarii Commentariis* (Paris: Gabriel Buon, 1560).

[41] Johannes Caesarius (ed.), *Introductio Jacobi Fabri Stapulensis in Arithmeticam; Ars supputandi Clichtovei; Epitome rerum geometricarum Bovilli* (Deventer: R. Pafraet, 1507).

Lefèvre, Clichtove, Bovelles, *Epitome compendiosaque introductio in libros Boetii, Praxis numerandi, Introductio in geometricen, perspectiva, Theoricen* (Paris, 1503)	*Introductio Stapulensis, etc.* (Deventer, 1507), ed. and abbrev. by Johannes Caesarius
dedication by Clichtove to Jean **Molinar**	
	poem by Hermann **Busch**
	poem by Johann **Caesarius**
	dedication by **Caesarius** to Henricus **Monacensis**
preface by **Lefèvre**	= preface by **Lefèvre**
	letter to the reader by **Caesarius**
Epitome Boetii by **Lefèvre**, commentary by **Clichtove**	= *Introductio in arithmeticam* by **Lefèvre**
dedication by Clichtove to Philippe **Prévost**	
Praxis numerandi by **Clichtove**	= *Ars supputandi* by **Clichtove**
dedication by **Bovelles** to Jacobo **Ramírez de Guzmán**, bishop of Catania	
	excerpt from *De musica* by **Augustine**
	prologue on geometry by **Caesarius**
Introductio in Geometriam by **Bovelles**	= *Epitome rerum geometricarum* by **Bovelles**
Liber de Quadratura circuli by **Bovelles**	
Liber de cubicatione sphere by **Bovelles**	
Perspectiva introductio by **Bovelles**	
dedication by **Lefèvre** to **Germain de Ganay**	
Insuper Astronomicon theoricen by **Lefèvre**	
	De quadratura circuli demonstratio by **Campanus**

Figure 4.1. Contents of two Fabrist compendia of mathematics: Lefèvre et al., *Epitome Boetii, etc.* (1503) and Johannes Caesarius (ed.), *Introductio Stapulensis, etc.* (Deventer, 1507).

was excerpted in a collection of short treatises on the liberal arts—even though the handbook did not include Clichtove's actual introduction to arithmetic.[42]

In a final twist of publishing irony, it was Caesarius' highly abbreviated version of Lefèvre, Clichtove, and Bovelles's introductory mathematics that was most widely disseminated, thanks to Oronce Fine (1494–1555). As the first professor of mathematics at François I's new Collège Royal, Fine is rightly seen as a proponent of the Fabrist interest in mathematics; certainly he helped Paris printers to republish a great number of their mathematical works, from 1515 to his eventual friendship with Bovelles in the 1540s.[43] It was Fine who selected Caesarius' abbreviation of the Fabrist introductions to supplement the 1535 Basel edition of Gregor

[42] Alexander Hegius, *Dialogi duo de sacro sancte incarnationis mysterio adiuncta pache inveniendi ratione, in quibus continetur ratio totius computi ecclesiastici et ferme totius sphere mundi, Ars supputatoria calcularis, Tractatulus de numero ad alium relato sive numerorum proportionibus*, ed. Jacobus Faber (Cologne: Heinrich Quentell, 1508), B3v.

[43] On Fine's relationship to Lefèvre's circle, see Chapters 5 and 7.

Reisch's bestselling *Margarita philosophica.* In this appendix, Fine included all the paratextual apparatus of prefatory letters, tables, and numbered enunciations.[44] By redaction and reproduction, the Boethian introductions of Lefèvre and Clichtove eclipsed Boethius entirely, making them authorities in their own right.

Ultimately, such examples show the move from manuscript to print in slow motion. Just how these new textbooks could enable generational changes in established disciplines, mixed mathematics and natural philosophy, we will see in Chapters 5 and 6. But already we can observe one big implication for the long-standing debate over the agency of print in early modern knowledge.[45] I suggest that print did change the landscape of knowledge. We should not exaggerate the *mechanical* benefits of reproduction. Instead, I would emphasize the new *human* roles of selecting, emending, commenting, and framing a text. The printing press expanded the range of experts involved in publishing a book: it now included more correctors and writers of paratext whose agendas sometimes competed with the original author. In Lefèvre's case, I have argued in this section, the wider authorial community was his students. In the next section, I turn to what difference these students made to his textbooks.

GENRE AS METHOD

As they experimented with the new possibilities of printed textbooks, Lefèvre, Clichtove, Bovelles, and their friends played every generic note in their pedagogical range: commentaries, epitomes, introductions, paraphrases, translations, and dialogues.[46] The proliferation of genres was no accident; they represented a range of practices that might help students know better. Particular genres therefore were intended to compress knowledge, to expand minds—to serve the *exercitationes ingenii* described in the last chapter. One measure of the cognitive importance that Lefèvre and his colleagues placed on genre is their familiar railing against sophistry, the disputation practices of the very logic their textbooks aimed to replace.

Another way of looking at what these textbooks were intended to do is through the history of method, but in a slightly different mode than is usual. The history of method has long been oriented around the cognitive modes in *regressus*, particularly as formulated by Paduan Aristotelians in the late sixteenth century and taken up as analysis and synthesis in the mathematical sixteenth century.[47] Some version

[44] Gregor Reisch, *Margarita philosophica*, ed. Oronce Fine (Basel: Henricus Petrus for Conrad Resch, 1535). Following editions that included the same works were published in 1583 (Basel), 1599 (Venice, 2 editions), and 1600 (Venice).

[45] For historiography, e.g. Sabrina Alcorn Baron, Eric N. Lindquist, and Eleanor F. Shevlin (eds), *Agent of Change: Print Culture Studies After Elizabeth L. Eisenstein* (Boston, MA: University of Massachusetts Press, 2007).

[46] These genres are categorized by Lines, 'Lefèvre and French Aristotelianism', 273–90.

[47] The most influential study is Neal W. Gilbert, *Renaissance Concepts of Method* (New York: Columbia University Press, 1960). Scholarship on method still tends to gravitate towards the Paduan Aristotelianism that Gilbert sought to decentre: e.g. Marco Sgarbi, *The Aristotelian Tradition and the Rise of British Empiricism* (Dordrecht: Springer, 2012).

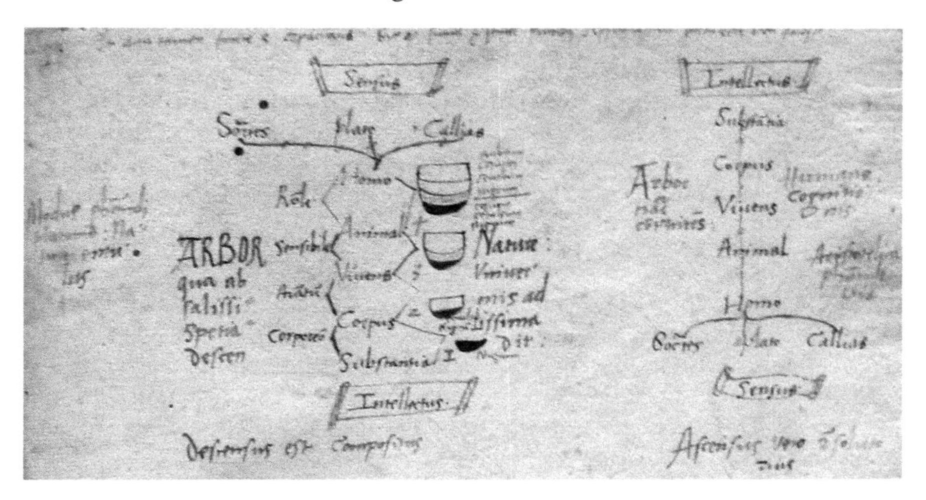

		Senses			Intellect	
	Socrates	**Plato**	**Callias**		Substance	
					I	
Plato's mode of		Human			Body	
philosophizing:	TREE of	Rational <	[White	TREE of human	I	Aristotelian
Aemulator	Nature by	Animal	Red	nature, of	Living thing	way of
of nature	which one	Sensible <	Blue	conceptualizing	I	philosophizing
	descends	Living thing	Black]	everything	Animal	
	from the most	Animate <			I	
	universal to	Body			Human	
	the most	Bodily <			I	
	specific	Substance		**Socrates**	**Plato**	**Callias**
		Intellect			Senses	
	Descent is			But ascent is		
	composition			resolution		

Figure 4.2. Aristotle's and Plato's modes of philosophizing: Lefèvre, *Libri logicorum* (1503), BHS K1047, endpage. Beatus contrasts the resolutive and compositive modes of philosophizing, portrayed as variants of the Porphyrian tree. The translation does not include the diagrams of substances, which fill in layers of colour for different composites: black for a substance with body, or bodily nature; blue represents the addition of life, or animate nature; red for the addition of soul, or sensible nature; and white for rational nature. The descent from universals is identified with Plato and the order of nature (left, moving from substance to the individual Plato and friends), and the ascent from particulars with Aristotle and the order of cognition (right, moving from individuals towards substance).

of these notions of method or modes of inquiry (*modus*) was already a familiar part of the medieval tradition, and Beatus Rhenanus inscribed in his logic textbook a lovely figure that contrasts the compositive method of Aristotle with the resolutive method of Plato (Figure 4.2).[48]

But *methodus* was not a tidy term and could be used in many ways. Over the course of the sixteenth and seventeenth centuries, the word *methodus* became one of the most popular words on textbook title pages. Too quickly historians of

[48] Lefèvre, *Libri logicorum*, BHS K 1047, endpages.

philosophy have seen early modern method as chiefly a matter of mind, rather than webs of authorities, genres, and practices. Method referred here not to a logical procedure but to a material artefact: the introductory book that set out a domain of study. Here I will explore how three Fabrist textbook genres could be linked to particular cognitive, methodical modes, first in the broader genre of introductions themselves, then in two specific kinds of introduction (paraphrase and 'elements').

(a) Introduction as Method. Throughout his works, Lefèvre refused to adopt any of the various schools of the 'Aristotelians', but urged readers to follow Aristotle himself. At first glance, this seems a commitment to follow the ancients alone in method, the standard humanist profession *ad fontes* with its dismissal of the warring schools. But we have seen from Beatus Rhenanus' library that such textual asceticism was no real option. Students had to master the commentators as well as the authors. To be mastered, knowledge had to be adulterated, revisualized, and condensed. It had to be made methodical.

It is the sense of *methodus* as the material tool for learning a discipline that is thematized in the textbooks of Lefèvre's circle. The first examples emerge in accounts of why they chose to write these brief introductions and epitomes. Lefèvre explained that Guillaume Gontier, who travelled with him to Italy, had urged him to write dialogues as a good way to introduce the *Metaphysics*. Dialogues could model how students should ask questions and formulate answers. Gontier also requested that Lefèvre preface the dialogues with a short *Isagoge*, 'where by artifice [the discipline] might be easily committed to memory'.[49]

Elsewhere Gontier explained that Lefèvre's introductions were a sort of *methodus* helping students quickly absorb the basics. Introductions should be short, so students would not be detained unduly in preliminaries such as logic:

> Let no one despise what does not immediately hold itself out to understanding, nor force by violent exposition what has been less understood; though let one regard as nothing what does not help both parts of philosophy [i.e. contemplative and practical], rightly and by way of preparation, as I say. Again, let no one condemn brevity. For whatever an art teaches ought to be brief; a little stream often sweetly quenches the thirst, but the whole sea cannot do that. It is madness to grow old in trivial[50] affairs when there is an easy exit available. Moreover, the intellect rejoices in brevity, as one who is aided by a little may be confounded by a lot.... [one will be] quickly brought to the port of the disciplines by this very short method [*methodus*], as if followed by the blowing wind and aided by oarsmen at their seats.[51]

[49] Lefèvre, *Introductio in metaphysica*, b1v (PE, 22): 'Admonuit me praeterea legentes monefacere, fronti praefixam Isagogen, quo facile memoriae mandetur in artificio esse constitutam'.

[50] A pun on the trivium of grammar, logic, and rhetoric.

[51] Lefèvre, *Introductiones logicales*, d6r–v (PE, 40): 'Sed nemo despiciat quod statim se non praebet intelligendum, neque violenta expositione extorqueat quod minus fuerit intellectum; verum nihil esse putet quod non ad utrasque philosophiae partes afferat praesidium, legitime dico atque subductitie. Nemo item brevitatem damnet. Nam quicquid ars praecipit breve esse debet; et fonticulus plerumque sitim sedat suaviter, id autem amplum mare praestare non potest; et in trivialibus senescere (ubi facilis ad exitum pateat via) dementiae est. Adde quod intellectus brevitate gaudet, utpote qui paucitate

This passage selects a distinct interpretation of the word *methodus*, linking it to brevity—even to the point of taking short cuts. This was a live option for fifteenth-century commentators on Peter of Spain's *Summule*, which began with the definition of logic as not only the 'art of arts' (mentioned earlier), but also as the 'path leading to the principles of all *methods*'.[52] Now, what did *methodus* mean for Peter of Spain? The influential late medieval logician Jean Buridan simply glossed *methodus* as *scientia*. Later logicians explicitly argued that method was not brief but the equivalent to *ars* or *scientia*—the habit of knowledge. Pierre Tataret, famous in Lefèvre's day as a Scotist teacher, put the matter clearly: 'Another way [of understanding *methodus*] is that it teaches the arguments to make from the principles and terms of each science—it should be understood this way. From this it is clear that *methodus* is not here properly taken to mean "brief pathway", but is taken for *scientia* itself.'[53]

In contrast, Lefèvre conspicuously chose the alternative interpretation of method as a brief pathway into an art. His view was conspicuous because the issue can be traced to a conspicuous passage of Aristotle, the first line of Aristotle's *Topics*; in fact, Peter of Spain's famous opening statement on logic and method paraphrased Boethius' translation of that passage. In his edition and commentary on Aristotle's logical works, Lefèvre glossed the word:

> *μέθοδος* properly means pathway, and is applied to a compendious teaching. Aristotle rightly calls his disciplines [i.e. books on each field of study] by this name. For they are very brief, quickly leading us to the understanding of the matter they deal with.[54]

Aristotle may not be brief, but Lefèvre's view is clear. A method or 'pathway' cuts with compendious brevity through a domain of study (*disciplina*). Lefèvre here chooses deliberately to explain method not as the abstract logical structure of a field of study, as contemporaries like Tataret did, but as the material route or procedure for introducing a discipline. It can only be in this sense that he presents Aristotle's own texts as methods. To frame method as the very stuff that makes up a brief, compendious textbook, Lefèvre drew on the sense of 'brief, compendious

iuvetur, confundatur autem multitudine.... hac methodo etiam quam brevissimo ad disciplinarum portus ocissime appellere, quasi aura flante secunda et quasi transtris remigibusque iuti.'

[52] Peter of Spain, *Tractatus*, 1. Early printed editions follow the majority of manuscripts: 'Dialectica est ars artium et scientia scientiarum ad omnium methodorum principia viam habens.'

[53] Commentators on Peter of Spain or Jean Buridan normally dealt with the first line of the *Summule*, which paraphrased the first line of Aristotle's *Topics*: 'Dialectica est ars artium [et] scientia scientiarum ad omnium methodorum principia viam habens.' Tataret comments: 'Tertio sciendum quod dialecticam habere methodum ad principia omnium scientiarum, potest dupliciter intelligi. Uno modo quod ipsa probet principia cuiuslibet scientiae, et sic non est intelligendum. Alio modo quod doceat formare argumentationes ex principiis et terminis cuiuscum scientiae, et sic est intelligendum. Ex quo patet quod methodus non capitur hic proprie per brevi via, sed capitur per ipsa scientia' (Tataret, *Expositio in summulas* (Paris: n.d. [1490s?], 2v). Compare also Johannes de Monte and George of Brussels at the same place.

[54] Lefèvre, *Libri logicorum*, 229r: '*μέθοδος* semitam proprie significat, transsumitur ad compendiariam disciplinam, quo nomine suas iure vocat Aristoteles disciplinas. Sunt enim brevissime, et cito nos ad rei de qua sunt ducentes cognitionem.'

way' that John of Salisbury had explained in his *Metalogicon* using the word *compendiaria*.[55]

As mentioned earlier, the vocabulary of methodical textbooks grew tremendously in the sixteenth century. Lefèvre's language of introductions was hardly the only source of this vocabulary in the sixteenth century.[56] But his case shows how brief methods grew more significant in an age of overflowing print. A glance at Eugene Rice's bibliography of Fabrist works reveals textbooks as *introductiones*, *epitomes*, *paraphrases*, and above all *artes*. In these titles, perhaps we can see the emphasis on brevity grow more insistent as books themselves proliferate. The terms are not applied offhandedly; the insistent emphasis on artful brevity captures an anxiety of the age.

When Lefèvre, Clichtove, Gontier, and others provide introductions to students for their use, they make clear one thing: these *isagoges*, *introductiones*, *paraphrases*, and *epitomes* are intended to serve as a means to facilitate access to the authorities, not replace them. They are intended to be 'gates', 'entrances', or 'pathways' to the disciplines, or, as Beatus might have said, *exercitationes ingenii*. Indeed, more than once Lefèvre compared their propaedeutic nature to medicines. His introduction to Boethius' *Arithmetica* was a remedy for the mind, for

> unless a mind has been properly prepared, it will gain nothing in the disciplines. For just as skilled physicians give potions, digestives, and strong medicines to strengthen one's health, so introductions are useful in all kinds of disciplines, in order that we might more easily attain understanding of the discipline, like a certain whole soundness of mind.[57]

In passages like these, Lefèvre develops the late medieval language of the philosophical *viae* or pathways into a concrete object. Late medieval university life is often depicted as a set of warring schools and *viae*. Since the nineteenth century these have been chiefly seen as conceptual schools: nominalism, realism, and so on. Maarten Hoenen has noted that in fact proponents of these *viae* were less concerned with conceptual consistency but rather chose their *via* as a pedagogical tool.

[55] John of Salisbury, *Metalogicon*, ed. Clement C. J. Webb (Oxford: Clarendon Press, 1929), 28: 'Est autem ars ratio que compendio sui naturaliter possibilium expedit facultatem. Neque enim impossibilium ratio prestat aut pollicetur effectum; sed eorum que fieri possunt, quasi quodam dispendioso nature circuitu compendiosum iter prebet, et parit (ut ita dixerim) difficilium facultatem. Unde et Greci eam *methodon* dicunt, quasi compendiariam rationem que nature vitet dispendium, et amfractuosum eius circuitum dirigat, ut quod fieri expedit, rectius et facilius fiat.' Cit. Gilbert, *Renaissance Concepts of Method*, 55–6.

[56] Though Agostino Nifo copies the definition of *methodus* almost word for word from Lefèvre, when he says that Aristotle 'per methodum brevem artem intelligit: nam licet μέθοδος graece, latine sit semita, transumitur tamen ad compendiariam artem, quae brevissima est, et cito nos ad rei cognitionem ducens'. Agostino Nifo, *Aristotelis Stagiritae Topicorum libri octo* (Venice: Girolamo Scoto, 1569), 5r. Compare Lefèvre above.

[57] Lefèvre, *Arithmetica, musica, etc*, h7v: 'Ita enim ferme comparatum est ut nisi mens rite praeparata fuerit, nullum in disciplinis capiat emolumentum. Ut enim periti medici potiones digerentiaque foribus praemittunt pharmacis quo firmiorem inducant valetudinem, ita quoque in omni disciplinarum genere operae pretium est introductiones praemittere, ut faciliorem assequamur disciplinae intelligentiam, velut quandam perfectam nostrae mentis sanitatem.'

Their real concern was to set out the best authorities and texts for teaching the curriculum. In fact, the philosophical positions of a *via* in one domain do not reflect positions in other domains; a *via* was only secondarily about doctrine or logical structure, and in the first place reflected the interpretive tradition or *processus* of Scotus, Aquinas, and so on.[58] As I have noted before, Lefèvre refused to associate himself with any of these schools, adopting instead the rhetoric that Jean Gerson had used to bypass the schools: 'Aristotle alone'. Yet his entire programme of textbooks was built around the same need met by the *viae*, the need to digest and mediate Aristotle, helping students through the arts course. The new Fabrist printed textbooks, as brief methods distinctive of the *cursus Fabri*, erased the *viae* only to replace them.

(b) Paraphrase as Analogy. Lefèvre's most distinctive introductions were paraphrases, beginning with his first printed work, the *Paraphrases on the Whole of Natural Philosophy* (1492). The actual practice of paraphrasing was nothing new; the Aristotelian paraphrases of the late antique philosopher Themistius had circulated since the thirteenth century, when students and teachers also began to circulate brief summaries of Aristotle's *libri naturales*.[59] But these were not conceived with deliberate thought to the genre of paraphrase. Medieval copies of Themistius' works usually were titled *commentaria*.[60]

In contrast, at the Collège du Cardinal Lemoine, paraphrase became a distinctive genre. The genre had received new attention among humanists who hoped to restore the particularities of ancient texts, a point Ermolao Barbaro made by retranslating Themistius in 1481. Beatus Rhenanus, we saw in the previous chapter, carefully situated paraphrasts such as Lefèvre within a framework of authorities and their genres, distinguishing paraphrase from introduction from commentary. On another end page in his logic textbook, Beatus copied a quotation from Barbaro's prefatory letter:

> 'Paraphrase is a genre of exercise among the rhetoricians. It is described by the Greeks in this way: it should reflect a narrative by analogy, or rather it is that which we make

[58] On the terminology of 'schools' and 'viae', see Maarten J. F. M. Hoenen, 'Via Antiqua and Via Moderna in the Fifteenth Century: Doctrinal, Institutional, and Church Political Factors in the *Wegestreit*', in *The Medieval Heritage in Early Modern Metaphysics and Modal Theory, 1400–1700*, ed. Russell L. Friedman and Lauge O. Nielsen (Dordrecht: Kluwer, 2003), 9–36. Hoenen's work explains well the important findings of Zenon Kaluza, *Les querelles doctrinales à Paris: Nominalistes et realistes aux confins du XIVe et du XVe siècles* (Bergamo: Lierluigi Lubrina, 1988).

[59] e.g. the various short *summae philosophiae naturalis*, often spuriously attributed to figures such as Robert Grosseteste and Albert the Great: Martin Grabmann, *Methoden und Hilfsmittel des Aristotelisstudiums im Mittelalter* (Munich: Verlag der Bayerischen Akademie der Wissenschaften, 1939); Neil Lewis, 'Robert Grosseteste's Notes on the Physics', in *Editing Robert Grosseteste: Papers given at the Thirty-Sixth Annual Conference on Editorial Problems, University of Toronto, 3–4 November 2000*, ed. Joseph Ward Goering and Evelyn Anne Mackie (Toronto: University of Toronto Press, 2003), 103–34; Andrew Cunningham and Sachiko Kusukawa, 'Introduction', in Cunningham and Kusukawa (eds), *Natural Philosophy Epitomised: Books 8–11 of Gregor Reisch's Philosophical Pearl (1503)* (Farnham: Ashgate, 2010).

[60] Themistius, *Commentaire sur le traité de l'ame d'Aristote: Traduction de Guillaume de Moerbeke*, ed. G. Verbeke (Leiden: Brill, 1973).

of others' writings . . . only keeping the same sense. One Themistius, the noblest of the peripatetic school made himself a paraphrast of Aristotle.' Thus writes Ermolao [Barbaro].[61]

Two features of this definition matter here. First, paraphrase is a 'genre of exercise among the rhetoricians', a turn of phrase that evokes the practices and *exercitationes ingenii* which Beatus deployed to master the arts cycle, and which Barbaro highlighted by quoting Quintilian. Genre is not inert text, but an exercise, a process. Second, Beatus here focuses on the term *proportio*, which I have translated as 'analogy'. As we shall see, paraphrase involves some of the cognitive processes associated with analogy.

Lefèvre adapted the genre of paraphrase gingerly, self-consciously. He certainly was responding to Quintilian's definition of the genre as the essential meaning of the original, rather than its narrative order.[62] Lefèvre had chosen a genre in which he could reorder the account, so long as he preserved Aristotle's fundamental sense. 'I did not judge it the paraphraser's task to follow the author's order in everything, changing nothing. Instead, keeping the sense of the letter and the author's intent, one should by every means make it clear and easy, and so far as possible cut off everything causing confusion.'[63] He defended this reordering on cognitive grounds: 'we followed this arrangement so that everything might be clearer and more easily remembered'.[64] So far, so uncontentious.

Paraphrase sits in an uneasy but fruitful proximity to analogy. Years ago Eugene Rice identified analogy as a mode of cognition that Lefèvre used to rise from Aristotelian physics to Platonic theology.[65] But the account of analogy as *ars artium* found in Beatus' notes, discussed in the previous chapter, suggests that the concept held deep wells of possibility to this circle of scholars. What possibilities they?

We already find the link of paraphrase to analogy in the prologue to the *Paraphrases philosophiae naturalis* of 1492, where Lefèvre said the meaning of Aristotle preserved in paraphrase was a certain *analogia* of the original. His prologue took on some of the same themes of Barbaro's letter about paraphrase, especially the stylish critiques of various schoolmen who strayed from Aristotle himself into *sophismata* and *suppositiones*. Noting that Plato condemned such sophistic,

[61] Beatus, K 1047, *Libri logicorum*, end page: '"Paraphrasis est exercitamenti apud rhetoras genus. Ea finitur a Grecis hoc modo: ut sit quae narrationi proportione respondeat. Aut sic, in qua vertimus aliorum scripta [. . .] modo sensu servato. Unus Themistius peripateticae sectae nobilissimus se paraphrasten Aristotelis facit." Hec Hermolaus.' Cf. *Paraphraseos Themistii,* trans. Ermolao Barbaro (Venice, 1481), a1v.

[62] Quintilian, *Inst. or.* 10.5.4.

[63] Lefèvre, *Paraphrases* (1492), b4v: 'Rati non esse Paraphrastis officium in omnibus autoris seriem sequi et nichil immutare. Verum id potius, seruata littere sententia et autoris mente eam modis quibus potest claram et facilem efficere, et omnem quo ad potest amputare confusionis occasionem.'

[64] Lefèvre, *Paraphrases* (1492), b4r: 'ne miraris si aliquot in locis Aristotelice littere non simus ordinem secuti: sed parum postposita interdum proposuimus. hoc enim consulto fecimus: quo omnia et clariora et memoratu promptiora rederentur'.

[65] On *analogia* as defining Lefèvre's work on Aristotle, see Eugene F. Rice Jr, 'Humanist Aristotelianism in France: Jacques Lefèvre d'Étaples and His Circle', in *Humanism in France*, ed. A. H. T. Levi (Manchester: Manchester University Press, 1970), 132–49, at 144.

false propositions, Lefèvre claimed that to remedy such inanities one must understand a hidden analogy:

> throughout the whole philosophy of Aristotle there lies hidden a certain secret Analogy, just as touch is spread out throughout the whole body. Without it our philosophy is inanimate (as a body without touch), and devoid of life. If God grants me the grace, I will uncover a little of it in commentaries, though not of course entirely. For I suppose its power and size is so great that it would be impossible to do [fully] . . .[66]

Lefèvre's remedy to sophistry is the stripped-down meaning of 'Aristotle himself' that he identifies as a 'secret analogy'. The comparison with the senses, drawn from Cicero, suggests that this analogy in Aristotle somehow illumines or communicates.[67] Is this analogy a secret teaching in Aristotle? Is this analogy simply the naked sense of Aristotle captured in Lefèvre's paraphrases? Or is it a specific method that Lefèvre believes to unify Aristotle's work, of the sort in Beatus' *cahier d'étudiant*? As we saw there, all of these terms were loaded with philosophical meaning: *analogia* could mean a mathematical proportion, a 'ratio of ratios'; or it could hold specific linguistic resonance with paraphrase and figures of speech. Lefèvre and his students, as we saw, made analogy—along with *similitudo* and like terms—the unity underlying both linguistic and mathematical modes of reasoning.

With this baggy definition of analogy, Lefèvre can present his paraphrase as much more than an approximation of Aristotle. Rather, analogy is the 'hidden essence' of Aristotle's reasoning present within the paraphrase. As a logical recomposition of Aristotle, paraphrase digs the essential meaning out of its original substrate, setting the hard nugget of truth into the open. In doing so, Lefèvre sees his paraphrase as uncovering the deeper hidden meaning of Aristotle's natural philosophy—and as he describes the methodological value of this exegetical task, his language takes on a vocabulary familiar from mathematics. Linked to analogy, paraphrase is a practice that does the work of *mathematical* method, enabling one to discover proportions between similar elements, just as when measuring a mathematical ratio. For Lefèvre teaches that paraphrase enables one to examine the proper 'principles' of natural philosophy, principles which, once heard, command immediate mental assent: axioms, *dignitates*, *communes*.[68]

66 Lefèvre, *Paraphrases* (1492), b2r: 'Id insuper te latere non debet per totam Aristotelis philosophiam abditam latentemque esse quandam secretam Analogiam perinde atque per totum corpus sparsus fususque tactus est [cf. Cicero, *De natura deorum* II.56]. Sine qua (ut sine tactu corpus) nostra philosophiae inanima est, vitaeque expers. Quam (si deus hanc michi largiatur gratiam) in commentariis aliquantulum detegam, non quidem omnino. Tanta enim eius virtus est et amplitudo ut id impossibile putem, sed quantum michi concessum fuerit.'

67 The lifelessness of philosophy without this analogy may have a precedent in John of Salisbury's twelfth-century account of logic, which adopts the same metaphor. John of Salisbury, *Metalogicon*, pp. 32, 37, 78, 79, 84, where he warns that without the 'vital organizing principle of logic', philosophy would rest 'lifeless and helpless'.

68 Lefèvre, *Paraphrases* (1492), b1v: 'Que publica intellectus et disciplinarum sunt limina, et ex aliorum propriorum principiorum luminibus nequaquam dinoscenda. Que si dignitates, proloquia, communesque scientie existant, talia sunt ut nos ipsis mente dissentire non putet, et que statim probamus audita.'

(c) **Elementator as Commentator.** The vocabulary of elementary principles—*dignitates, proloquia, communes, principia*—that Lefèvre evoked was drawn from the style of axiomatic reasoning of Euclid's *Elements*. This is somewhat puzzling at first, since in the same breath he rebuked the modern Aristotelians for forgetting Aristotle's own chief methodological postulate: the prohibition of metabasis, of mixing the unique principles proper to reasoning in each discipline. Yet Lefèvre simultaneously appears to apply a mathematical vocabulary (principles, axioms) to all the disciplines. The puzzle begins to unknot when we notice that Aristotle himself deliberately modelled method on the style of reasoning in geometry.[69] Likewise, Lefèvre saw the style of geometrical reasoning of Euclid's *Elements* as applicable across the disciplines, placing himself in a medieval tradition of *elementatores*.[70]

Lefèvre's view of Euclidean style depended on profound assumptions about the genre of elements and their authorship. We usually see Euclid as a model of systematic, axiomatic fixity, and medieval and Renaissance readers, encouraged by the likes of Aristotle, did often speak of Euclid this way. But they also saw Euclid's *Elements* in a subtly different way. For them, the text of Euclid was not one thing, the product of one mind. By 'Euclid' they meant the propositions or enunciations of the *Elements*—that is, they saw the text that Euclid had authored as only the propositions, without the supporting proofs.

This was how medievals encountered the text. Before the thirteenth century, the only translation of Euclid was a mere few dozen propositions and a handful of definitions from the first few books.[71] In this translation, attributed to Boethius, 'Euclid' meant a series of assertions about mathematical objects—no proofs support the propositions. By the thirteenth century, all the books of Euclid had been translated several times, and a thirteenth-century translation by Campanus

[69] Aristotle's account of the principles of disciplines took geometry as a paradigmatic case: *Analytica priora* 46a3–27; *Analytica posteriora* I.7 (71b19–25); *Metaphysica* I.2 (982b11–28); see also 1000a6–10, 1000b21–1001a3, 1060a27–36, 1075b13–14. At a general level, then, Aristotle and Euclid provided the Middle Ages with a common form of demonstration: e.g. Crombie, who argued that since the thirteenth century university philosophy, finding a methodological leader in Aristotle, had always loosely modelled itself on the axiomatic method of Euclid: Alistair C. Crombie, *Styles of Scientific Thinking in the European Tradition: The History of Argument and Explanation Especially in the Mathematical and Biomedical Sciences and Arts* (London: Duckworth, 1994), 1:31. For this reason, as Crombie knew, axiomatic reasoning in the Middle Ages can hardly be used an index of Neoplatonic influence. Cf. Jean-Luc Solère, 'L'ordre axiomatique comme modèle d'écriture philosophique dans l'Antiquité et au Moyen Âge', *Revue d'Histoire des Sciences* 56, no. 2 (2003): 323–45.

[70] The term *elementator* gives a name to what Henry Zepeda has called 'Euclidization'. See 'Euclidization in the Almagestum Parvum', *Early Science and Medicine* 20, no. 1 (2015): 48–76. The term was already used in the twelfth century, as mentioned by Richard Joseph Lemay, *Abu Ma'shar and Latin Aristotelianism in the Twelfth Century: The Recovery of Aristotle's Natural Philosophy Through Arabic Astrology* (Beirut: American University, 1962), 179. Peter Ramus described all those writing mathematics as a series of theorems as elementators (Robert Goulding, *Defending Hypatia: Ramus, Savile, and the Renaissance Rediscovery of Mathematical History* (New York: Springer, 2010), 63–6).

[71] Current scholarship is not definitive on whether Boethius really did translate these works. See Menso Folkerts, '*Boethius' Geometrie II: ein mathematisches Lehrbuch des Mittelalters* (Weisbaden: Steiner, 1970); David Pingree, 'Boethius' Geometry and Astronomy', in *Boethius: His Life, Thought and Influence*, ed. Margaret Gibson (Oxford: Oxford University Press, 1981). Certain they circulated as such: Gillian Rosemary Evans, 'The "Sub-Euclidean" Geometry of the Earlier Middle Ages, up to the Mid-Twelfth Century', *Archive for History of Exact Sciences* 16, no. 2 (1976): 105–18.

became the standard text into the Renaissance.[72] In this version too only the enunciations preserve the original Euclid, while the proofs or demonstrations that explain them are additions, usually constructed anew by the medieval translators. The *Elements*, as a body of propositions *and* proofs, was always experienced as a jointly-authored work.

This collective authorship shaped the cognitive style the *Elements* represented. One clue is in the word that medieval readers sometimes applied to these demonstrations: *commentaria*. By using the word 'commentary', the medieval proof draws on the broader remit of the Latin *probare* and *demonstratio* as 'testing' and 'explanation'. Thus, for Latin authors, the familiar QED, *quod erat demonstrandum*, properly terminates a Euclidean proof *as an interpretive note*, that is, as a scholium. If this marginalization of the proofs seems strange to us, it was not so in Lefèvre's time: Renaissance scholarship on Euclid's text centred on this problem. Until the late sixteenth century, consensus held that Euclid's proofs were in fact added by the late antique mathematician Theon of Alexandria.[73]

Lefèvre's own contributions to scholarship on Euclid engrained the tradition of joint authorship even deeper into Renaissance mathematics. In 1500, he published the early medieval translation of Boethius with a textbook on astronomy, reinforcing the idea that Euclid was really just the propositions. In 1517, he published the standard medieval text of Campanus alongside the new humanist one of Zamberti—Campanus largely supplied his own proofs to Euclid's propositions, while Zamberti stressed that he offered the ancient proofs of Theon. By setting alternative 'commentaries', Lefèvre's edition subtly maintained the multiple authorship of Euclid, titling the text 'commentaries of Theon', 'commentaries of Campanus', and 'commentaries of Hypsicles'.[74] This version had consequences throughout the sixteenth century reception of Euclid.[75] For medieval and Renaissance readers, down to Jacques Peletier and Peter Ramus, the *Elementa Euclidis* was just the propositions.[76]

This publishing history bears profound implications for the cognitive status of mathematical narrative. Modern mathematical texts still frequently follow a narrative that moves from axioms or common notions, to propositions, diagrams, and demonstrations or proofs. Above all, the narrative arc is suspended above demonstrations, which supply the bedrock of mathematical certainty. In contrast, this Renaissance tradition shifts the narrative weight to the propositions. Euclid's authority is only in the unchanging enunciations; they are the authoritative backbone of the work. Meanwhile, the demonstrations are the body that enfleshes that backbone as an optional gloss. In the Boethian tradition, in fact, demonstrations do not even appear; in other translations they are merely the commentary of

[72] Recent scholarship is summarized in H. L. L. Busard (ed.), *Campanus of Novara and Euclid's Elements*, vol. 1 (Franz Steiner Verlag, 2005), introduction.

[73] Goulding, *Defending Hypatia*, 151–4, shows that Theon's authorship of the proofs was popularized by Bartolomeo Zamberti in his new translation, mentioned in the preface and in the title: *Euclidis... elementorum libros XIII. cum expositione Theonis, etc.* (Venice: Joannes Tacuinus, 1505).

[74] Jacques Lefèvre d'Étaples, *Euclidis elementa*.

[75] See epilogue, Chapter 7.

[76] On these examples and their aftermath see Goulding, *Defending Hypatia*, 154–5 and chapter 6 more generally.

various authors. As a multi-authored work, Euclid exemplified a social reality that distinguished authority from mere explanation. Enunciations comprise the structure of the book, underwriting its prized authoritative systematicity. Demonstrations are useful explanations, but can be modified to suit new audiences; certainly they are not *necessary* to Euclid's integrity as a narrative.

In fact, the narrative style of elements applied beyond geometry to other mathematical disciplines. Euclid's *Elements* especially inspired Lefèvre's authorial and editorial efforts in arithmetic and music, dividing those disciplines into definitions, axioms, and propositions, and then deducing complex statements from simple ones.[77] In fact, Lefèvre's first mathematical work was an edition of the thirteenth-century number theory classic of Jordanus de Nemore, the *Arithmetica*, published in manuscript in 1493.[78] Jordanus had already seen himself as an elementator, using a Euclidean structure of definitions, axioms, and theorems to clarify the principles of arithmetic. Lefèvre reinforced the point by embellishing Jordanus' title to make it the ***Elementa** arithmetica*—furthermore, he retained all of Jordanus' enunciations while freely reworking most of the proofs, on a couple of occasions even adding excurses on theological and philosophical topics.[79] In 1496, the printed edition of the book further emphasized its Euclidean style, because Lefèvre added his own advanced treatise on music theory, the *Elementa musicalia*. The basic text was derived primarily from Boethius, but as he and Jordanus had done to arithmetic, Lefèvre reframed Boethius' original musical insights in Euclidean narrative style.[80]

By adopting the role of elementator, Lefèvre blurred the distinction between authoring, commentating, and editing. He did this for authorities such as Boethius and Jordanus, as well as for 'moderns' such as Nicholas of Cusa, revered in Lefèvre's circle as the modern example of the wisdom that mathematics should foster. In Lefèvre's preface to his major edition of Cusanus' *Opera omnia* (1514), Lefèvre praised Cusanus for his acute mathematical judgement, which led him to the heights of contemplative, spiritual insight.[81] Cusanus' name repeatedly emerges

[77] The now-standard division of theorems, problems, and propositions is a result of Proclus' influence on late sixteenth-century reception of Euclid, after he was widely available in Barozzi's Latin translation of 1560. See Thomas L. Heath (ed.), *The Thirteen Books of Euclid's Elements*, Vol. 1: *Books 1–2* (1908, 2nd edn 1925; New York: Dover Publications, 1956), 124–32. For example, the mathematical philologist of Urbino, Federico Commandino, explicitly drew his vocabulary from Proclus in his prolegomena to his edition, *Euclidis Elementorum libri XV. Unà cum scholiis antiquis* (Pisa: Iacobo Chriegher Germano, 1572), **1v. On the reception of Proclus and the shifting practices of demonstration, see, respectively, Eckhard Kessler, 'Clavius entre Proclus et Descartes', in *Les jésuites à la Renaissance: Système éducatif et production du savoir*, ed. Luce Giard (Paris: Presses universitaires, 1995), 285–308, and Paolo Mancosu, 'Aristotelian Logic and Euclidean Mathematics: Seventeenth-Century Developments of the Quaestio de certitudine mathematicarum', *Studies in History and Philosophy of Science* 23, no. 2 (1992): 241–65.

[78] As discussed above, in February 1494 Lefèvre described the *Elementa arithmetica* as having already been presented to Jean. Lefèvre, *Introductio in metaphysica*, a2r.

[79] This was first observed by B. B. Hughes, 'Toward an Explication of Ambrosiana MS D 186 Inf.', *Scriptorium* 26 (1972): 125–7.

[80] Lefèvre, *Elementa*. While the printer's title page did not use the term 'elements', within the book itself Lefèvre used the titles *Elementa arithmetica* (a2r) and *Elementa musicalia* (f1r).

[81] Nicholas of Cusa, *Opera*, preface to the reader (PE, 119).

in Lefèvre's introductions, commentaries, and even the notes of students; Lefèvre's valorization of intellectual vision and his use of figures owed a great deal to Cusanus' account of theoretical vision. But Cusanus's mathematics needed explanation. In his edition of Cusanus' mathematical works, Lefèvre deputized his 'devout brother and friend' Jacobus Faber of Deventer to add 'commentaries' (*commentarii*) on Cusanus' geometrical and arithmetical innovations. The monk obliged, and the treatises were published with extensive *annotationes*—all drawing out Cusanus' intuitive mathematics in Euclidean axioms and demonstrations, supplying Euclidean language where Cusanus had none.

In this context, then, the mathematical narrative line was carried through a collection of propositions. Such *elements* called for an act of commentary through demonstrations—understood as expansions, digressions, and elaborations, as well as proofs. The ongoing activity of demonstrating the propositions did not detract from the work's integrity in the propositions.

The spare introductions typical of Lefèvre's circle functioned in this mode of elementating. The *Elementa arithmetica* and the *Elementa musicalia* were both advanced treatises, a fact to which Lefèvre was quite sensitive. Without a crutch, Lefèvre suspected, most students would find the *Elementa arithmetica* itself simply too forbidding. To ameliorate the difficulty, he added to the first printed edition a condensed introduction to the basic terms and forms of Boethian mathematics—an *Epitome Boetii*. 'Aided by this work (unless I am mistaken), if one should betake himself to reading the books of Boethius's *Arithmetica*, he will find them much more open to the mind than they had been before.'[82] Later teachers agreed. In the 1510s the Paris-trained theologian Pedro Ciruelo republished Lefèvre's mathematical textbooks for the new university of Alcalá, and warned students planning to read the *Elementa musicalia* that, in his experience, the work was too difficult for complete novices and they would benefit from first reading Lefèvre's introductions to arithmetic.[83] As discussed earlier in this chapter, Lefèvre's *Epitome Boethii* was published with Clichtove's extensive commentary in 1503, and then in abbreviated form by Johannes Caesarius and Oronce Fine (with the *Margarita philosophica*). The spare introduction to classical number theory was also reprinted regularly in both Paris and Basel.[84]

Indeed, these epitomes functioned as collections of propositions—much like the 'secret analogy' hidden in Aristotle—which defined the fundamentals of a discipline. For what did Lefèvre's *Epitome Boethii* comprise? Fifteen pages of tables. First, a table (*formula*) of basic terms sets out a map of the discipline (see Figure 4.3). Second, a list of one-line definitions of those terms. Third, a set of numbered

[82] Lefèvre, Clichtove, and Bovelles, *Epitome*, 31v: 'Cuius ope (ni fallor) adiutus, si ad lectitandos divi Severini libros Arithmeticos se traduxerit, eos inveniet longe quam prius fuerant intellectui magis pervios.'

[83] Pedro Ciruelo, *Cursus quatuor mathematicarum artium liberalium* (Madrid: Michaelis de Eguia, 1528 [first edn Alcalá, 1516]), [a2r]: 'tamen experti sumus quod scholaribus introducendis esset nimium difficilis, tum propter operis exquisitissimam sublimitatem, tum et propter eorum ruditatem qui harum rerum sunt omnino expertes.'

[84] See note 39.

Jacobi Fabri Stapulensis Epitome in duos libros Arithmeticos diui Seuerini Boecij ad Magnificum dn̄m: Joannem Stephanum Ferrerium Episcopum Vercellensem.

1 Inter disciplinas mathematicas quibus neglectis omnis obscuratur disciplina: tanq̃ ceterarū parens/dux atq̃ domina: primum sibi vendicat Arithmetice locum que vt rite cognoscat quēadmodū et cetere certis eget adminiculis. inter que primo menti figenda est vniuersorū circa que versatur subiecta formula/mox singulorum diffinitiones/post quas numerorum affectiones/proprietatesq̃/post proprietates: quo ex loco vniuerse sunt monstrande.

Formula vniuersorum circa que negociat Arithmetica.

Numerus		Sesquiquintus	6.5.	Ante longior	15.		
Secundum se	2.	Superpartiens		Solidus			
Ad aliquid	4.2.	Superbipartiens	5.3.	Pyramis	4.		
Scōm figurā	3.	Supertripartiens	7.4.	Cubus 8.	latus 2.		
Numerus scōm se		Superquadripartiens	9.5.	Pyramis			
Par	2.	Superquintupartiens	11.6.	Trigona	4.		
Impar	3.	Supersextupartiens	13.7	Tetragona	5.		
Par		Multiplex superparticularis		Pentagona	6.		
Pariter par	4.	Duplus sesqualter	5.2.	Hexagona	7.		
Pariter impar	6.	Duplus sesquitertius	7.3.	Curta pyramis	29.	Sui tetragoni sūt	16.9.4.
Impariť par	12.	Duplus sesquiquartus	9.4.	Biscurta	25.	Sui tetragoni sunt	16.9.
Perfectus	6.	Triplus sesqualter	7.2.	Tricurta	41.	Sui tetragoni sunt	25.16.
Diminutus	4.	Triplus sesquitertius	10.3.	Laterculus	18.	Latera sunt	3.3.2.
Abundans	12.	Triplus sesquiquartus	13.4.	Asser	12.	Latera sunt	2.2.3.
Impar		Multiplex superpartiens		Cuneus	24.	Latera sunt	2.3.4.
Primus	3.	Duplus supbipartiens	8.3.	Nūerus circularis	25.		
Compositus	9.	Duplus suptripartiens	11.4.	Sphericus	125.		
Ad altū p⁹	9.25.	Duplus supq̄dripartiēs	14.5.	Parallelepipedus	12.	Latera sunt	2.3.2.
Numerus adaliq̄d		Triplus supbipartiens	11.3.	Medietas	6.4.2.		
Eq̄litas	10.10.	Triplus suptripartiens	15.4.	Arithmetica	6.4.2.		
Ineq̄litas	10.5.	Triplus supq̄dripartiēs	19.5.	Geometrica	9.6.4.		
Inequalitas		Numerus scōm figuram		Musica	6.4.3.		
maior ineq̄.10.5.		Linearis	2.	Arithmetica			
mior ineq̄l.5.10.		Planus	4.	Continua	6.4.2.		
Maior inequalitas		Solidus	8.	Disiuncta	7.5.6.4.		
Multiplex	2.1.	Planus		Geometrica			
Supptic'lar'	3.2.	Trigonus	3.	Continua	9.6.4.		
Suppartiēs	5.3.	Tetragonus	4.	Disiuncta	9.6.3.2.		
Multiplex		Pentagonus	5.	Quarta medie.	6.5.3.		
Duplus	2.1.	Hexagonus	6.	Quita medietas	5.4.2.		
Triplus	3.1.	Heptagonus	7.	Sexta	6.4.1.		
Quadrupl⁹	4.1.	Octogonus	8.	Septima	9.8.6.		
Quincuplus	5.1.	Ennagonus	9.	Octaua	9.7.6.		
Superparticularis		Decagonus	10.	Nona	7.6.4.		
Sesqualter	3.2.	Endecagonus	11.	Decima	8.5.3.		
Sesquiťti⁹	4.3.	Dodecagonus	12.				
Sesq̄quart⁹	5.4.	Altera parte longior	6.				

Figure 4.3. The table (*formula*) introducing Lefèvre's *Epitome Boetii* (1496), listing all the key terms of number theory from broadest headings down to narrowest subcategories. Note how each term is next to an example, so that a *numerus multiplex triplus* is shown as a ratio of 3:1. Lefèvre, *Elementa arithmetica, etc.* (1496), h8r. By kind permission of the Syndics of Cambridge University Library, Shelfmark: Inc.3.D.1.21.

propositions, which set out the narrative arc of Boethius' *Arithmetica*. Fourth, further tables which index these propositions of the *Epitome* to both Boethius' *Arithmetica* and in Jordanus' *Arithmetica*. The work was, in sum, a compendium of *tabulae* or lists, cross-referenced with care, serving as a kind of switchboard between authorities. A student could work his way back and forth through the whole discipline of number theory, connecting concepts to their explication at several levels, from simplest to most complex.

These shorter introductions, even those not explicitly labelled 'elements', still reflected a Euclidean style, both in their numbered propositions and in their flexible use of proof. As Aristotle put it in the *Analytica posteriora*, the *Elementa geometrica* was the paradigm of clarity for *all* disciplines because it clearly set out assumptions and definitions in the simplest, most intuitive terms. Similarly, the Fabrist introductions aimed to break down a discipline, an art, into its simplest and most intuitive terms, before rebuilding it into complex wholes.

PRINT AND METHOD

Such artful and methodical brevity, as the student Gontier put it, lay in the visual organization of the page. Lefèvre's textbooks addressed the same topics as earlier manuscript and even some early printed manuals, but on closer inspection their *form* differs. Unlike rival textbooks such as those by Pierre Tataret or Thomas Bricot, books printed by Lefèvre and his students—even those in large folio format—are printed in a single column, and crammed with paratexts from traditional prefatory letters to figures, tables, and diagrams, as historians have often noted.[85] Their significance is marked by how carefully printers reproduced them in later editions.

As an example, consider the book Lefèvre and Clichtove first printed together, the *Philosophiae naturalis paraphrases* (1492), which already deploys the typographic regime we find throughout these books. First, it begins with a synoptic table, memorably set in circles (Figure 4.4)—the same diagram Beatus described in his lecture notes on natural philosophy with the mnemonic NACAMILUT (Figure 3.1). The circles include examples too, much as we saw in the table introducing Boethius' *Arithmetica* (Figure 3.2), where Lefèvre illuminated the exotic Boethian language of proportions (e.g. *super quadripartient*) with an example ratio (e.g. 9:5).

[85] The figures and diagrams of these books were already noted by Cesare Vasoli, 'Jacques Lefèvre d'Etaples e le origini del "Fabrismo"', *Renascimento* 10 (1959): 238–40; Walter J. Ong, *Ramus, Method, and the Decay of Dialogue: From the Art of Discourse to the Art of Reason* (Cambridge, MA: Harvard University Press, 1958), 75–6. See also Augustin Renaudet, 'L'humanisme et l'enseignement de l'université de Paris au temps de la renaissance', in *Aspects de l'Universite de Paris* (Paris: Albin Michel, 1949), 135–55. For the judgement that Lefèvre's theism was 'dyed with the mysticism of the immeasurable and the infinite, built on the kind of metaphysics that runs to diagrams', see J. W. Brush, 'Lefèvre d'Etaples: Three Phases of His Life and Work', in *Reformation Studies: Essays in Honor of Roland H. Bainton*, ed. Franklin Hamlin Littell (Richmond, VA: John Knox Press, 1962), 125. On the apparatus of Lefèvre's paraphrase of *De anima*, see Florence Fournier, 'L'Enseignement d'Aristote à l'Université dans les années 1490: le rôle et la place de Jacques Lefèvre d'Étaples à travers l'étude de son cours et de son manuel de 1492 sur le De anima (livre I)', MA dissertation, Centre d'Études Supérieurs de la Renaissance, 2006. On the figural thinking of Bovelles in particular, see Anne-Hélène Klinger-Dollé, *Le* De sensu *de Charles de Bovelles: conception philosophique des sens et figuration de la pensée* (Geneva: Droz, 2016).

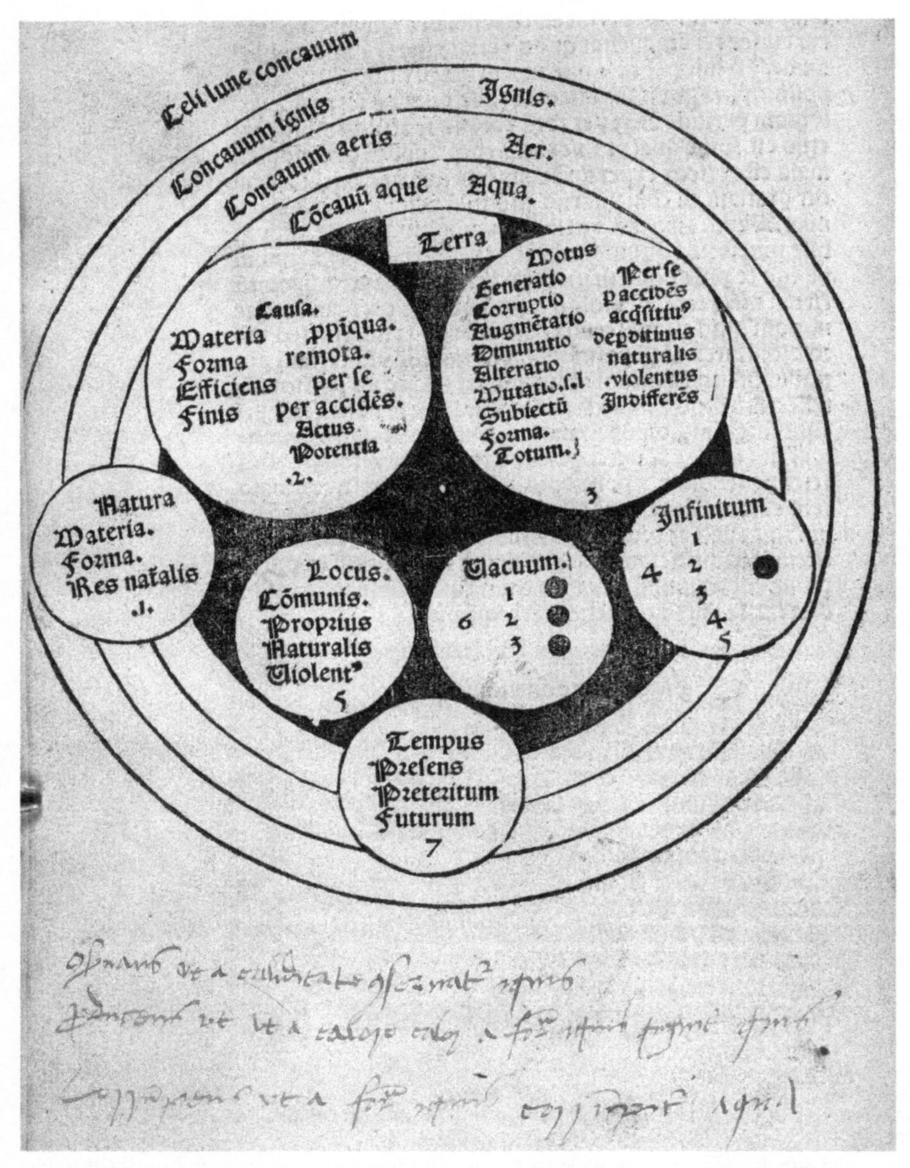

Figure 4.4. The *figura introductionis* which outlines the basic structure of sublunar physics in Aristotelian natural philosophy, with major headings and minor subdivisions. Lefèvre, *Paraphrases philosophiae naturalis* (1492), b2r. By permission of the British Library, Shelfmark IA.40121.

Such images exist quite apart from frontispieces, initials, and decorative illustrations. Fabrist textbooks incorporate an unusually high number of diagrams and tables which figure various arguments in the text itself, which can be called 'epistemic images' that 'express, prove, organize, or interpret' various arguments within the text.[86] In the next chapters I will address the particulars of certain images. But I want to conclude this chapter by focusing on the more general typographical structures that make these books a family, and I will argue that typography—the paratextual elements students most likely controlled—also belong to this category of 'epistemic image'. In brief, these books generally have three kinds of relevant typographical paratexts. First, synoptic tables of terms or propositions. Second, numbered theses and *conclusiones*. Third, marginal numbers which index theses both to synoptic overviews and to commentaries or authoritative texts (e.g. Jordanus, Boethius, Aristotle).

Such synoptic tables are a standard feature of Fabrist pedagogical works. The format varies. The simplest examples are lists of single terms, arranged in a hierarchy, as we find at the beginning of the paraphrases *De anima* (1492). The logical introductions (1497) include such a table at the beginning of each *ars* (each based on a book of Peter of Spain's standard logic textbook, the *Summule*). Thus each of the usual topics in logic, such as the *Art of Syllogisms* and the *Art of Places*, comprises a distinct set of definitions, listed in its own table. To a great extent, these tables performed the same cognitive function as the genres of introduction, paraphrase, and elements discussed in the last section. Lefèvre described the table prefacing the *Epitome Boethii* as a *formula* that fixes (*figere*) the entirety of arithmetic in the mind first, before passing on to definitions of 'singulars', then properties, and finally to demonstrations.[87] Such lists are not *simply* lists, but serve as a conceptual map on the threshold of a domain. Though presented only as a series, in columns, they arrange each discipline as a hierarchy of parts, proceeding from the largest relevant class to the smallest particular. Tables, whether synoptic lists of terms or lists of propositions, offer the sharpest reduction of a discipline—an *ars*—to its parts and internal hierarchy.

The second common feature of these books is a list of numbered theses (Figure 4.5). This list leads from the synoptic glance into the discursive word. Thus the circles of the synoptic *figura* of Aristotelian physics are no mere illustration; they are the subject of these theses, each explaining and unfolding the terms of the synopsis over the course of four pages. Then, each numbered thesis can be found back later, within the course of the paraphrase (Figure 4.6)—more on this in the next paragraph. The list of theses not only glosses the initial synopsis, but also relays the reader into the paraphrases themselves.

The third typical feature of Fabrist books is a series of numbers in the margin, each next to a proposition, which link the whole structure together, embedding each proposition within the architectonics of the discipline. The paraphrases on

[86] This definition of epistemic images is from Christoph Lüthy and Alexis Smets, 'Words, Lines, Diagrams, Images: Towards a History of Scientific Imagery', *Early Science and Medicine* 14, no. 1–3 (2009): 400–1. Lüthy and Smets here follow the pioneering John E. Murdoch, *Album of Science: Antiquity and the Middle Ages* (New York: Charles Scribner's Sons, 1984). Recent bibliography is outlined in Alexander Marr, 'Knowing Images', *Renaissance Quarterly* 69, no. 3 (2016): 1000–13.

[87] Lefèvre, *Elementa*, sig. b9r: 'primum sibi vendicat arithmetice locum que ut rite cognoscatur quemadmodum et cetere certis eget adminiculis inter que primo menti figenda est universorum circa que versatur subiecta formula, mox singulorum diffinitiones, post quas numerorum affectiones proprietatesque, post proprietates quo ex loco universe sunt monstrande.'

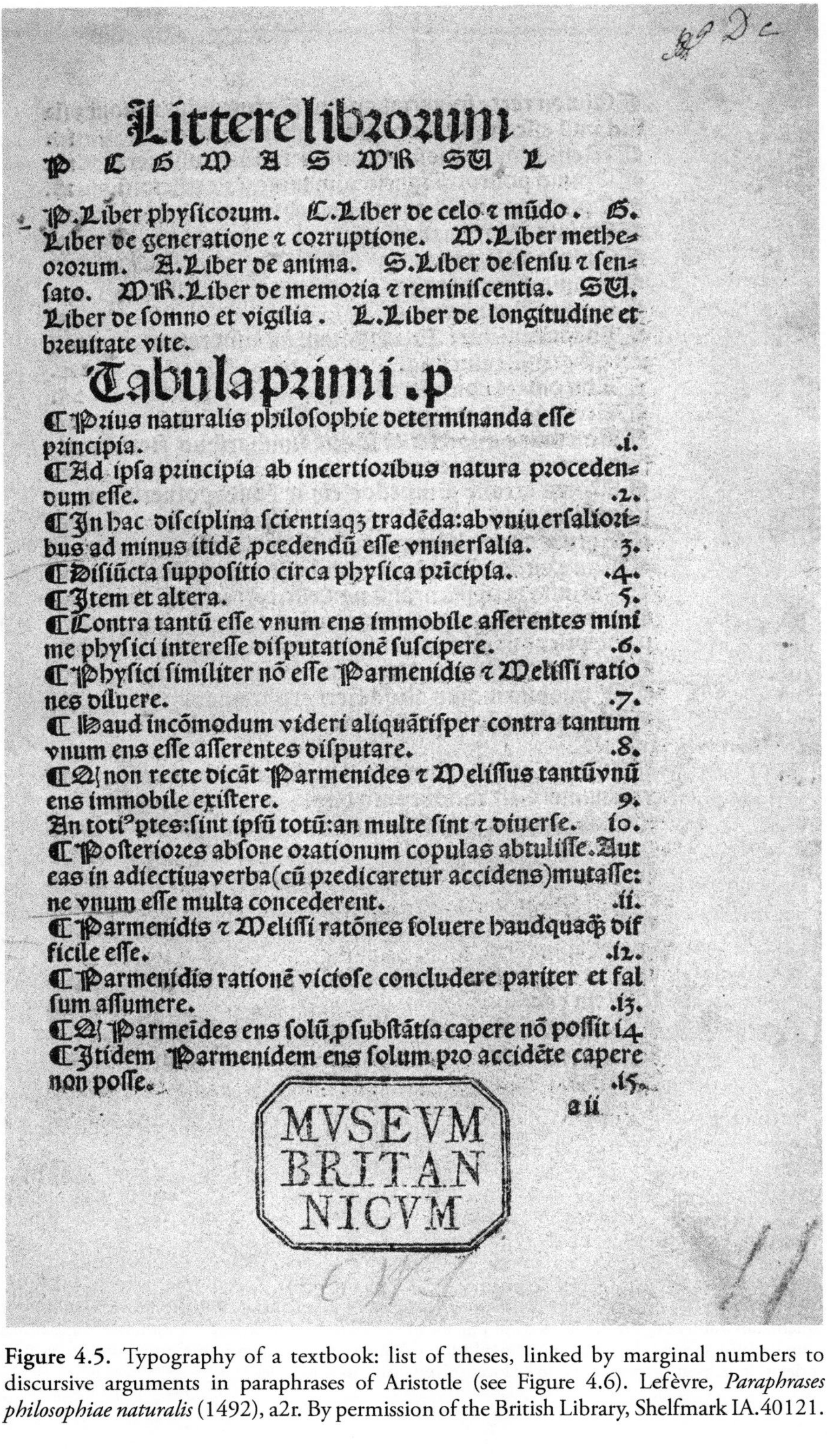

Littere librorum

P C G M A S MR SV L

P.Liber physicorum. C.Liber de celo ⁊ mūdo. G. Liber de generatione ⁊ corruptione. M.Liber metheororum. A.Liber de anima. S.Liber de sensu ⁊ sensato. MR.Liber de memoria ⁊ reminiscentia. SV. Liber de somno et vigilia. L.Liber de longitudine et breuitate vite.

Tabula primi .p

¶ Prius naturalis philosophie determinanda esse principia. .1.

¶ Ad ipsa principia ab incertioribus natura procedendum esse. .2.

¶ In hac disciplina scientiaqꝫ tradēda: ab vniuersalioribus ad minus itidē ꝓcedendū esse vninersalia. 3.

¶ Disiūcta suppositio circa physica pricipia. .4.

¶ Item et altera. 5.

¶ Contra tantū esse vnum ens immobile asserentes minime physici interesse disputationē suscipere. .6.

¶ Physici similiter nō esse Parmenidis ⁊ Melissi rationes diluere. .7.

¶ Haud incōmodum videri aliquātisper contra tantum vnum ens esse asserentes disputare. .8.

¶ Q̄ non recte dicāt Parmenides ⁊ Melissus tantū vnū ens immobile existere. 9.

An toti⁹ ptes: sint ipsū totū: an multe sint ⁊ diuerse. 10.

¶ Posteriores absone orationum copulas abtulisse. Aut eas in adiectiua verba (cū predicaretur accidens) mutasse: ne vnum esse multa concederent. .11.

¶ Parmenidis ⁊ Melissi ratōnes soluere haudquaq̄ difficile esse. .12.

¶ Parmenidis rationē viciose concludere pariter et falsum assumere. .13.

¶ Q̄ Parmeīdes ens solū ꝓ substātia capere nō possit 14.

¶ Itidem Parmenidem ens solum pro accidēte capere non posse. .15.

a ii

Figure 4.5. Typography of a textbook: list of theses, linked by marginal numbers to discursive arguments in paraphrases of Aristotle (see Figure 4.6). Lefèvre, *Paraphrases philosophiae naturalis* (1492), a2r. By permission of the British Library, Shelfmark IA.40121.

Jacobi Fabri stapulensis in Aristotelis octo Physicos libros Paraphrasis.

Primi tractatus primi physicorum note.

Principia/elementa/cause: Principiorũ sunt noia. ¶ Incertius/Ignotius. Notiora nobis/notiora ad nos. ¶ Cõfusiora/compositiora.

Iber physicorum Aristotelis octo partiales libros continet ¶ Primus quatuor tractatus. ¶ Primus tractatus libri proemiũ est: quo pacto i libris physicorum procedendũ sit ostendẽs. Tres cõtinens cõclusiones. ¶ Prima. Prius naturalis phie determinãda sunt principia. Patet. quia ea prius determinanda sunt ex quorum cognitõe rerum naturaliũ acquiritur noticia. At ex pricipiis naturalis phie: rerũ naturalium acquiritur noticia. Quoniamquidem itelligere et scire cõtingit circa oẽs sciãs quarum sunt pricipia/elemẽta aut cause/ex horum cognitiõe. Arbitramur enim oẽs tunc vnumquodque cognoscere/cũ principia/elemẽta et causas cognoscãt. Sũt igitur prius naturalis phie determinanda principia. ¶ Secunda ad ipa pricipia ab incertioribus natura procedẽdũ ẽ. ¶ Patet quia a notioribus nobis ad minus nota procedendũ est. Incertiora autẽ natura: nobis sunt notiora certioribus natura. Nõ enim oportet eadẽ nobis et natura similiter nota eẽ. Sũt enim cõpositiora nobis prius manifesta simpliciora vero posterius. Ad ipa igitur pricipia ab incertioribus natura procedendum est. Etenim ingenita est nobis via ex nobis notioribus et certioribus: in notiora et certiora natura procedere. ¶ Tercia In hac disciplina ab vniuersalioribus ad minus vniuersalia procedere expediet. ¶ Ratio quia a notioribus ad minus nota procedendũ est. In disciplinis autẽ vniuersaliora notiora sunt minus vniuersalibus. ¶ Primo quia vle totũ est intellectus (cõprehẽdit enim sub se partes multas) sicut cõpositum: totũ est sensus. Cõpositũ autẽ notius est sensui suis partibus/ita et vle notius erit itellectui suis partibus. ¶ Tertio quia vniuersaliora cõfusiora sunt: sue enim diffinitões quasi indistincte importãt: distingũt et in particulas diuidũt. Cõpositi-

Figure 4.6. Typography of a textbook: a numbered thesis. Lefèvre, *Paraphrases philosophiae naturalis* (1492), b5r. By permission of the British Library, Shelfmark IA.40121.

Aristotle's *Physics* already present the format followed throughout all the later works. Each paraphrase begins with a few *notae* which alert the reader to the most important topics below. Then each paraphrase itself is constructed as a series of *conclusiones*, supported by brief arguments and sometimes corollaries. Thus Aristotle is reduced to essential propositions, each a brick in the architecture of the whole.

These three features allow the text to expand or contract as needed. In the *Textus de sphaera* (1495), Lefèvre opened with an 'index of the book', in which the numbered propositions again offered a panorama of the whole narrative Lefèvre then unfolded. In this case, the index or concordance was especially useful, because Lefèvre offered much *more* than Sacrobosco's own text, as I discuss in Chapter 5, including further tables on the distances between the spheres, tables of right ascensions, and tables of longitudes and latitudes based on the new-found *Cosmographia* of Ptolemy. The astronomy contrasts with Lefèvre's logical introductions, where tables *shorten* rather than expand on the commented text. One such logical text is his *Introductiones in suppositionem, etc* (1496), meant to let students bypass the standard university logic text of Peter of Spain. Lefèvre offered the entire work as a list of propositions—these were, he submitted, as much introduction to scholastic *sophismata* and *sorites* as anyone would need.[88] Such tables are concordances that let one co-locate the entirety of the text in mind at once. Lefèvre's edition of Jordanus' *Arithmetic* and his own advanced treatise on music theory is accompanied by the most elaborate example of this sort. This 'formula of properties', like the index in Sacrobosco, is a kind of concordance between propositions concerning the properties of numbers found in Jordanus *Elementa arithmetica*, linking them back to their original sites in Boethius's *Arithmetica*.

* * *

Such printed paratexts imply certain cognitive practices. Synoptic tables of terms or propositions offered an immediate view of a discipline's minimal parts, arranged in a hierarchy. On the few occasions Lefèvre and Clichtove explain this paratextual regime, they use the vocabulary of the mathematically inspired axiomatic genres. In his prologue to the *Paraphrases philosophiae naturalis* of 1492, Lefèvre set out his new style of textbook while criticizing those who misapply methods by mixing up their proper principles, axioms, common notions (*communes*), and postulates (*dignitates*).[89] Aristotle had said that the basic elements of a discipline should be immediately available to the mind; axioms for a given discipline should be apprehended all at once.[90]

Panoptic tables offered just this sort of visual threshold into a field of study: an *accessus ad auctores*. Such tables take a prominent role in Lefèvre's introductions or epitomes as thresholds into a text that themselves are thresholds into the *artes*.

[88] See note 18. [89] See discussion in the section 'Music' in Chapter 5.

[90] Prologue to the paraphrases of Aristotle's *Physics* in Lefèvre, *Paraphrases* (1492), b1r–2r: 'paucis annotatum firma mente teneto. Notas capitum quibus preponuntur prenotioni quid est subseruire. Interiectasque interim sed raro dignitates atque positiones. He siquidem vt plurimum proprium in littera locum obtinent'. See also Clichtove's commentary on this methodological statement in the 1502 edition of the same work.

They offer a structure by which to grasp, at the outset, the shape of the text to follow. They help the student imagine some knowledge that they have not yet mastered—they are a pedagogue's answer to Plato's famous conundrum in the *Meno* of whether we can learn what we do not know. On questioning a slave boy, Socrates suggested that mathematical knowledge is innate to the soul, and learning is therefore recollection. Lefèvre countered this view by drawing on the venerable analogy of the intellect as the mind's eye, offering mathematical propositions as examples of how the axioms and universals of mathematics are immediately available to the mind in a moment of illumination.[91] Set at the outset of a work, panoptic tables offer a flash of clarity evocatively modelled on mathematical argumentation.

In this axiomatic mode, the numbered theses and *conclusiones* set out the skeleton of assumptions, assertions, and corollaries that structure a domain. As Lefèvre put it at the beginning of his first paraphrases, the *notae* he added to the beginning of chapters supply missing 'postulates and assumptions' and identify main arguments 'in their proper place with a letter'.[92] Lefèvre explained the mode of the *Introductio in Ethicen* (1497) as divided similarly into minimal *elementa* aimed to delineate and stimulate conformity to the discipline. 'The art, I say, is set in order in questions, elements, and precepts for what to do... Elements set [the art] in order and resolve it.... If explicit elements are not there, questions should be resolved using the elements of similar questions, either explicitly or by understanding an easy analogy.... The numbers added to the side [of the text] designate the books of Aristotle's *Ethics* which the heads of the introductions are introducing.'[93] Propositions refer back to the synoptic table, at the beginning of the work, which allow the student to survey the whole of the book at a glance. Marginal numbers cast the reader ahead, to compare Aristotle's own words. The index or table serves as more than a finding aid—though certainly it is that. It enables the mind to move back and forth, from part to whole, to toggle swiftly between close argumentation and large-scale visualization. In this movement, such tables provided a way of re-forming the philosophical arts—literally giving them renewed shape. These books show how the impulse to *elementate* resulted in practices of visualizing disciplines, decomposing a discipline into its smallest constitutent parts and then recombining them.

[91] Lefèvre, *Libri logicorum*, 177v: 'Verum quedam talia sunt que cognitis terminis (attentione mentis adhibita) statim cognoscimus, perinde ac apertis fenestris et revelatis ciliis statim lumen cognoscimus.' Lefèvre here responds to Plato's *Meno* as cited by Aristotle in the *Posteriora analytica*, book 1, 71a29–30.

[92] Lefèvre, *Paraphrases* (1492), b1v: 'paucis annotatum firma mente teneto. Notas capitum quibus preponuntur prenotioni quid est subservire. Interiectasque interim sed raro dignitates atque positiones. He siquidem ut plurimum proprium in littera locum obtinent.' See also Clichtove's commentary on this methodological statement in the 1502 edition.

[93] Lefèvre, *Introductio in Ethicen Aristotelis* (Paris, 1534), 2r, where the book's structure is presented as integral to its art: 'Ars inquam quaestionibus, elementis praeceptisque officiorum digesta.... Elementa digerunt, atque dissolvunt.... Et si expressa non adducuntur elementa, ex superioribus elementis similium quaestionum, quaestiones sunt diluendae, vel expressis, vel facili analogia cognoscendis.... Et numeri ad latus adiecti, libros moralium Aristotelis (ad quos introductionum capita introducunt) designant.'

Formula proprietatū ex Boetio reperiendarū/ atq; ex Jordano demonstrandarum.

¶Di uus Se ue ri nus Bo e ti us ¶Jor da nus.

Numerorum proprietates.	Caput.	Liber.	Propositio.	Liber.
Numerus.				
1	7	1	2	1
2	7	1	2	1
Numerus par.				
1	5	1	2	7
2	5	1	2	7
3	46	2	10	7
4	46	2	12	7
Numerus impar.				
1	5	1	3	7
2	46	2	11	7
3			10	7
Numerus pariter par.				
1	9	1	31	7
2	9	1	29	7
3	9	1	32	7
4	9	1	54	7
5	9	1	25	4
6	9	1	26.40	2.7
Numerus pariter impar.				
1	10	1	33	7
2	10	1	34	7
3	10	1	35	7
4	10	1	35	7
5	10	1	2	1
6	10	1	3	1
Numerus impariter par.				
1	11	1	37	7
2	11	1	38	7
3	11	1	40	7
Numerus perfectus				
1	20	1	0	0
2	20	1	60	7
Numerus diminutus et abundās				
1	0	0	55	7
2	0	0	55	7
Numerus primus et compositus				
1	0	0	1	3
2	0	0	2	3
3	17	1	25	7
4	17	1	25	7
Numerus ad alterum primus.				
1	17	1	12	3
2	18	1	15	3

Figure 4.7. Epitome of arithmetic. The *formula* of core propositions collated from Boethius and Jordanus shares a goal with the table from Ramus shown in Figure 4.8: to epitomize a discipline. Lefèvre, *Elementa arithmetica, etc.* (1496), i5r. By permission of the British Library, Shelfmark C.106.a.8.

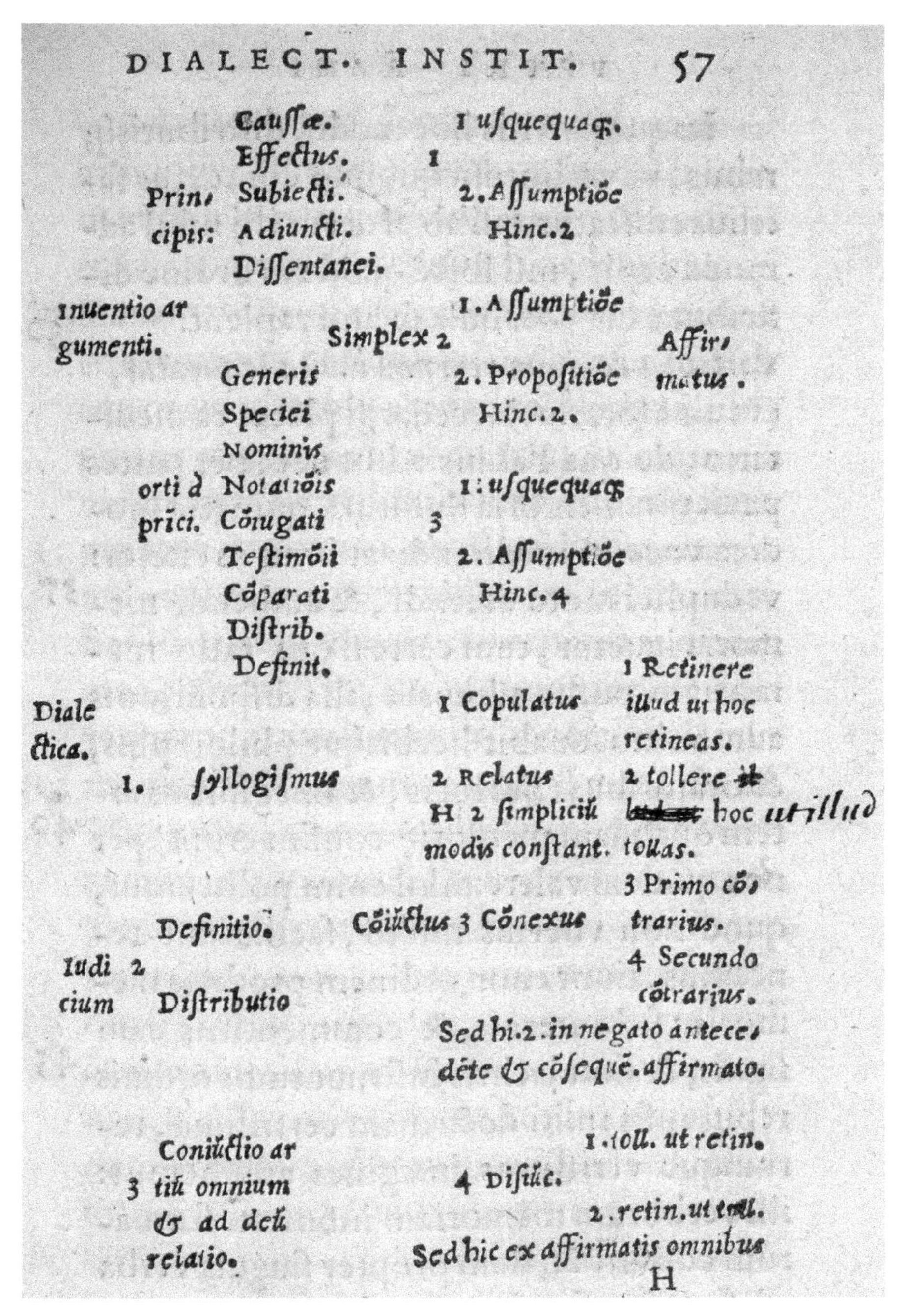
DIALECT. INSTIT. 57

Inuentio argumenti.
Principis: Caussæ. Effectus. Subiecti. Adiuncti. Dissentanei.
orti à prīci. Generis Speciei Nominis Notatiōis Cōiugati Testimōii Cōparati Distrib. Definit.

Simplex 2
1 usquequaq. 1 2. Assumptiōe Hinc. 2
1. Assumptiōe 2. Propositiōe Hinc. 2.
Affirmatus.
1: usquequaq. 3 2. Assumptiōe Hinc. 4

Dialectica.
1. syllogismus
1 Copulatus
2 Relatus
H 2 simplicib modis constant.
Cōiūctus 3 Cōnexus
4 Disiūc.

1 Retinere illud ut hoc retineas.
2 tollere ~~it~~ ~~[illegible]~~ hoc ut illud tollas.
3 Primo cōtrarius.
4 Secundo cōtrarius.
Sed hi. 2. in negato antecedēte & cōseqūē. affirmato.

1. toll. ut retin.
2. retin. ut toll.
Sed hic ex affirmatis omnibus

Iudicium 2
Definitio.
Distributio

3 Coniūctio artiū omnium & ad deū relatio.

H

Figure 4.8. Epitome of dialectic: Peter Ramus, *Dialecticae institutiones* (Paris: Jacques Bogard, 1543), 57r. By permission of the British Library, Shelfmark C.106.a.8.

Lefèvre's modest project of introductions as thresholds to the arts therefore served as the first step to the most ambitious encyclopaedic projects of the later Renaissance. Most fundamentally, this movement back and forth between simplicity and depth answers the twin challenge outlined in the previous chapter: how to balance the encyclopaedic ideal of knowledge *and* the need for methodological precision in each discipline. The mode is that of collation, a back-and-forth movement between compressed and expanded accounts, between concise elements and their discursive explanations. This mode deepens the significance of Lefèvre and his little band to the long tradition of methodical reflection that carries through Peter Ramus to René Descartes, and more generally through the German encyclopaedists. Fabrist tables belong in this story; they offer the same cognitive mapping of later dichotomous charts that proceed in a direction from *maxima* to *minima*. The student was expected to proceed from universals to singulars—an expectation that was the explicit foundation of Ramus' *Institutiones dialecticae* (1543), which (not incidentally) included a similar table, not yet divided by the printer's art into branching trees (Figures 4.7, 4.8). In print, such maps of the disciplines lead to Christophe de Savigny's remarkable *Tableaux accomplis de tous les ars liberaux* (Paris, 1587), where the whole of learning fits into a few elaborate plates.[94] One might compare the process of *enumeratio* that Paolo Rossi found in memory treatises, which characterized Peter Ramus' method, and which in turn was where Descartes started his methodological instructions in his *Regulae ad directionem ingenii*: 'We shall be following this method exactly if we first reduce complicated and obscure propositions step by step to simpler ones, and then, starting with the intuition of the simplest ones of all, try to ascend through the same steps to a knowledge of all the rest.'[95]

In this chapter I have examined the first comprehensive efforts to use the printing presses to reorder the university textbook. For decades, Lefèvre was famous for authoring reliable, up-to-date introductions to university learning, which helped to navigate the pressures of university learning described in the last chapter. Of course, Lefèvre did not create the printed textbook out of nothing. He drew on Porphyry, Boethius, and Themistius, whose epitomes and paraphrases had long served students. Still, Lefèvre books were the first distinctive creatures of their kind in print.

In fact, Lefèvre only invented the printed textbook with a great deal of help. His students selected them for print, editing, amending, and augmenting them, sometimes even writing sections. This was partly in the nature of the textbook, an extension of glossing practices that had long characterized the classroom. But mathematical genres turn out to be particularly extreme cases of such joint

[94] An edition of Savigny is included in Steffen Siegel, *Tabula: Figuren der Ordnung um 1600* (Berlin: Akademie Verlag, 2009).

[95] René Descartes, *The Philosophical Writings of Descartes*, Vol. I (Cambridge: Cambridge University Press, 1984), 20 (AT 10:379), for Rule 7 on *enumeratio*. See Paolo Rossi, *Logic and the Art of Memory: The Quest for a Universal Language*, trans. Stephen Clucas (1983; London: Continuum, 2000), 123–9, as well as Wilhelm Schmidt-Biggemann, *Topica universalis: eine Modellgeschichte humanistischer und barocker Wissenschaft* (Hamburg: Meiner, 1983), 293–7.

authorship. Precisely because they focused on the enunciation alone as the authoritative centre, with demonstrative commentary optional, these texts—like Euclid's *Elements*—invited further authors, accruing further layers of text as they went.

Textbooks, after all, always served other texts, at least in principle. Introductory manuals were not an end in themselves, but a means to engage Aristotle, Euclid, Jordanus, and so on. In the oscillation between the table at the outset and the textual argument given later in the manual, introductions as a whole were intended as a large synoptic table, as an icon, a threshold that welcomes the reader beyond itself into the greater profundities of the art itself. In practice, however, print seems to have encouraged a kind of superficiality. The tables and numbers themselves served as pins on which to hang new commentary. The old paratext shuffled to the centre of the page, crowded by new paratext. Repeated editions turned some students into authorities in their own right; the paratexts they devised to support Lefèvre's introductions were themselves printed as surveys of a discipline, removed even further from the original.

Textbooks have often been ignored in the history of science because of an assumption that creativity happens elsewhere. The work of commentary and introduction—digesting, revisualizing, explaining, abbreviating—seems to confirm old canards rather than stimulate new knowledge. The role of print has been especially suspect. The brilliance and weakness of these books meet in a reliance on visual reasoning: to use the ordered page, one proceeds by visualizing connections without making them verbally explicit. The driving theme of Walter Ong's study of Ramus famously presented such a visualizing impulse as the decay of focused, rational discourse.[96] The reasoning is familiar to any teacher: if students take short cuts, they may never learn what it takes to tackle more difficult, profound topics. In short, we do well to wonder whether Lefèvre's new printed textbooks meant anything good for mathematical culture.

[96] Ong, *Ramus, Method, and the Decay of Dialogue.*

5

The Senses of Mixed Mathematics

By the seventeenth century intellectual culture had become mathematical, so that in 1674 the Jesuit architect Claude-François Milliet de Chales could proclaim that mathematics had displaced theology as the queen of the sciences.[1] Once medieval universities had taught chiefly the quadrivium of Boethius: arithmetic and geometry, music and astronomy. Now, in thirty books spread over three volumes, De Chales provided what he called a 'world of mathematics', surveying whole swathes of technical domains under the rubric of mixed mathematics: magnetics, stone-cutting, hydraulics, pyrotechnics, conic sections, and the method of indivisibles. The tidy quadrivium had fractured into an all-encompassing set of disciplines, and new philosophers from Descartes to Christian Wolff threatened to make mathematics the arbiter of all knowledge. This did not happen all at once. The beginnings of fault lines are found in the early printed books of Lefèvre's circle. In this chapter, we shall see how these books fostered practices that eroded old disciplinary barriers in two mathematical disciplines central to Renaissance culture: music and cosmography.

The transformation of the medieval quadrivium into the mathematical philosophies of the seventeenth century is complex, and no one explanation will do. Most have looked to the world of mathematical practice, to courtly experts such as John Dee or Galileo, army engineers, the Italian abbacisti and German Reckenmeister.[2] Others trace the collapse of the medieval quadrivium to pressures from recovered ancient mathematics and commentators such as Pappus, Hero of Alexandria, and Proclus.[3] Both explanations are important, and I have no wish to replace them.

[1] Claude-François Milliet de Chales, *Cursus seu mundus mathematicus* (Lyons: Officina Anissoniana, 1674), dedication.

[2] A representative range of studies includes Mario Biagioli, 'The Social Status of Italian Mathematicians', *History of Science* 27, no. 1 (1989): 41–95; Eric H. Ash, *Power, Knowledge, and Expertise in Elizabethan England* (Baltimore, MD: Johns Hopkins University Press, 2004); Stephen Johnston, 'The Identity of the Mathematical Practitioner in 16th-Century England', in *Der 'Mathematicus': Zur Entwicklung und Bedeutung einer neuen Berufsgruppe in der Zeit Gerhard Mercators*, ed. Irmgarde Hantsche (Bochum: Brockmeyer, 1996), 93–120; Karin J. Ekholm, 'Tartaglia's Ragioni: A Maestro d'abaco's Mixed Approach to the Bombardier's Problem', *British Journal for the History of Science* 43, no. 2 (2010): 181–207; Alexander Marr, *Between Raphael and Galileo: Mutio Oddi and the Mathematical Culture of Late Renaissance Italy* (Chicago: University of Chicago Press, 2011).

[3] The classic statement is Paul Lawrence Rose, *The Italian Renaissance of Mathematics* (Geneva: Droz, 1975). Studies using this approach include Eckhard Kessler, 'Clavius entre Proclus et Descartes', in *Les jésuites à la Renaissance: Système éducatif et production du savoir*, ed. Luce Giard (Paris: Presses universitaires, 1995), 285–308, and Domenico Bertoloni Meli, 'Guidobaldo Dal Monte and the Archimedean Revival', *Nuncius* 7, no. 1 (1992): 3–34.

But engineers grew in prestige and Proclus was read by a growing audience of bureaucrats and gentlemen whose attitudes and literacy in mathematics were shaped by university books.[4] By examining the first wave of those books in Lefèvre's circle, we will see that they did not simply transmit the medieval quadrivium unchanged, but in the process *reimagined* the disciplines.

In fact, the instability of these disciplines was already in the DNA of the medieval quadrivium. The apparent solidity of the quadrivium depended on a few key distinctions. As Boethius explained, only arithmetic and geometry are 'pure' mathematics, respectively considering number and lines on their own.[5] The others are impure: music matches arithmetical reasoning to sound, and astronomy uses geometry to trace the heavens' movements. It was easy to think these distinctions well founded and authoritative, because of what Aristotle said at *Physics* II.2, the basis for late medieval discussions of 'mixed' or 'subaltern' sciences.[6] In his paraphrase of the passage, Lefèvre described how the mathematician properly considers magnitudes and figures by intellectually abstracting them from all accidents, matter, and even movement.[7] But, he continued, the musician, optician, and astronomer (*musicus, perspectivus, astrologus*) follow a different method.[8] They do use mathematical explanations (*demonstrationes*), but apply them to physical phenomena of sounds, sights, and the moving heavens. In such formulations, the antiquity of the quadrivium and a few attendant disciplines such as optics lulled Renaissance thinkers into supposing that these mixed mathematics formed a tidy, principled group. The group turned out to be unstable. The fracturing had already begun among medieval philosophers such as Roger Bacon and Jordanus de Nemore—the one imagined mathematics as the basis of all forms of philosophy,

[4] e.g. Mordechai Feingold, *The Mathematician's Apprenticeship*; Dear, *Discipline and Experience: Science, Universities and Society in England, 1560–1640* (Cambridge: Cambridge University Press, 1984), Antonella Romano, *La contre-réforme mathématique: Constitution et diffusion d'une culture mathématique jésuite à la Renaissance (1540–1640)* (Rome: École française de Rome, 1999).

[5] See Boethius' statement of these dependencies in *Arithmetica* I. A modern introduction and translation, to be used with care, is Michael Masi, *Boethian Number Theory: A Translation of the De Institutione Arithmetica* (Amsterdam: Rodopi, 1983).

[6] Some argue that 'mixed mathematics' only arose as a term around 1600: e.g. Gary I. Brown, 'The Evolution of the Term "Mixed Mathematics"', *Journal of the History of Ideas* 52, no. 1 (1991): 81–102. However, the notion of a *scientia mixta*—for which the main examples were harmonics, optics, and astronomy—was common in late medieval philosophy: e.g. relevant sections of Steven J. Livesey, 'William of Ockham, the Subalternate Sciences, and Aristotle's Theory of Metabasis', *British Journal for the History of Science* 18, no. 2 (1985): 127–45; Jean-Marc Mandosio, 'Entre mathématiques et physique: note sur les "sciences intermédiaires" à la Renaissance', in *Comprendre et maîtriser la nature au Moyen Âge: Mélanges d'histoire des sciences offerts à Guy Beaujouan* (Geneva: Droz, 1994), 115–38.

[7] Jacques Lefèvre d'Étaples and Josse Clichtove, *Totius philosophiae naturalis paraphrases, adiecto commentario* (Paris: Henri Estienne, 1502), 23v: 'Mathematicus autem ea a re naturali, materia et motu cogitatione et intellectu abstrahit.' Lefèvre even paraphrases Aristotle's jab at Plato for making mathematical objects too real, not merely produced by the mind: 'haec latuerunt Platonem ideas introducentem, quas non modo cogitatione et intellectu sed re ipsa a materia se iungebat, faciebat enim Physicam magis Mathematica abstractam, cum ea minus abstracta sit.'

[8] Lefèvre and Clichtove, *Totius philosophiae naturalis paraphrases*, 23v: 'Tertio quia physicus et Mathematicus ut musicus, Perspectivus, et Astrologus (qui proprius ad physicam accedere videntur) non eodemmodo demonstrant, quia Musicus, Perspectivus, et Astronomus demonstrationes ab Arithmetica et Geometria accipiunt.'

the other systematically constructed a new mixed science of weights.[9] In the fourteenth century, studies of physical change known as the 'latitude of forms' earned the title of mixed sciences too.[10] The process accelerated around 1500 as cosmography, ballistics, pneumatics, and so on laid claim to the title.[11] Such disciplines were considered subaltern because they borrowed from other disciplines, promiscuously mixing one discipline's explanatory method with another's objects. They were mixed because they occupied an in-between place in Aristotle's taxonomy of knowledge, in which the overarching methodological restriction was the *prohibition of metabasis*, of using principles of one discipline to reason in another. But if numbers and lines can be used to reimagine one discipline, why should they not apply to others?

Lefèvre's books of cosmography and music are ideal cases for observing how these in-between forms of mathematics were remixed in the making of mathematical culture. Both of Lefèvre's books repackaged the main texts of the medieval quadrivium; both books inaugurated disciplines that put mathematics at the heart of European culture: cosmographers and musicians could be artisan practitioners, polite courtiers, or erudite scholars alike. In the Renaissance, cosmography came to offer the mathematical basis for navigation, mapping, and erudite learning.[12] Even as a theoretical enterprise, music became a site of experiments with mathematical instruments, theories of sound, and courtly competition.[13] The Fabrist books contributed to these trajectories by aiming to create a community of expert readers—not practitioners who defined their careers around mathematics, but lovers of the liberal arts, and specifically 'lovers of mathematics' with the experience to discern mathematical knowledge.[14]

Learning to wield these disciplines required a certain expertise and good judgement about how mathematical ideas, spun of the finest abstractions our minds can make, enmesh the physical world that scrapes our senses so unevenly. Lefèvre's books on cosmography and music theory present their most distinctive insights precisely at this juncture of mind and sense. This standpoint enriches our understanding of early modern scientific novelty and sensory experience. Our historiographies often

[9] David C. Lindberg, 'On the Applicability of Mathematics to Nature: Roger Bacon and His Predecessors', *British Journal for the History of Science* 15 (1982): 3–25; Jens Høyrup, 'Jordanus de Nemore, 13th Century Mathematical Innovator: An Essay on Intellectual Context, Achievement, and Failure', *Archive for History of Exact Sciences* 38, no. 4 (1988): 307–63. More generally, see H. L. L. Busard, 'Über die Entwicklung der Mathematik in Westeuropa zwischen 1100 und 1500', *NTM Zeitschrift für Geschichte der Wissenschaften, Technik und Medizin* 5, no. 1 (1997): 211–35.

[10] See Chapter 6.

[11] Brigitte Hoppe, 'Die Vernetzung der Mathematisch ausgerichteten Anwendungsgebiete mit den Fächern des Quadriviums in der Frühen Neuzeit', in *Der 'mathematicus' zur Entwicklung und Bedeutung einer neuen Berufsgruppe in der Zeit Gerhard Mercators*, ed. Irmgarde Hantsche (Bochum: Brockmeyer, 1996), 1–33.

[12] The literature is large, but see Adam Mosley, 'The Cosmographer's Role in the Sixteenth Century: A Preliminary Study', *Archives internationales d'histoire des sciences* 59, no. 2 (2009): 423–39; Jean-Marc Besse, *Les Grandeurs de la terre: aspects du savoir géographique à la renaissance* (Paris: ENS Editions, 2003).

[13] e.g. Rebecca Cypess, *Curious and Modern Inventions: Instrumental Music as Discovery in Galileo's Italy* (Chicago: University of Chicago Press, 2016).

[14] On the topos of 'lovers of mathematics' see Oosterhoff, 'Lovers in Paratexts'.

regard the visual as a marker of new knowledge. To be sure, much of our evidence is printed, and I shall show how print, as a technology of 're-imaging', catalysed the process of mixing mathematics.[15] Lefèvre, Clichtove, and Oronce Fine's astronomical books include images to illustrate scientific theories and practices—frontispieces, iconography, and so on—of the sort that played a powerful role in legitimating mathematical disciplines, defending them as reliable and useful.[16] Tables, diagrams, and lists worked together to constitute the very genres of mathematics. But learning to use these tools required other senses beyond sight. By keeping in view the classroom context and the paratexts discussed in the previous chapters, we can understand how vision cooperated with other senses, especially touch and hearing, to cultivate mathematically discerning readers. These books became part of the experience needed to extend mathematics imaginatively to new domains.

THE MIND'S EYE AND THE SENSES

A student learned mathematical verities in Lefèvre's circle with the goal of sharpening his intellectual vision, to train his mind's eye. This might tempt us to interpret Fabrist textbooks chiefly as an example of the Western preoccupation with vision as the primary sense of understanding.[17] The print textbooks by Lefèvre and Bovelles mark an early high point in the visual possibilities of print, with their wide range of diagrams, tables, figures, and even allegorical images.[18] Lefèvre's works rarely include images that are not tables, numbers, lines, or circles. Even the more representational images of Bovelles, such as the famous *palestrites studiosus* (the scholarly athelete), are schematic, inscribed with lines and labels (Figure 5.1). At first, Fabrist books seem a paradigmatic proof of Walter Ong's view in which visual, diagrammatic reasoning eclipsed oral discourse after the advent of print. But the twin oppositions of manuscript/print, sight/other senses, fail. In previous chapters I have already built on the post-Ong consensus that manuscript and print cultures shared practices. Over the course of this chapter, I will go further, suggesting that actually learning mathematics required readers to use vision in tandem with other senses.

[15] This chapter therefore contributes to the growing literature on images in scientific cultures, surveyed by Alexander Marr, 'Knowing Images', *Renaissance Quarterly* 69, no. 3 (2016): 1000–13.

[16] Volker R. Remmert, *Picturing the Scientific Revolution: Title Engravings in Early Modern Scientific Publications*, trans. Ben Kern (Philadelphia, PA: Saint Josephs University Press, 2011); Inga Elmqvist Söderlund, *Taking Possession of Astronomy: Frontispieces and Illustrated Title Pages in 17th-Century Books on Astronomy* (Stockholm: Center for History of Science at the Royal Swedish Academy of Sciences, 2010).

[17] Stuart Clark, *Vanities of the Eye: Vision in Early Modern European Culture* (Oxford: Oxford University Press, 2007), 9–20.

[18] Rebecca Zorach and more recently Anne-Hélène Klinger-Dollé have remarked on the insistently visual character of their thought, and their frequent recourse to mathematical figures. Rebecca Zorach, 'Meditation, Idolatry, Mathematics: The Printed Image in Europe around 1500', in *The Idol in the Age of Art: Objects, Devotions and the Early Modern World*, ed. Michael Wayne Cole and Rebecca Zorach (Farnham: Ashgate, 2009), 317–42; Anne-Hélène Klinger-Dollé, *Le* De sensu *de Bovelles: conception philosophique des sens et figuration de la pensée* (Geneva: Droz, 2016). See Chapter 4, note 85.

Figure 5.1. The 'studious athlete' in Bovelles, *Liber de intellectu, etc.* (1511), 60v. Note the lines between *lectio*, *scriptio*, and vision, which together with hearing and speaking fill the scholar's starry imagination. By kind permission of the Syndics of Cambridge University Library, Shelfmark: Acton.b.sel.32.

This does not deny the profound importance of vision in Lefèvre's circle. Throughout their books, the abundance of figures and a schematic style reflects a pervasive pedagogical struggle to make intellectual realities visible. The Fabrists worked so hard to reveal unseen things because of how they understood the vocation of the Christian scholar. Already in the *Textus de sphaera* (1495) Lefèvre turned to the Apostle Paul's famous line that 'the invisible things of [God] from the creation of the world are clearly seen, being understood by the things that are made'.[19] For Lefèvre, this topos justified much more than an abstract natural theology, the idea that the attributes of God—his wisdom and power—can be found in creation. It also justified the faithful scholar's vocation of striving to see hidden matters. In this early work of astronomy, Lefèvre hinted at less licit motivations, with the curiosity of magicians, citing them on the four cardinal points of the heavens which relate to particular heavenly powers: God, intelligences, the blind, and the evil. For years Lefèvre busied himself with the *invisibilia* of the magicians, particularly in the Cabbalistic and Pythagorean sections of the treatise *De magia naturali*. The theme even emerged in his commentary on the Psalms of 1509. There he used the Pauline passage to explain the central theme of the Psalms—that of the blessed man (*beatus*)—as a matter of rising from the visible things of creation to the invisible things of God. Tellingly, he explained that every discipline from logic to physics to mathematics should be used, not simply for their own sakes, but as instruments of 'supramundane' learning. Lefèvre alluded to 1 Corinthians 13, praying that 'although now we are imperfect, tracing through a mirror, by enigma, the *supramundana disciplina* rises from the traces of things to you [O God].'[20] By this time, Lefèvre had publicly distanced himself from natural magic, but not from all forms of occult knowledge: at several points in his Psalms commentary he returned to Cabbalistic lines of reasoning about *invisibilia*.[21] Throughout his life, Lefèvre navigated the line between curiosity and studiousness, between legitimate and illegitimate modes of making visible the hidden, secret things of God and nature.[22]

The hidden *invisibilia* should be made manifest because this was what faith was supposed to do.[23] Lefèvre was acutely conscious of the responsibility. In 1510 he published a treatise *De trinitate* by the Victorine mystic Richard of St Victor, which begins with phrase that 'the just shall live by faith', a phrase found both in the Old Testament prophet Habakkuk (2:4) and in the New Testament book of Hebrews

[19] The biblical passage is Romans 1:20, cit. *Textus de sphera* (1495), a8r.

[20] Jacques Lefèvre d'Étaples, *Quincuplex Psalterium* (Paris: Henri Estienne, 1509), 174v: 'Ergo non rationalis, non naturalis, non mathematica disciplina beatum efficit, sed supramundana—que dum adhuc imperfecti sumus et in speculo et enigmate vestigamus, ex rerum vestigiis ad te surgit, divinorum eloquiorum pia sedulaque perscrutatione adiuta, que et nobis et secundum nos verissime theologia est.'

[21] Brian P. Copenhaver, 'Lefèvre d'Étaples, Symphorien Champier, and the Secret Names of God', *Journal of the Warburg and Courtauld Institutes* 40 (1977): 189–211, at 197–9.

[22] For more on this process in the *Quincuplex*, see Richard J. Oosterhoff, 'Why Marginalia Still Matter: Finding a Voice for Humility in Google Books', *NDIAS Quarterly* (Fall 2014): 6–11.

[23] Bernard McGinn, *The Harvest of Mysticism in Medieval Germany* (New York: Crossroad, 2006), 457, points out that the Western Christian tradition is nearly unified in agreement that the vision of God—the *theoria Theou*—is the ultimate goal of the human life.

(10:38). Commenting on this line, Lefèvre went directly to the famous definition of faith in Hebrews: 'faith is the substance of things hoped for, the evidence of things not seen.'[24] Lefèvre found the Greek here especially illuminating because it used words central to Greek philosophy, so he quoted it in full: '*Ἔστιν δὲ πίστις ἐλπιζομένων ὑπόστασις, πραγμάτων ἔλεγχος οὐ βλεπομένων.*' Thus *ὑπόστασις* becomes *fundamentum* or ground of understanding, and *ἔλεγχος* is *argumentum* or rational argument. Lefèvre chose to render faith in terms of its cognitive bases, as the basic reasoning that leads to a trustworthy understanding of *res invisibilia*.[25] Similar points would crop up in his influential commentary on Hebrews a couple of years later, in ways that would become important to Martin Luther's emphasis on *sola fide*.[26] The vocation of faithful intellectuals was to make invisible things manifest.

Lefèvre, Bovelles, and others in their circle partly owed this intellectualist account of faith to a distinct strand of late medieval devotional culture. Historians now generally recognize that late medieval religious practice was shot through with the material traces of the divine, with tasting and seeing God's acts in the world in the Eucharist as well as through relics and visions.[27] As I sketched in Chapter 2, the Fabrists were profoundly influenced by some of the most important Northern traditions here, including university reformers such as Jean Gerson and Jean Mombaer, both inspired by the *devotio moderna*. More specifically, they drew on the intellectualist tradition of Nicholas of Cusa and Ramon Lull, who were particularly conscious of the power of images in constructing knowledge.

Bovelles developed these views forcefully in a suite of treatises that begin with the intellect and the senses and lead to his celebrated *De sapiente*.[28] Here the *mediated* nature of knowledge emerges in a Cusan and Lullian view of man as a microcosm who uses his imagination to model the whole external world within himself. Imagination mediates between the senses and the mind, leading the outside world inward, transforming sensory stimulus into intellectual concepts (*species*). The underlying point of reference is Aristotle's *De anima* III.3, which prompted most late medieval philosophers to explain cognition as a matter of images (*phantasmata*)

[24] Hebrews 11:1, KJV.

[25] Jacques Lefèvre d'Étaples (ed.), Richard of St. Victor, *De superdivina Trinitate theologicum opus hexade librorum distinctum. Commentarius artificio analytico* (Paris: Henri Estienne, 1510), 5r: 'Est igitur fides eorum quae sperantur idest invisibilium et eternorum bonorum fundamentum... Sed et fides est *elenchus*, id est argumentum rerum invisiblium, hoc est eternarum quae (ait) non videntur.'

[26] Jacques Lefèvre d'Étaples, *Epistolae ad Rhomanos, Corinthios, Galatas, Ephesios, Philippenses, Colossenses, Thessalonie, Timotheum, Titum, Philemonem, Hebraeos. Epistolae ad Laodicenses, Senecam. Linus de passione Petri et Pauli* (Paris: Henri Estienne, 1512), 253r–v. Shortly after teaching from Lefèvre's *Quincuplex Psalterium* in 1513, Martin Luther lectured on Romans and reimagined Christian practice around the phrase from Habbakuk and Romans, that 'the just shall live by faith'. On Luther's annotations on Lefèvre's *Quincuplex Psalterium* see Guy Bedouelle, *Le Quincuplex Psalterium de Lefèvre d'Étaples: un guide de lecture* (Geneva: Droz, 1979), 226–39.

[27] Caroline Walker Bynum, *Christian Materiality: An Essay on Religion in Late Medieval Europe* (New York: Zone, 2011); Jeffrey F. Hamburger, 'The Visual and the Visionary: The Image in Late Medieval Monastic Devotions', *Viator* 20 (1989): 161–82; Jeffrey F. Hamburger and Anne-Marie Bouché (eds), *The Mind's Eye: Art and Theological Argument in the Medieval West* (Princeton, NJ: Princeton University Press, 2005).

[28] Bovelles, *Liber de intellectu etc.*

and imagination (*phantasia*). But philosophers disagreed about what happened to those images, and whether the highest kinds of intellection also involved such images—by thinking of God, for example, was one making an image of God and thereby sinning against the Ten Commandments?

Bovelles adopted an approach from Lull and Cusanus that swells the role of mathematical intuition in the imagination's operation. Though Lullian ideas permeate his work, he rarely cites the Catalan mystic. One of the few times he cites Lull explicitly is on imagination as the interior common sense—the sense that coordinates the soul's various inputs. 'For [Lull] says that in the imagination itself sensible species are thus separated and become purer. From there they pass directly to the intellect, denuded of all matter and stripped of their former sediments of accidents.'[29] Bovelles's own account adopts the optimistic view that the imagination purifies images.[30] For him, the imagination has two parts: one accepts phantasmata from without, another actively judges them and transmits them to the intellect.[31] The active, fertile imagination can construct and conjecture new images that approximate the external world.[32] This active, imaginative power belongs to man's divine, immortal nature, creating inwardly just as God created outwardly. It is this cognitive power that puts man in God's image, so that man is an 'earthly god'.[33] And because it is an actively *purifying* power, Bovelles can answer the worry of error, mis-imagined realities—which ultimately raises the spectre of idolatry in imagining proxies for God. Bovelles here picks up a strand in the late medieval Albertist tradition (including Cusanus and his teacher Heymeric de Campo) on the possibility of image-ing God through mathematical analogies.[34] Operating broadly within a 'symbolic' theology inherited from Dionysius the Areopagite, transmitted through twelfth-century students of Boethius, this tradition offered a *theologia geometrica* committed to the idea that mathematical figures offered the best tools for carrying finite minds to 'see' infinite realities.[35] The tension between

[29] *De intellectu*, chapter V.9: 'Hoc autem medium (ut et Remundus Lullius sentit) est sensus interior, qui communis sensus dicitur sive imaginatio et phantasia. Ait enim in ipsa imaginatione sensibiles huiusmodi species desecari et evadere puriores, indeque transire cominus ad intellectum, nudatas omni materia et prisca accidentium eluvie spoliatas.'

[30] Compare Leen Spruit, *Species Intelligibilis: From Perception to Knowledge*, Vol. II: *Renaissance Controversies, Later Scholasticism, and the Elimination of the Intelligible Species in Modern Philosophy* (Leiden: Brill, 1995), 39–45.

[31] *De sensu*, chapter VI. In another case (30r) he describes the imagination as a glass sphere encapsulating the intellect; the senses trace images on its outside, as if projecting on a screen, while the intellect views those images through the imagination, thus purified of immaterial qualities.

[32] On this point, consistent throughout his life, Bovelles closely follows Nicholas of Cusa, *De coniecturis*. See Emmanuel Faye, 'Nicolas de Cues et Charles de Bovelles dans le manuscrit "Exigua pluvia" de Beatus Rhenanus', *Archives d'histoire doctrinale et littéraire du moyen âge* 65 (1998): 415–50, at 423.

[33] *De sapiente*, 129r: 'Quod sapiens omnium finis et ut terrenus quidam deus'.

[34] Katharine Park, 'Albert's Influence on Late Medieval Psychology', in *Albertus Magnus and the Natural Sciences*, ed. James A. Weisheipl (Toronto: Pontifical Institute of Mediaeval Studies, 1980), 522–35. For historiographical context, see David Albertson, 'Mystical Philosophy in the Fifteenth Century: New Directions in Research on Nicholas of Cusa', *Religion Compass* 4, no. 8 (2010): 471–85.

[35] On Heymeric's *theologia geometrica* and its influence on Cusanus, see Ruedi Imbach, 'Das Centheologicon des Heymericus de Campo und die darin enthalten Cusanus-Reminiszenzen', *Traditio*

the urge to visualize and the fear of idolatry therefore resolved in mathematics as a devotional practice.

Without lionizing the imagination, Lefèvre nevertheless made mathematics the genre suitable for imagining God in his remarkable commentary on the twelfth-century Richard of St Victor's *De trinitate* (1510)—the one in which he described faith as the *fundamentum* and *argumentum* of unseen things. In the preface Lefèvre made clear how mathematics transcended other forms of reasoning (notably logic) in his account of how to seek out God in three modes, each corresponding to a cognitive level. The first belonged to imagination, which presented the *idola* of the brute senses. Then at the second level, in a 'servant space' (*mediastinum locum*), reason traced out the reasons of things. Since this level is 'proper' to humans, it is tempting to remain here, at the level of the 'empty little fantasies' (*inanes phantasiolas*) propounded by logicians blinded by their own 'barbarous' sophistries; Lefèvre decried in particular the '*calculatoria* which yet perverts all calculation of reason'—a potshot at the fourteenth-century *ars calculatoria* of Richard Swineshead and others.[36] Instead, Lefèvre urged readers to move up one further mode of abstraction: 'just as we rose from imagination to reason, so the right order of movement is to go beyond reason to understanding.'[37] These prefatory comments do not show just how to make this mental motion; but all becomes clear as soon as we turn to the ensuing commentary. While Richard of St Victor's text is unchanged, the bulk of the edition is Lefèvre's commentary, wholly arranged in the Euclidean axiomatic mode. It begins with common notions and postulates. Each section of commentary is structured as a proof, referring back to those starting principles. The whole of the book is an interlinked web of 'elementating', in which complex arguments are built stepwise out of simple axioms and propositions. Early on Lefèvre explicitly disavowed the goal of a natural theology to convert unbelievers. 'Faith is among theologians what an understanding of principles is among philosophers.'[38] Cusanus had used mathematical figures, but never constrained his texts to follow the strict format of axioms, enunciations, and demonstrations. For Lefèvre, the Euclidean genre explores the invisible implications of the given principles of faith. In brief, mathematical reasoning helps the eyes of faith to 'see' the Trinity.

But, persuaded by all of the visual evidence before us, it is possible to read the primacy of sight too absolutely. Despite the elevation of the mind's eye, *all* the senses here can serve intellectual vision. Bovelles and Lefèvre do describe intellectual processes chiefly by analogy with sight. Nevertheless, this implies no contrast between

39 (1983): 466–77; David Albertson, *Mathematical Theologies: Nicholas of Cusa and the Legacy of Thierry of Chartres* (Oxford: Oxford University Press, 2014), 222–49.

[36] This passage is discussed by Rice, PE, 227. Such criticism of the *calculatores* is surveyed by Carlo Dionisotti, 'Ermolao Barbaro e la Fortuna di Suiseth', in *Medioevo e Rinascimento: Studi in onore di Bruno Nardi*, 2 vols (Florence: Sansoni, 1955), 1:219–53, and Chapter 6 below.

[37] Lefèvre, preface to Richard of St. Victor, *De trinitate*, 2r: 'Et ut ex imaginatione ad rationem conscendimus, ita proprius motionis ordo est ex ratione ad intelligentiam moveri.'

[38] Lefèvre, commentary to Richard of St. Victor, *De trinitate*, 5r: 'Fides igitur apud theologos, quod intellectus principiorum apud philosophos.'

intellectual vision and touch, for example. The imagination purifies input from *all* the physical senses. Indeed, at one point Bovelles elaborates a hierarchy of the senses in which he make the unusual choice of placing hearing at the top.[39]

In fact, if we consider this evidence for what it says about the actual experience of learning, the aural and the haptic are constantly below the surface. The studious athlete of Bovelles (Figure 5.1) holds open a book that displays the sort of visual culture which permeates these books, but as the culmination of *De sensu* the image as a whole is a laboured reflection on the range of senses in learning. Lines from his hands lead to the scholar's eyes, filling his starry imagination, while speech from his mouth reverberates in his ears. Reading or *lectio* is here a silent, solitary affair, for it is associated with the eyes, an association deepened by the rich visual imagery on the open book's pages—they resemble something like Bovelles's own pages. But the book's white margins call out for *scriptio*, the pen readied for action in the scholar's other hand. The pen is surely what Bovelles has in mind when he describes the role of hands in the accompanying poem. Reading calls for writing, a matter of touch. Far from nullifying the other senses, vision puts the other senses to work.

The same dynamic is found in Lefèvre's own suggestions for how to go about the business of mathematical reasoning. He summarized the goal of intellectual insight with particular clarity in his contribution to the astronomical genre of *theoricae*, textbooks that used geometry to model planetary motions. The very name *theorica* self-consciously referred to abstract visualizations. As Lefèvre himself explained it, this part of astronomy 'contains within it an altogether pure and free contemplation (which they call a *theorica*)'.[40] The ultimate value of such astronomy, as he presented it, was the fact that astronomical modelling provides an inward glimpse of the divine:

> For this part of astrology is almost entirely imaginative and creative. And in just the same way that the wisest, best craftsman produces by craftsmanship of the divine mind the true heavens and true motions, so too our mind always emulates its father (when the blemish of ignorance is wiped away), offering within itself the figured heavens and figured motions, and the certain images [*simulacra*] of the true motions, in which it apprehends truth as the traces of the creator's divine mind. Thus the astronomer's mind, when it diligently makes the heavens and their motions, is like the maker of things when he created the heavens and their motions.... By this our mind declares itself to be divine, the companion and likeness of the immortal nature, so that it retains the sole right in heaven of apprehending that it exists among immortals, just as if it abided in the regions of the immortals themselves. For who doubts this [understanding] comes from kinship with that immortal nature?[41]

[39] Klinger-Dollé, *Le De sensu*, part III (titled 'une philosophie des médiations sensibles'), and also Thomas Frangenberg, '"Auditus Visu Prestantior": Comparisons of Hearing and Vision in Charles de Bovelles's "Liber de Sensibus"', in *The Second Sense: Studies in Hearing and Musical Judgement from Antiquity to the Seventeenth Century*, ed. Charles Burnett, Michael Fend, and Penelope Gouk (London: Warburg Institute, 1991), 71–94.

[40] Lefèvre, Clichtove, and Bovelles, *Epitome*, 97r (PE, 113): '[Haec pars astrologiae] que prorsus sinceram et liberalem in se continet contemplationem (quam theoricen appellant)'.

[41] Lefèvre, Clichtove, and Bovelles, *Epitome*, 97r: 'Nam haec astrologiae pars tota ferme imaginaria effictrixque est. Et haud secus quam rerum sapientissimus optimusque opifex veros caelos et veros

The ability to figure worlds in the mind is what sets humanity in God's own image. What matters in mathematics is the experience of that inward making—of intuition based on visualization. Above all, Lefèvre's description of astronomical insight focuses on its dynamic character. The value of mathematics lies not in concluding statements, or even a static vision of the order of things; its value lies an inner performance. As with Bovelles's account of the imagination's operation, the goal is not to prove certain statements, but to help students experience the world in a way that will define them. A *sapiens* is someone marked by a certain set of intellectual experiences.

For Lefèvre's students, such experiences involved more than passively viewing printed texts. In fact, he counselled using additional diagrams and three-dimensional models such as aequatoria: 'it will be especially helpful if diagrams, and descriptions of lines and surfaces are put before your eyes, or one of those representations which are somehow solid—a better equipped version will be of service to the less experienced students, for such diagrams could not be printed suitably.'[42] Lefèvre thus anticipates that the printed text will not be enough, and recommends tools. His recommendations aim toward mental experience, but conjure up a context in which the student wrestles with the text with a pen in hand, while hearing an oral explanation: 'For this reason those who possess the particular disposition of being quicker to follow the understanding of those speaking, or equally those who are especially skilled at astronomical calculation; such people will better construct [*fingere*] them in the mind.'[43]

Our evidence for mathematical teaching in Lefèvre's circle is static, visual, and printed—but expects the student to require a dynamic, aural, and haptic experience. A student learned to 'see' mathematically by experiencing the performance of a specimen problem, supported by a pen, oral instruction, and imitation. In the remaining sections of this chapter, I trace emerging shifts in mathematical culture through Lefèvre's textbooks. In each case, a multisensory experience draws the reader towards understanding.

motus divinae mentis opificio producit, mens nostra sui semper aemula parentis (cum ignorantiae labes pluscum detergitur) effictos caelos effictosque motus intra se profert, verorumque motuum simulacra quadam, in quibus ut in vestigiis divinae mentis opificii deprehendit veritatem. Est igitur astronomi mens, cum caelos caelorumque motus gnaviter effingit, similis rerum opifici caelos caelorumque motus creanti [...] qua in re mens nostra se declarat esse divinam et immortalis naturae sociam atque affinem, ut quae sola ius in caelo ipsum apprehendendi cum immortalibus retinet, haud secus ac si in ipsorum immortalium regionibus degeret. Id enim quis dubitat ex immortalis naturae cognatione illi obtingere?'

[42] Lefèvre, Clichtove, and Bovelles, *Epitome*, 97r: 'Et proderit tum maxime si diagrammata, linearumque et superficierum descriptiones subiecte erunt oculis, aut eorum que dicuntur solide quedam representationes, quod instructiorum rudioribus munus erit providere, non enim commode excudi potuerunt.'

[43] Lefèvre, Clichtove, and Bovelles, *Epitome*, 97r: 'Quocirca qui mentem ad rite effingendum melius habent affectam, promptiores sunt ut dicendorum consequantur intelligentiam, pariter et qui supputationum maxime astronomicarum sunt industrii.'

COSMOGRAPHY

Cosmography was a child of the Renaissance, as a discipline that grew up around Ptolemy's *Geographia* after its discovery around 1400.[44] Cosmography borrowed its principles and techniques from other disciplines, mostly optics and astronomy. Although medieval geographical works such as Pierre d'Ailly's *Ymago mundi* (*c.*1410) mentioned the fact that earthly maps bear a mathematical relationship to the cosmic sphere—it views earth from a heavenly vantage point—they went little further, largely keeping to ethnographic and qualitative accounts of the climatic zones, their temperatures, and their inhabitants. By contrast, Ptolemy had written his *Geographia* as a manual for projecting a sphere onto a flat surface: that is, as the techniques for mapping the longitudes and latitudes of the heavenly sphere onto earth. These techniques fed a new spectrum of cosmographical genres. On the luxurious end, we find the maps of cosmographers such as Sebastian Münster, Abraham Ortelius, and Gerardus Mercator. On the humbler end of the spectrum lay handbooks that explained the mathematics linking heavenly and earthly coordinates. Such cosmographic handbooks were retrofitted onto a standard medieval introduction to astronomy: Sacrobosco's *Sphere*.

Lefèvre and his students grafted Ptolemy's cosmography onto the medieval textbook in 1495. As we found in Chapter 2, since the thirteenth century the *Sphere* had been the first—and often only—introduction that students had for astronomy. In manuscript the *Sphere* had changed little; it bore a standard stock of diagrams and often circulated with commentaries by luminaries such as Grosseteste and Cecco d'Ascoli.[45] At first, printed editions of Sacrobosco simply conserved the medieval text and images, even though the *Sphere* was one of the first basic university arts texts to be printed regularly.[46] The first editions after the *editio princeps* of 1472 were printed in quarto format, without the illustrations that graced many of the manuscripts—though spaces were left so that readers could add diagrams later.[47] Then in 1478, in Venice, the printer Franciscus Renner first added several

[44] The word was occasionally used by medieval authors such as Cassiodorus, Bernard of Sylvestris, and others, but did not consistently refer to a single discipline. Adam Mosley, 'Early Modern Cosmography: Fine's Sphaera Mundi in Content and Context', in *The Worlds of Oronce Fine*, ed. Alexander Marr (Donington: Shaun Tyas, 2009), 114–37. On the early discovery and naming of Ptolemy's work as *Cosmographia*, see James Hankins, 'Ptolemy's Geography in the Renaissance', in *Humanism and Platonism in the Italian Renaissance*, 2 vols (Rome: Edizioni di storia e letteratura, 2003), 1:457–68.

[45] Various examples are given in Richard J. Oosterhoff, 'A Book, a Pen, and the Sphere: Reading Sacrobosco in the Renaissance', *History of Universities* 28, no. 2 (2015): 1–54, at 41 n. 17.

[46] The thirty-four editions published before 1500 are distributed roughly according to the overall pattern of early European printing, with eight editions at Paris, compared with thirteen at Venice, while six at the next most prolific, Leipzig—other locations only published one: *Italy*: Bologna (2); Brescia (1); Ferrara (1); Milan (1); Venice (13). *France*: Paris (8). *Germany*: Leipzig (6); Cologne (1); Strassburg (1). These figures taken from Jürgen Hamel, *Studien zur 'Sphaera' des Johannes de Sacrobosco* (Leipzig: Akademische Verlagsanstalt, 2014), who notes two Paris editions not listed in the ISTC.

[47] A copy of a 1472 Venice edition of the *Sphere* at Cambridge University Library (Inc.4.B.3.8) has some figures drawn in. The text references an image, but does not specify whether it has been included: 'this is found by an equidistant line, as appears in the figure below or is made on that page' (*et hoc*

images.[48] In 1482 Erhart Ratdolt added the image of an armillary sphere balanced on a hand, thereby setting a new trend in the *Sphere* genre. When he printed the book again in 1485, Ratdolt added many more woodcuts, bringing the total to twenty-six, if we include the frontispiece.[49] The first Paris editions follow the same format: Wolfgang Hopyl himself published the first in 1489, to be followed in 1493 with separate editions by Georg Middelhus and Antoine Caillaut. No printed *Sphere* included commentary and no printed *Sphere* included images beyond those already in manuscript—before 1495.[50]

That year, Lefèvre and his students published the *Textus de sphera*, an edition of Sacrobosco. Although the *text* this edition served was deeply traditional, the *book* (text, paratexts, layout, etc.) as a whole innovated in three ways: it was larger; it developed a distinctive new visual regime; and it included a comprehensive commentary. First, the Fabrist book was set in large folio format. From the *editio princeps* of 1472[51] until the early 1490s every edition of the *Sphere* included the *Theoricae novae* of Peurbach and the *Contra Cremonensia* of Regiomontanus. Altogether they comprised only twenty-four quarto leaves. In contrast, Hopyl published the *Textus de sphera* in full folio format, and Lefèvre's commentary and visual programme augmented Sacrobosco to fill all twenty-four large leaves, without the additional works. The larger format may imply an effort to raise the book's status: it was surely more expensive than predecessors. Whether intended or not, the larger format enabled certain reading practices, providing white space that later readers filled with annotations and diagrams.[52]

The book's second, more elaborate innovation was a striking new visual regime, from *mise-en-page* to tables and diagrams. As was usual, the authoritative text was set in larger type, with commentary in smaller type. But in a move unusual for a folio volume, Hopyl set the text in a single column, a format common to all of Lefèvre's works. Up to this point, folio university books normally showed the close-printed double columns that reflected manuscript priorities—scribes found them easiest to copy.[53] The single column accented the linear organization by chapters, each followed

inveniturper lineam aeque distantem sicut apparet in figura subiecta sive facta in isto folio). Thanks to Roger Gaskell for this reference.

[48] Most importantly, these included the famous circles of the heavenly spheres (a2v), a half-circle with lines representing the seven habitable zones (d2v), and an image of the solar and lunar eclipses (d5v). Most such images, particularly those on the circularity of the earth, were present in the manuscript tradition: e.g. Ambrosiana H 75 Sup., fols 6r–17v.

[49] On these, Gilbert R. Redgrave, *Erhard Ratdolt and His Work at Venice* (London: Chiswick Press, 1894), figure immediately following page 16.

[50] This judgement is based on my own examination of most editions published before 1500; but for confirmation see Hamel, *Studien zur 'Sphaera'*; Jürgen Hamel, 'Johannes de Sacroboscos Sphaera', in *Gutenberg-Jahrbuch*, ed. Stephan Füssel, 81 (Weisbaden: Harrassowitz Verlag, 2006), 113–36; Jürgen Hamel, 'Johannes Sacrobosco: Handbuch der Astronomie, Kommenierte Bibliographie der Drucke der "Sphaera" 1472 bis 1656', in *Wege der Erkenntnis: Festschrift für Dieter B. Herrmann zum 65. Geburtstag*, ed. Dietmar Fürst, Dieter B. Herrmann, and Eckehard Rothenberg (Frankfurt am Main: Harri Deutsch Verlag, 2004), 115–70.

[51] That year two editions were published, one in Ferrara and another in Venice. It is not clear which was first.

[52] See the section 'A Reader' in this chapter, and Oosterhoff, 'A Book, a Pen, and the Sphere'.

[53] The short lines of double columns require less wrist movement of the scribe.

by commentary, accentuating the extraordinary amount of white space of unprinted page. The spacious effect was augmented by the many woodcut images that surrounded and broke up the text. The typography also reprises the same distinctive paratextual apparatus described in the last chapter, still experimental in the 1490s: title page, index, numbering of commentary, use of titles and extra-forme material, and so on (Figure 5.2). The book opened onto the kind of table that characterized Lefèvre's other introductions. Each numbered proposition was keyed to an *argumentum* or *topos*, identified by small numbers in the margin, within Lefèvre's commentary, not Sacrobosco's own text—an example of the paratextual commentary becoming more authoritative (see Chapter 3). This visual structuring of Fabrist commentary was replicated in all sixteenth century editions of the *Textus de sphera.*

The visual economy of the *Textus de sphera* was particularly schematic, even by the diagrammatic standards of the genre. Even the frontispiece—often more pictorial than other woodcuts—exudes this austerity: a large woodcut of Ptolemy communing with the muses Astronomia and Urania, seated below an armillary sphere (Figure 5.3).

The frontispiece integrated two images already present in earlier printed editions of the *Sphere*. However, in earlier cases the images served different purposes. The armillary sphere had been elsewhere in the book, at the appropriate place for illustrating the various circles of the heavens from the tropics to the equator. The older image of Ptolemy and the muses had offered a contrast between the ordered austerity of the superlunary world, distinct from the organic fertility of the rabbits and greenery below. Ptolemy, Astronomia, and Urania represented the liberal arts meditating on the heavens above, with their feet planted on the earth below. The new Paris edition of 1495 brings these pictures together, armillary sphere with a simplified version of Ptolemy and the muses (Figure 5.4). Much less detail connects the muses to earth, and the grass and rabbits are nearly squeezed out of the woodcut. Ptolemy and the muses still meditate on the heavens, but now the stars are merely a backdrop to the instrument above their heads, as they carefully observe the armillary sphere. The iconography is schematic, prioritizing instruments—and using them for calculation.

The diagrams found throughout the *Textus de sphera* deliberately induct the reader into the performance of calculations, coming closer to specimen diagrams. This is a striking development within the *Sphere*, which had been a genre that presented a preliminary *qualitative* description of astronomy, leaving actual quantitative skills to more advanced works. In this book, the visual rhetoric of the diagrams facilitate quantitative readings. Lefèvre's frontispiece already evacuates words from the original versions of these images. Instead, a key of letters deftly coordinates text and diagram. Unlike all other commentaries on the *Sphere*, images throughout the book are uncluttered by words.[54] In the same way, the edition prioritized geometrically clean lines. Images in other editions of Sacrobosco are cartoonesque, exaggerating certain features to help students visualize cases on a scale ordinarily unavailable to observers, such as the sphericity of the world. The most famous is

[54] e.g. Lefèvre, *Textus de sphera* (1495), inter alia at a7v, and the diagram of the spheres on c2r and c8r.

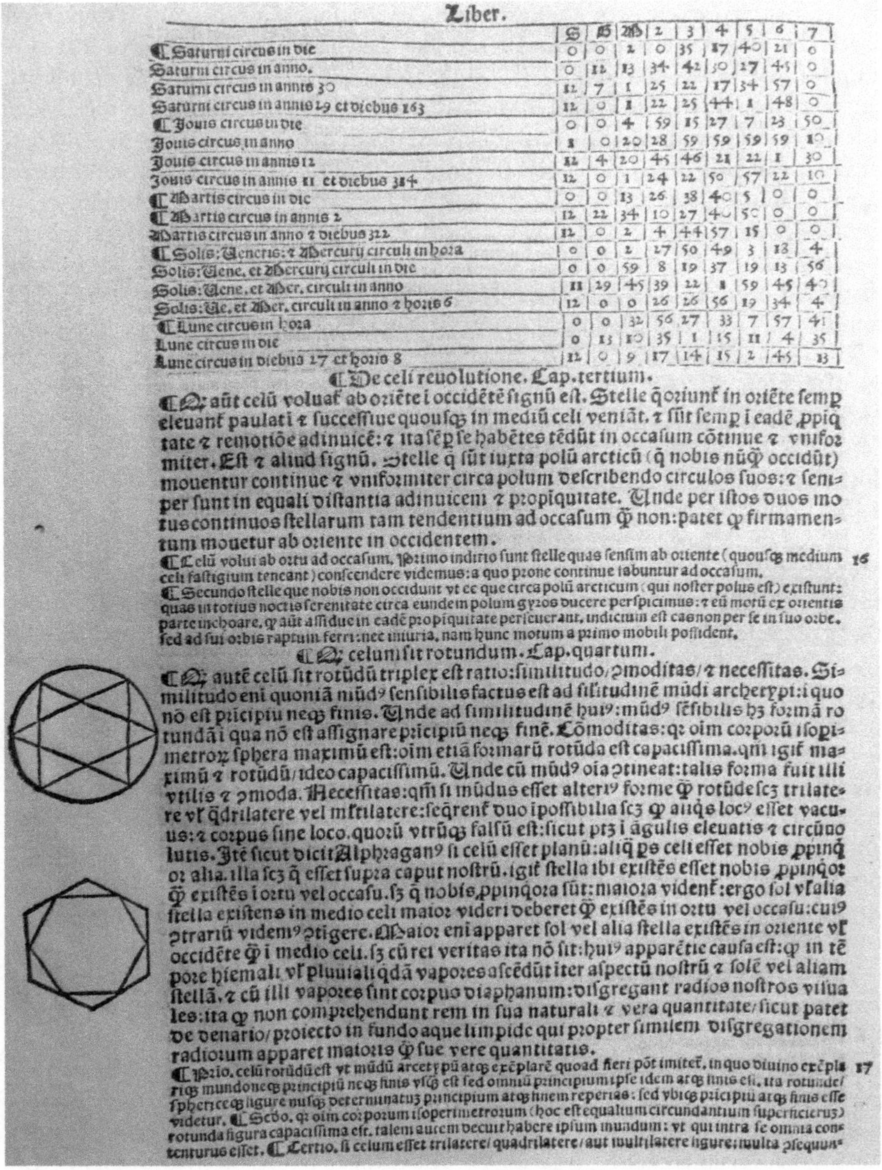

Liber.

	S	G	M	2	3	4	5	6	7
¶ Saturni circus in die	0	0	2	0	35	17	40	21	0
Saturni circus in anno.	0	12	13	34	42	50	27	45	0
Saturni circus in annis 30	12	7	1	25	22	17	34	57	0
Saturni circus in annis 29 et diebus 163	12	0	1	22	25	44	1	48	0
¶ Jouis circus in die	0	0	4	59	15	27	7	23	50
Jouis circus in anno	1	0	20	28	59	59	59	59	10
Jouis circus in annis 12	12	4	20	45	46	21	22	1	30
Jouis circus in annis 11 et diebus 314	12	0	1	24	22	50	57	22	10
¶ Martis circus in die	0	0	13	26	38	40	5	0	0
¶ Martis circus in annis 2	12	22	34	10	27	40	50	0	0
Martis circus in anno ⁊ diebus 322	12	0	2	4	44	57	15	0	0
¶ Solis: Veneris: ⁊ Mercurij circuli in hora	0	0	2	27	50	49	3	13	4
Solis: Vene. et Mercurij circuli in die	0	0	59	8	19	37	19	13	56
Solis: Vene. et Mer. circuli in anno	11	29	45	39	22	1	59	45	40
Solis: Ve. et Mer. circuli in anno ⁊ horis 6	12	0	0	26	26	56	19	34	4
¶ Lune circus in hora	0	0	32	56	27	33	7	57	41
Lune circus in die	0	13	10	35	1	15	11	4	35
Lune circus in diebus 27 et horis 8	12	0	9	17	14	15	2	45	13

¶ De celi reuolutione. Cap. tertium.

¶ Q; aūt celū voluat ab oriēte i occidētē signū est. Stelle q̄ oriunt in oriēte semp eleuant paulati ⁊ successiue quousq; in mediū celi veniāt. ⁊ sūt semp i eadē ꝓpiqtate ⁊ remotiōe adinuicē: ⁊ ita sēp se habētes tēdūt in occasum cōtinue ⁊ vniformiter. Est ⁊ aliud signū. Stelle q̄ sūt iuxta polū arcticū (q̄ nobis nūq̄ occidūt) mouentur continue ⁊ vniformiter circa polum describendo circulos suos: ⁊ semper sunt in equali distantia adinuicem ⁊ propiquitate. Unde per istos duos motus continuos stellarum tam tendentium ad occasum q̄ non: patet ꝙ firmamentum mouetur ab oriente in occidentem.

¶ Celū volui ab ortu ad occasum. Primo inditio sunt stelle quas sensim ab oriente (quousq; medium celi fastigium teneant) conscendere videmus: a quo prone continue labuntur ad occasum. 16

¶ Secundo stelle que nobis non occidunt vt ee que circa polū arcticum (qui noster polus est) existunt: quas in totius noctis serenitate circa eundem polum gyros ducere perspicimus: ⁊ eū motū ex orientis parte inchoare. ꝙ aūt assidue in eadē propiquitate perseuerant. indicium est eas non per se in suo orbe. sed ad sui orbis raptum ferri: nec iniuria. nam hunc motum a primo mobili possident.

¶ Q; celum sit rotundum. Cap. quartum.

¶ Q; autē celū sit rotūdū triplex est ratio: similitudo/ ɔmoditas/ ⁊ necessitas. Similitudo eni quoniā mūdꝰ sensibilis factus est ad silitudinē mūdi archetypi: i quo nō est pricipiu neq; finis. Unde ad similitudinē huiꝰ: mūdꝰ sēsibilis hꝫ formā rotundā i qua nō est assignare pricipiū neq; finē. Cōmoditas: q: oim corporū isoperimetrorꝫ sphera maximū est: oim etiā formarū rotūda est capacissima. qm̄ igit maximū ⁊ rotūdū/ ideo capacissimū. Unde cū mūdꝰ oia ɔtineat: talis forma fuit illi vtilis ⁊ ɔmoda. Necessitas: qm̄ si mūdus esset alteriꝰ forme q̄ rotūde scꝫ trilatere vł q̄drilatere vel mltilatere: seq̄rent duo ipossibilia scꝫ ꝙ aliq̄s locꝰ esset vacuus: ⁊ corpus sine loco. quorū vtrūq; falsū est: sicut ptꝫ i āgulis eleuatis ⁊ circūuolutis. Itē sicut dicit Alphraganꝰ si celū esset planū: aliq̄ ps celi esset nobis ꝓpinq̄or alia. illa scꝫ q̄ esset supra caput nostrū. igit stella ibi existēs esset nobis ꝓpinq̄or q̄ existēs i ortu vel occasu. sꝫ q̄ nobis ꝓpinq̄ora sūt: maiora vident: ergo sol vł alia stella existens in medio celi maior videri deberet q̄ existēs in ortu vel occasu: cuiꝰ ɔtrariū videmꝰ ɔtigere. Maior eni apparet sol vel alia stella existēs in oriente vł occidēte q̄ i medio celi. sꝫ cū rei veritas ita nō sit: huiꝰ apparētie causa est: ꝙ in tēpore hiemali vł pluuiali q̄dā vapores ascēdūt iter aspectū nostrū ⁊ solē vel aliam stellā. ⁊ cū illi vapores sint corpus diaphanum: disgregant radios nostros visuales: ita ꝙ non comprehendunt rem in sua naturali ⁊ vera quantitate/ sicut patet de denario/ proiecto in fundo aque limpide qui propter similem disgregationem radiorum apparet maioris q̄ sue vere quantitatis.

¶ Prio. celū rotūdū est vt mūdū arcetypū atq; exēplarē quoad fieri pōt imitet. in quo diuino exēplarīq; mundo neq; principiū neq; finis vsq; est sed omniū principium ipse idem atq; finis est. ita rotunde/ sphericeq; figure nusq; determinatū principium atq; finem reperias: sed vbiq; pricipiū atq; finis esse videtur. ¶ Scdo. q: oim corporum isoperimetrorum (hoc est equalium circundantium superficierū) rotunda figura capacissima est. talem autem decuit habere ipsum mundum: vt qui intra se omnia contenturus esset. ¶ Tertio. si celum esset trilatere/ quadrilatere/ aut multilatere figure: multa ɔsequun- 17

Figure 5.2. A typical page in Lefèvre's *Textus de sphera*: table, text, commentary (in smaller type), with diagrams and section numbers both printed outside the forme. Lefèvre, *Textus de sphera* (1495 [here 1516]), a5v. © Fitzwilliam Museum, University of Cambridge.

the image of the ship bearing two observers peering at a port; the one at the mast's foot, as Sacrobosco explains, cannot see the port over the earth's bulge, while the man standing at the mast's height can (Figure 5.5). An image drawn to scale would not fit on the page, of course, so—following Ratdolt's 1485 edition—a kind of cartoon was used to make the point.[55] Another example (also in Figure 5.5) was based on Sacrobosco's description of how one sees different stars depending on where one is on the earth's globe (different stars at different longitudes), which resulted in a cartoon of stick figures walking around a circle that represented the earth, not to scale. Lefèvre's *Textus de sphera* eschewed such images, instead representing heavenly objects or movements with perfect circles, triangles, or other purely geometrical figures.

This commitment to astronomy's distinctive *mathematical* objects is especially evident in the edition's most innovative feature: tables. The number of woodcut diagrams alone is less than we find in Ratdolt's luxurious edition from 1485 (although this is more than most other incunabula editions): twenty-two woodcuts plus a frontispiece. But the number jumps when we also count the twenty-three tables that fill the book. Some of these tables simply present common values, such as the distances between the various planetary orbits; such tables substitute pictorial diagrams with numerical values, cohering with the book's overall schematic approach. Other tables give ascensions based on observations. Still other tables cue students on the use of astronomical tables, so they will know how to calculate specific locations of heavenly objects (I will discuss this further below). Finally, some tables locate cities on earth by their longitudes and latitudes. In this last category of table, astronomy becomes cosmography.

The new regime of paratexts offered readers the techniques of calculation needed to move beyond simply assenting to astronomical propositions to practising the rudiments of astronomy for themselves. Other versions of Sacrobosco helped students gain some of the visual skills needed to make sense of Ptolemaic planetary models.[56] But just how much Lefèvre's textbook pushed towards practice can be sensed already at the book's opening pages, where a short *introductoria additio* offers definitions from geometry and arithmetic (Figure 5.6). From geometry, readers must know the definition of a circle, a half-circle, and the differences between right, curved, acute, and obtuse angles. Some Venetian Sacroboscos included similar primers; but this was more. Besides establishing the dependence of astronomy on geometry, these details attempt to draw together the chaotic education in the quadrivium with which a student would approach the *Sphere*. From arithmetic, examples helped readers to practise the techniques of sexagesimal addition and

[55] Evidently, the image was not always effective; in at least one case, an engraver did not understand the crucial difference between the men at each end of the ship's mast, and so in 1501 Giovanni Sessa published an edition where both figures could see the port. See Isabelle Pantin, 'L'illustration des livres d'astronomie à la renaissance: l'évolution d'une discipline à travers ses images', in *Immagini per conoscere: Dal Rinascimento alla Rivoluzione scientifica*, ed. Fabrizio Meroi and Claudio Pogliano (Florence: Leo S. Olschki, 2001), 10.

[56] For examples, see Kathleen M. Crowther and Peter Barker, 'Training the Intelligent Eye: Understanding Illustrations in Early Modern Astronomy Texts', *Isis* 104, no. 3 (2013): 429–70.

Figure 5.3. Frontispiece from Sacrobosco, *Sphaera mundi* (Venice: Ottaviano Scoto, 1490). © Fitzwilliam Museum, University of Cambridge.

Figure 5.4. Frontispiece from Lefèvre, *Textus de sphera* (Paris: Estienne, 1516), a3v, reusing the woodblock from the first edition of 1495. © Fitzwilliam Museum, University of Cambridge.

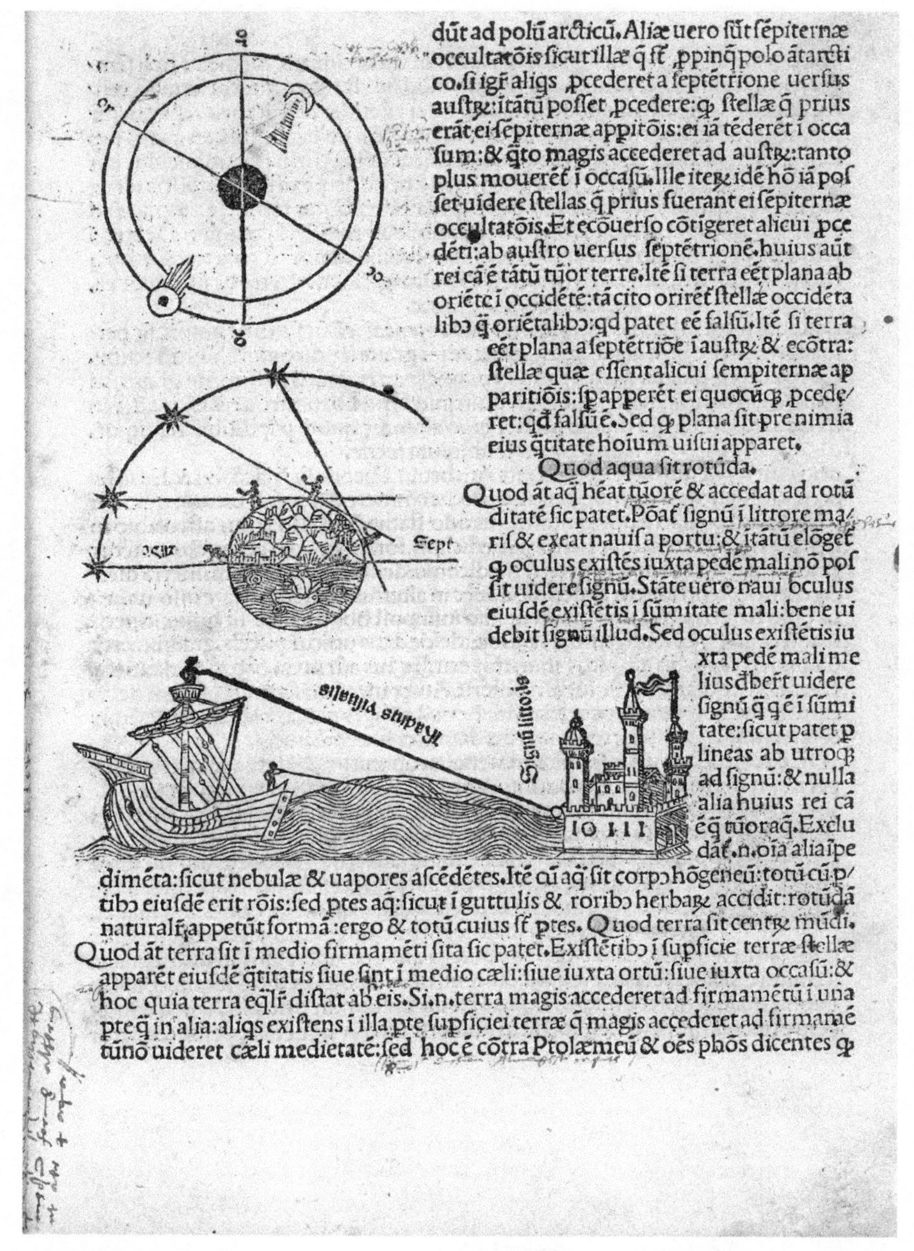

dũt ad polũ arcticũ. Aliæ uero sũt sépiternæ
occultatõis sicut illæ q̃ sũt ꝑpinq̃ polo átarcti
co. Si igr̃ aliqs ꝑcederet a septẽtrione uersus
austꝝ: itãtũ posset ꝑcedere: ꝙ stellæ q̃ prius
erãt ei sépiternæ appitõis: ei iã téderét i occa
sum: & q̃to magis accederet ad austꝝ: tanto
plus moueréť i occasũ. Ille iteꝝ idẽ hõ iã pos
set uidere stellas q̃ prius fuerant ei sépiternæ
occultatõis. Et ecõuerso cõtigeret alicui ꝑce
déti: ab austro uersus septẽtrioné. huius aũt
rei cã é tãtũ tũor terre. Ité si terra eét plana ab
oriẽte i occidẽté: tã cito oriréť stellæ occidẽta
libꝫ q̃ oriẽtalibꝫ: qd patet eé falsũ. Ité si terra
eét plana a septẽtriõe i austꝝ & ecõtra:
stellæ quæ essent alicui sempiternæ ap
paritiõis: sp appereťt ei quocũqꝫ ꝑcede/
ret: qđ falsũ é. Sed ꝙ plana sit pre nimia
eius q̃titate hoĩum uisui apparet.

Quod aqua sit rotũda.

Quod ãt aq̃ hẽat tũoré & accedat ad rotũ
ditaté sic patet. Põaf signũ i littore ma/
ris & exeat nauis a portu: & itãtũ elõgeť
ꝙ oculus existẽs iuxta pedẽ mali nõ pos
sit uidere signũ. Stãte uero naui oculus
eiusdẽ existẽtis i sũmitate mali: bene ui
debit signũ illud. Sed oculus existẽtis iu
xta pedẽ mali me
lius đbeť uidere
signũ q̃ q é i sũmi
tate: sicut patet ꝑ
lineas ab utroqꝫ
ad signũ: & nulla
alia huius rei cã
é q̃ tũor aq̃. Exclu
dať .n. oĩa alia ĩpe
diméta: sicut nebulæ & uapores ascẽdẽtes. Ité cũ aq̃ sit corpꝰ hõgeneũ: totũ cũ p/
tibꝫ eiusdẽ erit rõis: sed ptes aq̃: sicut i guttulis & roribꝫ herbaꝝ accidit: rotũdã
naturalr̃ appetũt formã: ergo & totũ cuius sť ptes. Quod terra sit centꝝ mũdi.
Quod ãt terra sit i medio firmaméti sita sic patet. Existẽtibꝫ i supficie terræ stellæ
apparẽt eiusdẽ q̃titatis siue sint i medio cæli: siue iuxta ortũ: siue iuxta occasũ: &
hoc quia terra eq̃lr̃ distat ab eis. Si .n. terra magis accederet ad firmamẽtũ i una
pte q̃ in alia: aliqs existens i illa pte supficiei terræ q̃ magis accederet ad firmamẽ
tũ nõ uideret cæli medietaté: sed hoc é cõtra Ptolæmeũ & oẽs phõs dicentes ꝙ

Figure 5.5. The exaggerated realism of Sacrobosco, *Sphaera mundi* (1490), a8r. © Fitzwilliam Museum, University of Cambridge.

subtraction, skills necessary to record and calculate planetary positions. The primer taught readers just enough geometry and arithmetic actually *to calculate* astronomically: 'These things have been added concerning the physical structure of abacus [i.e. arithmetic], not because they are enough of an introduction for abacus and astronomical calculation, but so that they might aid calculation and those skilled in counting who are studying this discipline of astronomy—without this skill they will find themselves driven from the temple of the quadrivium and never reap any benefit from it.'[57] The new diagrams and tables reframe the entirety of the medieval textbook; the *Textus de sphera* transforms Sacrobosco into a book of practical mathematics that teaches readers how to do.

The book's third striking innovation was its commentary, in which Lefèvre similarly stressed visual skills alongside competencies in calculation. The practical origin of astronomical theory is reinforced through a material history of the core object of Ptolemaic astronomy, the sphere. Lefèvre begins by correcting some history. Some think the 'inventor' of the sphere was Archimedes, the mathematician whom the Roman general Marcellus had hoped to save when attacking Syracuse. But Lefèvre emphasizes that Archimedes came from a lineage originating with Perdix, the nephew of the craftsman Daedalus, inventor of the steel compass, the iconic tool of mathematical practitioners.[58] Then, 'our Euclid clearly showed the utility of the compass for making a large, more useful sphere'.[59] Euclid thus precedes Archimedes in antiquity as well as in mechanical prowess, and so is the rightful inventor of the sphere. Lefèvre completes his material history by sending readers to the artisan's workshop in order to find a 'marvellously efficacious description' of Euclid's meaning:

> This is indeed a description of marvellous efficacy, which clearly teaches (insofar as sensible matter can take it) how to make an artificial sphere. Its use and right understanding is worth its weight in gold to artisans in our time who wish to fashion figures with a lathe in metal or another material. So having taken a compass of thin steel or iron, a semicircle is inscribed on some line which is then cut out from the arc to the diameter, and moreover the diameter in between as well; then it is fit for cutting and dividing, and you have a tool very much suited for turning a sphere, just as a compass is for turning circles.[60]

[57] Lefèvre, *Textus de sphera*, a3r: 'Haec de abaci physica ratione adiecta sunt, non quia ad abacum, astronomicumque calculum sufficienter introducant, sed ut calculum calculique peritos consulant, qui hoc astronomico instituto sunt informandi. Sine qua numerandi peritia ex adytis quadrivii se cognoscant explosos, nullum unquam ex eo fructum suscepturi. Et sit semper oculis tum docentium, tum discentium subiecta materialis sphaera.'

[58] Lefèvre quotes Ovid, *Metamorphoses* 8.547–9.

[59] Lefèvre, *Textus de sphera*, a4r: 'Hoc est qui ferra, circinumque repperit, quid ergo noster Euclides, qui usum fabricande longe quidem utilioris sphere, dilucide monstravit.' Unless otherwise specified, following citations to Lefèvre's *Textus* are to this edition.

[60] Lefèvre, *Textus de sphera*, a4r: 'Et hec profecto mire efficacie descriptio est, que aperte docet (quantum sensibilis materia recipere valet) artificialem constituere spheram, cuius utilem commodamque intelligentiam nostre tempestatis artifices multi auri pondo comparare deberent, qui metalo, ligno, aut alia materia figuras torno exprimere volunt. Si itaque in levi calybe aut ferro, sumpto circino supra quancunque lineam semicirculus educatur qui ab arcu ad diametrum usque excavetur, quin immo et medium diametri interstitium, et mox ad arcum circumferentiamque excavatur ut ea ex parte ad scindendum secandumque fiat aptus, exurget instrumentum tornandis spheris (haud secus quam circinus circulis) aptissimum.'

Figure 5.6(a). The *introductoria additio* in Lefèvre, *Textus de sphera* (Paris: Estienne, 1500), BHS K 1046c, a2v–a3r. Lefèvre's primer explains the geometrical objects needed to understand astronomy, as contemporary editions of Sacrobosco sometimes also did (*right*). Here Lefèvre goes further, explaining how to perform sexagesimal arithmetic, which is the subject of Beatus Rhenanus' extensive notes on the facing page (*left*).

¶ Non nulle ad sequentia note

Irculus est figura plana vna quidē circūducta linea cōtenta: in cuius medio pūctus est: a quo omnes recte linee ad circundātem lineam educte / ad inuicē sunt equales ¶ Figura plana est cuius mediū non subsultat / egrediturue ab extremis. ¶ Circūferentia circuli est linea circulū cōtinens: hoc est: est linea illa ad quam omnes recte linee a centro circuli eiecte / adinuicē sunt equales: que et ambitus circuitus: curuatura qz: et circulus nonnunq̄ dicitur. ¶ Centrū circuli est punctus ille: a quo omnes recte ad lineā circulū cōtinentē educte / ad inuicem sunt equales. ¶ Dimidius circulus est figura plana diametro circuli et medietate circūferentie cōtenta. ¶ Diameter circuli: est quecunqz linea recta per centrū circuli trāsiens vtrinqz ad circūferētiā circuli eiecta. ¶ Linea recta est a puncto ad punctum extensio breuissima. ¶ Solidum: corpus longitudine / latitudine / altitudineqz dimensum. ¶ Altitudo / crassities / profunditas. ¶ Angulus est duarum linearum mutuus contactus: est enim figure particula a linee cōtactu in amplitudinem surgens. ¶ Angulus rectus est angulus ex linea supra lineā cadente: et vtrinqz altrinsecus duos adinuicem equales angulos faciente causatus: vt angulus a d b. et angulus a d c. ¶ Quem si due recte linee continēt: angulus rectilineus nominatur: si autem eum linee curue continent / angulus curuus / sphericalisqz dicetur. Linea curua: circūferētia: aut circunferentie portio est. ¶ Angulus obtusus est angulus qui est recto maior: vt angulus e d b: continet enim angulum rectum a d b: et insuper angulum e d a. ¶ Angulus acutus est angulus recto minor: vt angulus e d c. Continet enim angulus rectus a d c: angulum e d c: et insuper angulum a d e: et anguli recti: equales: normalesqz dicuntur. Obtusi autem et acuti: obliqui: inequalesqz. ¶ Integrum est res tota: aut rei pars: que sexagenaria partitione non prouenit. ¶ Minutum est sexagesima pars integri. ¶ Secundū est sexagesima pars minuti. ¶ Tertiū est sexagesima pars secundi: ⁊ ita deinceps secundum naturalē numeri semperqz vnitate crescentē multitudinē. ¶ Dies partitur in 24 horas: hora in sexaginta minuta: minutū in 60 secunda: secundū in 60 tertia: et ita deinceps secundum naturalem numerorū seriem. Quo fit vt hora: secunda cōtineat 3600: et tertia 216000. ¶ Signum est duodecima pars circuli. ¶ Gradus est tricesima pars signi: at triginta duodecies multiplicata: 360 reddūt: quo fit vt iterum recte diffiniat gradus esse tricētesima sexagesima pars circuli. Itē et gradus: ptes circuli nūcupātur. ¶ Frangit ergo circulus in duodecim signa: et signū in 30 gdus: et gradus in sexagita minuta: et minutū in sexagita secūda: et secūdū i 60 tertia: et hoc pacto deīceps. ¶ Aiaduerte tamē in hac fractione sexagenaria: si frāgit hora fragmēta illa: minuta horaria: secunda: et tertia horaria dicunt. Et si frāgit signū: dicunt minuta: secūda: tertia signi: ⁊ ita deinceps. ¶ Abaci physica ratio in sexagenaria collectione (que fit addendo) atqz sexagenaria mutuatione (que fit distrahendo) intelligit: in qua sumopere curandū est: vt integra: similia sub similibus integris collocētur. et similes minutie sub similibus: vnius eiusdemqz denoīationis minutiis: suis quidē interuallis distincte. Minutie sūt minuta: secūda: tertia: quarta: et ita deinceps: et in eisdē interuallis spaciisqz denaria collectione aut mutuatione: que vulgaris est: vtendū est: et est a tenuiorib⁹ minutiis: collectionis / distractiōisqz inchoādus labor: verbi causa: volo in vnū colligere: hoc ē simul addere duos pmos subiecte formule numeros: quor vnus supior: et alter iferior collocat: aut miorē a maiori subducē: addo subducoqz vt sbiecta mōstrat formula

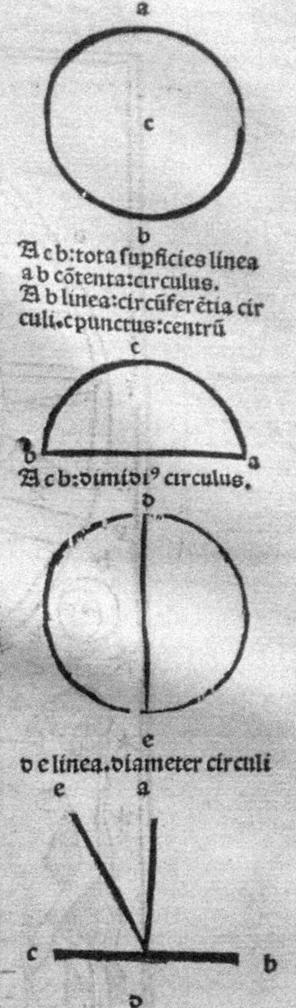

¶ Hoc pacto fit physica additio.

Sig	Grad	Minu.	Scda	Tertia
0	54	48	37	20
0	50	36	39	42
1	45	25	17	02

¶ Tertius numerus subter: ex duobus superioribus additione colligitur.

¶ Hoc pacto fit physica distractio.

Sig.	Gra.	Mi.	Se.	Ter.
0	54	48	37	20
0	50	36	39	42
0	04	11	57	38

¶ Tertius numerus subter: ex duobus superioribus distrahendo relinquitur.

Hec de abaci physica ratione adiecta sunt: non quia ad abacū: astronomicūqz calculū sufficienter introducti sunt: sed vt calculū calculiqz peritos cōsulant: qui hoc astronomico īstituto sūt īformādi: sine qua numerandi peritia ex additis quadriuij se cognoscāt explosos nullū vnq̄ ex eo fructū suscepturi. Et sit semp ocu... tam docentiū: tum discentiū subiecta materialis sphera. Sed nūc pricipale institutū aggrediamur.

Figure 5.6(b). Continued

Lefèvre assured readers that 'it was this use that Euclid brings to us, which he meant when he said that a sphere is the revolution of half a circle'.[61] So with geometry one can create a metal jig, a blade shaped like a concave half-circle, using a compass. Applied to rough matter spinning on a lathe, this tool makes a perfect sphere, as an image showed (see Figure 5.7). The pedagogy of astronomy draws here on the tactile world of daily life, visualizing the artisan's tools alongside the mathematician's instruments in students' imaginations.

The practical orientation of the commentary is reinforced by regular references to instruments, beginning already with the primer on how to do sexagesimal arithmetic, where Lefèvre suggested that calculation should involve an instrument for visualization: 'So let the "material sphere" [i.e. the armillary sphere] always be before the eyes of teachers and students alike.'[62]

In fact, the commentary marks the movement from instruments to tables and back again. Instruments are useful as aids for visualizing the heavenly motions. But they are not always precise enough and Lefèvre suggests tables as a more accurate alternative:

> And although inspection of the sphere with your eyes will somewhat help you to understand correctly this and what follows, nevertheless you will scarcely find one made with such artful ingenuity that it can show plainly enough the divisions of the arcs of ascensions, small and large. For that reason, to set everything out more clearly, one will often consult tables of ascensions. Nor is it the task of the present introduction to bear the burden of demonstrating how it is necessary that the zodiac circle's ascensions be unequal, and other things that follow of that sort. For in each discipline, it is worthwhile to treat only those things which are effective for understanding or comprehending well.[63]

Visualization will help one understand the general outlines of planetary theory, but not to clarify the precise quantities of heavenly change. That requires tables. Tables cannot model the underlying principles ('demonstrating how it is necessary'), but they are 'effective for understanding well'. With these tables, Lefèvre operates at a level of quantification far beyond Sacrobosco's manual. The original *Sphere* defined terms such as 'solstice', 'equator', and the parts of the zodiac, but never showed how to calculate their movements. In contrast, throughout his commentary, Lefèvre offered his readers a step-by-step tutorial in how to use tables to compute basic values. The core skill of the practising astronomer was to plot the locations of stars and planets in the night sky based on their rising times. Sacrobosco had

[61] Lefèvre, *Textus de sphera*, a4r: 'illamque intendebat [Euclides] cum diceret spheram esse transitum dimidii circuli'.

[62] Lefèvre, *Textus de sphera*, a3r: 'Et sit semper oculis tum docentium, tum discentium subiecta materialis sphaera.'

[63] Lefèvre, *Textus de sphera*, c2r: 'Et quamvis ocularis sphaerae inspectio, ad haec et sequentia rite intelligenda nonnihil afferat praesidii, vix tamen tanto ingenio tamque fabrefactam invenias, quae arcuum ascensionum, tum parvorum tum magnorum discrimina satis aperte monstret. Quapropter, ut dilucidius omnia pateant, saepius ascensionum tabulae consulendae erunt. Neque presentis introductionis officium, pondus demonstrationis sustinet, quo pacto signiferi circuli in utroque horizonte ascensionum inaequalitatem esse necesse est, et caetera id genus sequentia. In unaquaque enim disciplina, operae pretium ducendum est illa sola tractari, quae in ea bene cognosci deprehendique valeant.'

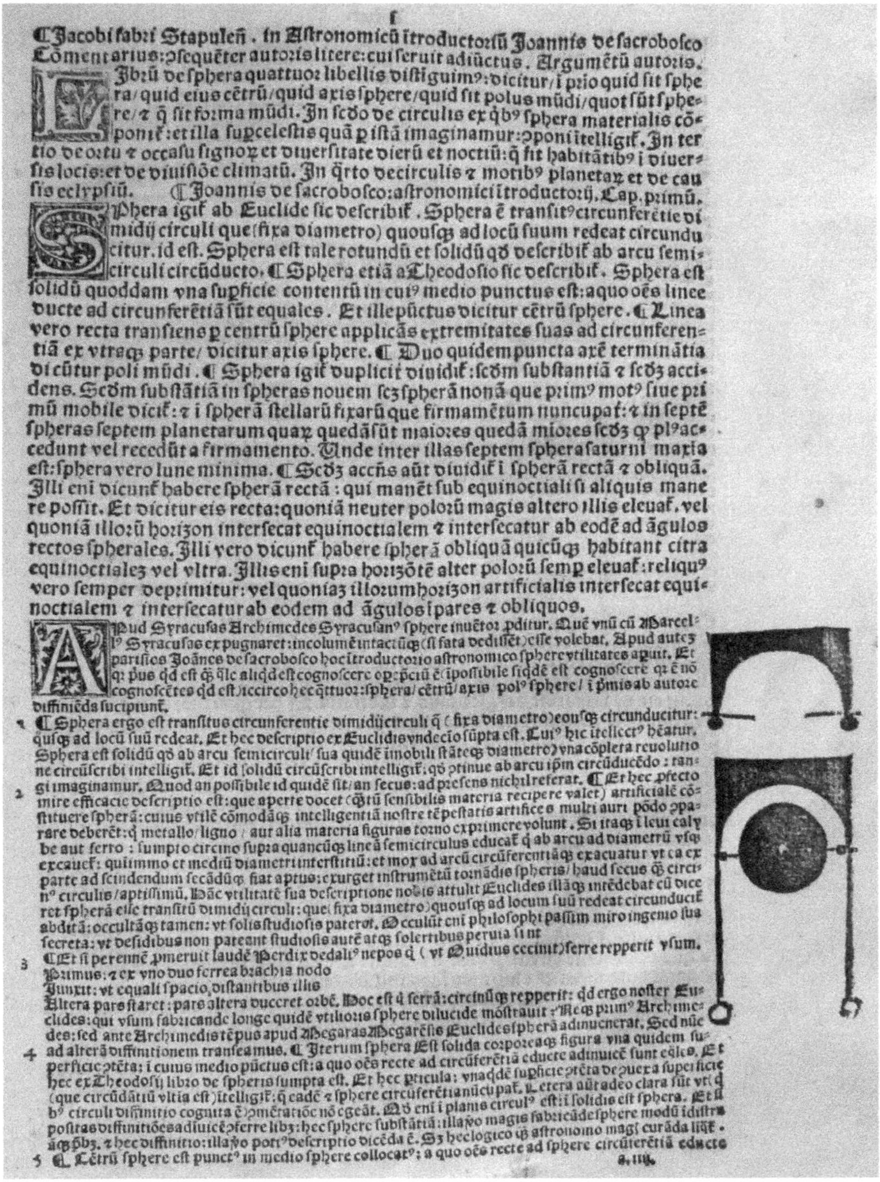

f

¶ Jacobi fabri Stapulen̄. in Astronomicū itroductoriū Joannis de sacrobosco Cōmentarius: ꝯsequēter autoris litere: cui seruit adiūctus. Argumētū autoris.

Librū de sphera quattuor libellis distiguimꝰ: dicitur/ i prio quid sit sphera/ quid eius cētrū/ quid axis sphere/ quid sit polus mūdi/ quot sūt sphere/ & q̄ sit forma mūdi. In scđo de circulis ex q̄bꝰ sphera materialis cōponit: et illa supcelestis quā p istā imaginamur: ꝯponi itelligit. In tertio de ortu & occasu signoꝝ et diuersitate dierū et noctiū: q̄ fit habitātibꝰ i diuersis locis: et de diuisiōe climatū. In q̄rto decirculis & motibꝰ planetaꝝ et de causis eclypsiū. ¶ Joannis de sacrobosco: astronomici itroductorij. Cap. primū.

Sphera igit ab Euclide sic describit. Sphera ē transitꝰ circunferētie dimidij circuli que (fixa diametro) quousq; ad locū suum redeat circunducitur. id est. Sphera est tale rotundū et solidū qđ describit ab arcu semicirculi circūducto. ¶ Sphera etiā a Theodosio sic describit. Sphera est solidū quoddam vna supficie contentū in cuiꝰ medio punctus est: a quo oēs linee ducte ad circunferētiā sūt equales. Et ille pūctus dicitur cētrū sphere. ¶ Linea vero recta transiens p centrū sphere applicās extremitates suas ad circunferentiā ex vtraq; parte/ dicitur axis sphere. ¶ Duo quidem puncta axē terminātia dicūtur poli mūdi. ¶ Sphera igit dupliciē diuidit: scđm substantiā & scđz accidens. Scđm substātiā in spheras nouem scz spherā nonā que primꝰ motꝰ siue primū mobile dicit: & i spherā stellarū fixarū que firmamētum nuncupat: & in septē spheras septem planetarum quaꝝ quedā sūt maiores quedā miores scđz q̄ plꝰ accedunt vel recedūt a firmamento. Unde inter illas septem sphera saturni maxia est: sphera vero lune minima. ¶ Scđz accn̄s aūt diuidit i spherā rectā & obliquā. Illi eni dicunt habere spherā rectā: qui manēt sub equinoctiali si aliquis manere possit. Et dicitur eis recta: quoniā neuter polorū magis altero illis eleuat. vel quoniā illorū horizon intersecat equinoctialem & intersecatur ab eodē ad āgulos rectos spherales. Illi vero dicunt habere spherā obliquā quicūq; habitant citra equinoctialez vel vltra. Illis eni supra horizōtē alter polorū semp eleuat: reliquꝰ vero semper deprimitur: vel quoniaz illorum horizon artificialis intersecat equinoctialem & intersecatur ab eodem ad āgulos ipares & obliquos.

Apud Syracusas Archimedes Syracusanꝰ sphere inuētor pditur. Quē vnū cū Marcellꝰ Syracusas expugnaret: incolumē intactūq; (si fata dedisset) esse volebat. Apud autez parisios Joānes de sacrobosco hoc itroductorio astronomico sphere vtilitates aguit. Et q: pus qđ est q̄ q̄le aliqđ est cognoscere op: pciū ē (ipossibile siqđ ē est cognoscere q: ē nō cognoscētes qđ est) iccirco hec q̄ttuor: sphera/ cētrū/ axis polꝰ sphere/ i pmis ab autore diffiniēda suscipiunt.

1 ¶ Sphera ergo est transitus circunferentie dimidij circuli q̄ (fixa diametro) eousq; circunducitur: quusq; ad locū suū redeat. Et hec descriptio ex Euclidis vndecio sūpta est. Cuiꝰ hic itellectꝰ hēatur. Sphera est solidū qđ ab arcu semicirculi/ sua quidē imobili stāteq; diametro) vna cōpleta reuolutione circūscribi intelligit. Et id solidū circūscribi intelligit: qđ ꝯtinue ab arcu ipm circūducēdo: tangi imaginamur. Quod an possibile id quidē sit/ an secus: ad presens nichil referat. ¶ Et hec pfecto

2 mire efficacie descriptio est: que aperte docet (q̄tū sensibilis materia recipere valet) artificialē cōstituere spherā: cuius vtilē cōmodāq; intelligentiā nostre tēpestatis artifices multi auri pōdo ꝯparare deberēt: q̄ metallo/ ligno/ aut alia materia figuras torno exprimere volunt. Si itaq; i leui calybe aut ferro: sumpto circino supra quancūq; lineā semicirculus educat q̄ ab arcu ad diametrū vsq; excauet: quimmo et mediū diametri interstitiū: et mox ad arcū circūferentiāq; exacuatur vt ea ex parte ad scindendum secādūq; fiat aptus; exurget instrumētū tornādis spheris/ haud secus q̄ circinꝰ circulis/ aptissimū. Hāc vtilitatē sua descriptione nobis attulit Euclides illāq; intēdebat cū diceret spherā esse transitū dimidij circuli: que (fixa diametro) quousq; ad locum suū redeat circunducit abditā: occultāq; tamen: vt solis studiosis pateret. Occulūt eni philosophi passim miro ingenio sua secreta: vt desidibus non pateant studiosis autē atq; solertibus peruia sint.

3 ¶ Et si perennē pmeruit laudē Perdix dedaliꝰ nepos q̄ (vt Ouidius cecinit) serre repperit vsum.
Primus: & ex vno duo ferrea brachia nodo
Iunxit: vt equali spacio distantibus illis
Altera pars staret: pars altera duceret orbē. Hoc est q̄ serrā: circinūq; repperit: qđ ergo noster Euclides: qui vsum fabricande longe quidē vtilioris sphere dilucide mōstrauit? Neq; primꝰ Archimedes: sed ante Archimedis tēpus apud Megaras Megarēsis Euclides spherā adinuenerat. Sed nūc

4 ad alterā diffinitionem transeamus. ¶ Iterum sphera Est solida corporeaq; figura vna quidem superficie ꝯtēta: i cuius medio pūctus est: a quo oēs recte ad circūferētiā educte adinuicē sunt eq̄les. Et hec ex Theodosij libro de spheris sumpta est. Et hec pticula: vna qđē supficie ꝯtēta de ꝯuexa superficie (que circūdātiū vltia est) itelligit: q̄ eadē & sphere circūferētia nūcupat. Cetera aūt adeo clara sūt vt (q̄bꝰ circuli diffinitio cognita ē) ꝯmēratiōe nō egeāt. Qđ eni i planis circulꝰ est: i solidis est sphera. Et illas duas diffinitiōes ad inuicē ꝯferre libz: hec sphere substātiā: illa ꝙ magis fabricāde sphere modū idiistrat... hec diffinitio: illa ꝙ potiꝰ descriptio dicēda ē. Sz hec logico ꝙ astronomo magis curāda liqt.

5 ¶ Cētrū sphere est punctꝰ in medio sphere collocatꝰ: a quo oēs recte ad sphere circūferētiā educte

a. iiij.

Figure 5.7. Illustration of a sphere on a lathe, with a semi-circular cutting tool. Lefèvre, *Textus de sphera* (Paris: Estienne, 1516), a4r. © Fitzwilliam Museum, University of Cambridge.

explained the difference between 'direct' and 'oblique' risings of heavenly bodies. One observes direct risings from the equator, where the zodiac moves directly overhead so that rising times vary evenly. Oblique risings are more complicated, since as the observer moves further from the equator, the belt of the zodiac rises over the horizon at an angle, which varies the time at which segments of the zodiac rise. As a good pedagogue, Lefèvre first produced a table for the first, simpler scenario, from which one might deduce the length of any one sign when given another. To do this, one had to know the relative lengths of each sign (at that latitude), and accordingly Lefèvre provided another table for those values. He then added six rules for performing these calculations, each clarified with a concrete example and a supplementary table for easy memorization or reference. These did not constitute a complete working set of astronomical tables; they were simplified specimens of the ephemerides and calendars that comprised tables depicting the locations of celestial bodies at certain times, with the canons detailing their use. But through such specimen tables, the book became a primer in the basic techniques of using astronomical tables. A student who mastered the *Textus de sphaera* would have skills necessary to approach the advanced canons and tables that were the astronomer's (or astrologer's) stock in trade.

Why might one want to calculate the risings of zodiacal signs? In 1500, the chief answer was surely to cast a horoscope. This raises the possibility that Lefèvre was preparing students for those practical occupations that relied on astrological expertise, such as medicine or courtly counsellors. At some points in the book Lefèvre seemed quite friendly to astrology, alluding to the secrets of the *magi* and the occult knowledge of Pythagoreans.[64] And in fact he included a table of zodiacal ascensions that could be useful to a would-be astrologer, calibrated for 48 degrees latitude, for Paris. Lefèvre advised against using the Alphonsine tables, 'for they are not precise'; instead he sent the reader to Regiomontanus.[65] Lefèvre worked through calculations for correlating celestial and earthly time—a skill necessary for casting horoscopes as well as dialling in general. Sacrobosco had simply explained that natural days and natural hours depend on the movement of the sun, on the equatorial plane, so that at the equator it is always equinox. Solar, 'artificial', days get shorter in winter, of course, the further north one goes. But on the slant plane of the ecliptic, even in winter, Sacrobosco noted, one might correct

[64] For instance, on the material explanation of the sphere, Lefèvre explained why Euclid's language was not terribly clear. He observed that philosophers often put the kinds of phenomena one might see in daily objects and experiences into such obscure terms, 'obscure and hidden, so that it might be clear only to the diligent. For philosophers hide their secrets everywhere by their wondrous ingenuity, so that they might not be open to the indolent, but rather lead the way to the diligent and skilled.' Lefèvre, *Textus de sphera*, a4r: 'Hanc utilitatem sua descriptione nobis attulis Euclides, illamque intendebat cum diceret spheram esse transitum dimidii circuli, que (fixa diametro) quousque ad locum suum redeat circumducitur: abditam, occultamque tamen, ut solis studiosis pateret. Occulunt enim philosophi passim miro ingenio sua secreta, ut desidibus non pateant, studiosis autem atque solertibus praevia sint.' The topos, rooted in the secrecy associated with Pythagoras and his school, was often repeated in learned magic, that the ingeniosi might only teach things to those worthy of such secrets.

[65] Lefèvre, *Textus de sphera*, c4v: 'Caveant tamen abaciste adducta in hoc ultimo commentario per ascensiones tabulis alphonsinis adiectas numerando perquerere, nam precise non sunt; sed potius per tabulas ascensionum Joannis Nurenbergi ubilibet, et in omni altitudine poli que sexagesimum gradum non transcendit si placitum fuerit computent.'

the 'artificial' length of day and night by calculating from the known rate of zodiacal risings. Sacrobosco left this observation in principle; Lefèvre put it into practice. Completing the task was no mean feat, since it involved accounting for the different speeds of each section of the zodiac, which are more pronounced at higher latitudes like Paris. He succeeded, however, in punctuating the section with a chart for converting the celestical time of ascensions into terrestrial 'unequal' hours and minutes.[66]

Yet, as we have seen elsewhere, even though Lefèvre adroitly wielded the commonplaces of practical utility, he often retreated from making mathematics subservient to them. Here too Lefèvre remained coy about the value of astrology, vaguely warning off those who would misapply his teaching: 'a great many false things might be taken up if not always reasoning about ascensions from the fourth rule; these are facile matters, to do more with common and logical fantasy than astronomical contemplation.'[67] The usefulness Lefèvre advertised was that these tables would offer 'contemplation' or intellectual understanding.[68]

In fact, this seems to be more than mere teacherly prudence about revealing the full powers of a dangerous subject. For the student who had learned to use such astronomical tables was prepared to tackle a second subject—a subject that was more closely suited to the ideals of universal learning that Lefèvre espoused. This second domain was cosmography.

Lefèvre lavished some of his most sophisticated commentary in the *Textus de sphera* on cosmographical measurement. A discussion of the horizon even took on the language of mathematical demonstration. After giving some definitions, we have a proof in classical geometrical style. It began with the formulaic promise of demonstration: '*que hoc pacto demonstratur*... Let there be a circle *a b d*. The solstitial colure, which by definition runs through both the poles of the zodiac and the poles of the universe. And let line *a* be the ecliptic and *b* the equator, and point *c* the pole of the zodiac.'[69] The proof ended with *quod est propositum*, a variant of the QED. All this mathematical rigour, however, was not primarily for astronomical purposes. Certainly it explained standard astronomical topics in Sacrobosco: colures, tropics, the equator, and various points of solar declination. But the ultimate benefit of this knowledge emerged as Lefèvre walked students through the calculations needed to determine these locations, showing how these corresponded to a table of twenty-four deviations from the latitude of Paris (48 degrees). Thus the point of the exercise was to show how the heavenly sphere maps onto the earth.

[66] Lefèvre, *Textus de sphera*, c5v.

[67] Lefèvre, *Textus de sphera*, c4r: 'Alioquin falsa plerunque sumerentur nisi semper ascensionum ratione ex quarte principio habita; et hec facilia sunt, et in quibus potius communis logicaque phantasia est, quam astronomica contemplatione.'

[68] Lefèvre, *Textus de sphera*, c4r: 'Instancia quam autor diluit non est cognitu difficilis, nec ex tabula et superioribus diffinitionibus veritatem elicere difficile; modo intellexeris ubi in littera vocabula hec oritur, oriebatur, oriuntur, peroriuntur habentur; horum loco aptissime esse intelligenda orta est, orta erat, orte sunt, per orte sunt.'

[69] Lefèvre, *Textus de sphera*, b6v: 'Sit circulus a b d colurus solstitiorum, qui ex diffinitione per polos zodiaci pariter et polos mundi transit. Et sit linea a eclyptica et linea b equator et punctus c polus zodiaci'.

The motive for including these tables was explicit: 'this has great importance for the *Cosmographia* of Ptolemy and *Geographia* of Strabo.'[70] Both authors had been rediscovered in the fifteenth century. While Strabo recognized the distinction between the mathematical and qualitative descriptions of the earth, he himself focused on qualitative and ethnographical descriptions, tying those associations to the word *geographia*. In contrast, Ptolemy had become *princeps* of measuring the earth as well as the heavens; as his *Cosmographia* became more important, the cosmographer emerged as the practitioner who applied mathematical principles to the earth.[71] The cosmographer approached the earth from the perspective of the heavens—and with the tools of mathematics more than those of astrological, historical, or ethnographic analysis.[72] Ptolemy instructed the reader on how to construct a map, projecting the coordinates of the heavenly sphere on a plane. Such a projection depended on knowing the locations of cities, so the heart of Ptolemy's *Cosmographia* was a series of topographical tables, giving longitudes and latitudes of rivers, mountains, and cities from the farthest reaches of the Roman Empire, from the Pillars of Hercules to the Ganges.

Therefore the culmination of Lefèvre's commentary is a series of tables derived from Ptolemy's *Cosmographia*. All of the cities listed can be found in Ptolemy's own list of ancient cities, and he likely selected them from one of the several editions of the first Latin translation by Jacopo Angeli (*c.*1406), first printed in 1462.[73] The *Textus de sphera* of 1495 grafted Ptolemy's *Cosmographia* onto the medieval astronomical handbook.

This gave Lefèvre's version a unique place among commentaries on the *Sphere*, and for decades it remained a distinctive authority in the genre. Already in 1499, a Venetian printer set Lefèvre alongside commentaries by Cecco d'Ascoli and Francesco Capuano. The volume was the first in a new vogue for omnibus collections on spherical astronomy. Regularly reprinted and augmented, by 1531 such collections had swelled to include no less than sixteen texts, by authorities such as Michael Scot, Regiomontanus, Campanus, Pierre d'Ailly, and Robert Grossteste.[74]

[70] Lefèvre, *Textus de sphera*, b7r: 'non parvum ad Cosmographiam Ptolemaei et Geographiam Strabonis habet momentum.'

[71] M. Milanesi, 'Geography and Cosmography in Italy from the XVth to the XVIIth Century', *Memorie della Società Astronomia Italiana* 65 (1995): 443–68. See especially Milanesi's own work listed in his bibliography, on the relative status of the disciplines. Good resources on Ptolemy's reception include Zur Shalev and Charles Burnett (eds), *Ptolemy's Geography in the Renaissance* (London: Warburg Institute, 2011); Patrick Gautier Dalché, *La Géographie de Ptolémée en Occident (IVe–XVIe siècle)* (Turnhout: Brepols, 2009).

[72] Note chapters by Mosley and Besse in Alexander Marr (ed.), *The Worlds of Oronce Fine. Mathematics, Instruments and Print in Renaissance France* (Donington: Shaun Tyas, 2009).

[73] There is one major difference: many of the values of longitude and latitude differ from the earlier editions of Ptolemy's *Cosmographia*. Evidently Lefèvre took his values from sources he trusted more than Ptolemy, but it remains an open question which sources those were. Were these more recent, updated editions, circulating perhaps in manuscript? Or did someone in his circle recalculate these values?

[74] See Venetian editions from 1500, 1508, 1515, 1518, and the high point in Joannes Sacrobosco, *Spherae Tractatus* (Venice: Lucantonio Giunti, 1531). This trajectory was noted by Peter Barker, 'The Reality of Peurbach's Orbs: Cosmological Continuity in Fifteenth and Sixteenth Century Astronomy', in *Change and Continuity in Early Modern Cosmology*, ed. Patrick J. Boner (Berlin: Springer, 2011), 7–32, at 17–19.

The book developed particular status in Paris, where its editions charted the trajectory of mixed mathematics. Already in 1500, Henri Estienne had reprinted the *Textus de sphera* along with fragments of Boethius' *Elementa geometriae* and Bonetus de Lattes's short treatise on a ring-shaped 'astrolabe' or dial. Estienne printed it at least three more times, in 1507, 1511, and 1516. When Simon de Colines assumed leadership of the Estienne dynasty in 1521, he printed a new edition of the *Textus de sphaera* (note the new, humanistic spelling). He commissioned a new frontispiece by Oronce Fine, launching the young man's career as the most important mapmaker and mathematical practitioner of Renaissance France.[75] With this new edition of the older typographical achievement Colines advertised his plan to build on the reputation of Estienne; Fine's frontispiece heralded that transition by elegantly improving upon the old frontispiece. It is not clear whether Fine also recrafted all the diagrams, but likely Fine was the editor who added new marginal notes and expanded old ones to include references to newly-popular mathematical authors from antiquity, such as the *Sphaera* of Proclus. Colines reprinted this version of Lefèvre's *Textus de sphaera*—with the Boethian geometry and the ring astrolabe—at least five more times, to 1538.

Lefèvre's collage of medieval and ancient sources became a model for new handbooks in the budding genre of cosmography. The most illustrious example is Oronce Fine's own contribution to the genre, published first as *De mundi sphaera, sive cosmographia*, as part of his compendium of mathematical studies, the *Protomathesis* (1532).[76] The very title declared the origins of the new genre of cosmography in the old astronomical textbook. Indeed, Fine placed himself not only within the university tradition of astronomy, but specifically in the tradition of Lefèvre's *Textus de sphera*. The degree to which Fine's *Cosmographia* aimed to replace Lefèvre's commentary has not been appreciated. But even a superficial comparison makes the point. Fine addressed all the topics in Sacrobosco's handbook, but he emphasized precisely those cosmographical calculations that Lefèvre found so important—adding, moreover, a whole book on hydrographia, the nautical equivalent of geography. A concrete example of their similarities is the tables of ascendants. Precisely as Lefèvre had done, Fine provided first a table of ascensions as measured at the equator, and then a second set of tables for the latitude of Paris.[77]

Fine even emulated the pedagogic sensibility that defined Lefèvre's approach. I have tried to demonstrate the extent to which Lefèvre helped students learn mathematical skills, partly through material aids and partly through specimen problems, enabling students to witness and even perform techniques on their own.

[75] Isabelle Pantin, 'Altior incubuit animus sub imagine mundi: L'inspiration du cosmographe d'après un gravure d'Oronce Finé', in *Les méditations cosmographiques à la Renaissance* (Paris: Presses de l'université Paris-Sorbonne, 2009), 69–90. On the significance of this moment for Colines, see the entry in Schreiber and Veyrin-Forrer, *Simon de Colines*.

[76] More generally, see the chapters by Jean-Marc Besse, Adam Mosley, and Pascal Brioist in Marr (ed.), *The Worlds of Oronce Fine*.

[77] Oronce Fine, *De Mundi sphaera, sive Cosmographia, libri V. ab ipso authore denuo castigati, & marginalibus (ut vocant) annotationibus recens illustrati: quibus tum prima Astronomiae pars, tum Geographiae ac Hydrographiae rudimenta pertractantur* (Paris: Michel de Vascovan, 1555). Direct ascendants at 24r–v; oblique (at 48 degr. lat.) on 27r–v.

Throughout Fine's works one can find a similar rhetoric of enabling readers to perform mathematical techniques. Nowhere is this clearer, however, than in the *Cosmographia*, where Fine simply copied the specimen tables that Lefèvre used in the *Textus de sphera* to teach readers, step by step, how to use such tables and how to understand the relationship between celestial and terrestrial time. The borrowing specifies with precision that Fine was Lefèvre's student in print, if not in person. The *Textus de sphera*, mingling Sacrobosco and the *Cosmographia* of Ptolemy, had forged the new mixed mathematics of cosmography.

This chapter's last section will consider what readers made of this. Certainly Oronce Fine's *Cosmographia* was widely influential in shaping the discipline, cited by André Thevet, Robert Recorde, James Cheyne, Christoph Clavius, Antonio Possevino, and Barozzi.[78] The fact that Fine's cosmography was in large measure a rewriting of Lefèvre's commentary on a medieval textbook suggests something surprising about the formation of the new mathematical sciences. The 'cosmographical eye' is often taken to impose its mathematical grid on the world, exemplifying a twin Foucauldian regime: a moment of rupture and disciplinary control impressed from the top down. But such an account of the modern gaze simply has not looked hard enough at the actual processes of forming new disciplines. Instead, we find students engaged in paratextual authorship, impressing a gradual shift in practices; we find Lefèvre (and later Oronce Fine) seeking to coax and elicit understanding from their students and other readers through a rich range of sensory experiences.

MUSIC

Lefèvre also put the relation of mind and senses at the heart of his music theory, in a move that made him an authority among Renaissance theorists such as Francesco Gaffurio, Heinrich Glarean, Gioseffo Zarlino, and seventeenth-century musical authorities such as Marin Mersenne and the Jesuit Biancani.[79] The work in question was Lefèvre's *Elementa musicalia*, which was published in 1496 with the *Elementa arithmetica*, taking up the fundamental topic of tuning musical

[78] See Mosley, 'Early Modern Cosmography', 135–6; S. K. Heninger, Jr., 'Oronce Finé and English Textbooks for the Mathematical Sciences', in *Studies in the Continental Background of English Literature: Essays Presented to John L. Livesey*, ed. D. B. J. Randall and G. W. Williams (Durham, NC: Duke University Press, 1977), 171–85.

[79] Marin Mersenne, *Traité de l'harmonie universelle* (Paris: Fayard, 2003), 32; Biancani, *De mathematicarum natura dissertatio*, 59r. Studies that mention Lefèvre include Claude V. Palisca, *Humanism in Italian Renaissance Musical Thought* (New Haven, CT: Yale University Press, 1985), 224, 233, 242, 273; Ann E. Moyer, *Musica Scientia: Musical Scholarship in the Italian Renaissance* (Ithaca, NY: Cornell University Press, 1992), 153, 220, 271; Nan Cooke Carpenter, *Music in the Medieval and Renaissance Universities* (1958; New York: Da Capo Press, 1972), 141, 149–50. Many historians have mentioned Lefèvre's work in passing as significant for using a geometrical technique for equal division of the tone, but none have examined how or why he did so: Palisca, *Humanism*, Peter Pesic, 'Hearing the Irrational: Music and the Development of the Modern Concept of Number', *Isis* 101, no. 3 (2010): 501–30, at 507–8; Oscar João Abdounur, 'Ratios and Music in the Late Middle Ages: A Preliminary Survey', in *Music and Mathematics in Late Medieval and Early Modern Europe*, ed. Philippe Vendrix (Turnhout: Brepols, 2008), 23–69, at 50; David Paul Goldman, 'Nicholas Cusanus' Contribution to Music Theory', *Rivista internazionale di musica sacra* 10 (1989): 308–38.

intervals.[80] In this book, we can see geometrical techniques applied to arithmetical ratios—the beginning of the end for classical ratio theory, in which the archetypic case was music.

The human ear in most cultures finds pleasure in certain vibrations that overlap and match each other enough to 'sound together', to be consonant. A musical ear finds this consonance in notes or pitches spaced at particular intervals: the octave, the perfect fifth, and the perfect fourth. How to determine the distance between two musical pitches had been a core topic of musical theory since antiquity. Pythagoreans set the terms when they discovered that these consonant intervals matched the sounds made by different string lengths: an octave interval is the difference in pitches of one string half the length of another (at the same thickness and tension). It is in the harmonious ratio 2:1. A perfect fourth is made by strings in the ratio 4:3; a perfect fifth is in the ratio 3:2.

But what about the other intervals? Can we find a system for specifying the distances between intervals? One option was to stick with mathematics, devising ratios to measure the difference between string lengths for each pitch. Pythagoreans took up this second option. Applying mathematics, they set the tone as the difference between the fifth and the fourth: 3:2/4:3 = 9:8. The key to the system's integrity was that all intervals then fit together as ratios of whole numbers. Thus the Pythagorean system was five tones (9:8), filled in with two 'minor semi-tones' of 256:243, yielding this scale:

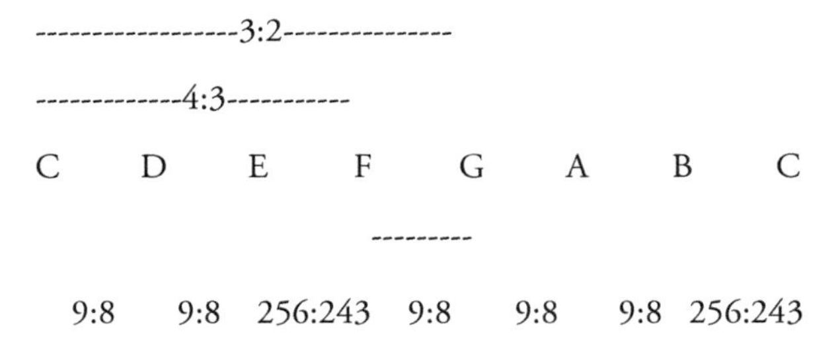

The alternative option was to use the ear, fitting six intervals into the octave by trial and error, depending on the sense of hearing to measure the spaces between pitches. The great spokesman for this option was Aristoxenus of Tarantum, who noticed a problem in the Pythagorean system. Its fatal flaw was that ancient ratio theory had no way to divide a tone (9:8) evenly in half—the result would be an irrational ratio—and therefore the 'semi-tones' were in fact slightly smaller than half a tone. The system of five tones of 9:8 and two 'minor' semitones of 256:243 could slightly distort

[80] The *Elementa musicalia* was first published with the *Elementa arithmetica* and the *Epitome in libros arithmeticos divi Severini Boetij. Rithmimachie ludus qui et pugna numerorum appellatur* (Paris: Johann Higmann and Wolfgang Hopyl, 1496). This whole volume was republished at Paris in 1514; the *Elementa musicalia* was published separately in Paris by Guillaume Cavellat in 1551 and 1552, and Pedro Ciruelo included it in his *Cursus quattuor mathematicarum artium liberalium*, published at Alcalá in 1516 and Madrid in 1528 [according to the colophon; the title page reads 1526].

certain intervals, making them sound harsh. Aristoxenus argued for dropping the mathematical ratios altogether, instead fixing the spaces between pitches by ear.[81]

The biggest change in theories of tuning before the eighteenth century was a gradual synthesis of Aristoxenus' sense-based perspective within mathematical efforts to understand sound. There were consequences in all three major shifts in mathematical culture that I noted in the Introduction. The first has to do with sound, conceived as a magnitude: Aristoxenus undermined the idea that sound was naturally divided into discrete intervals, leaving it a continuous magnitude.[82] Second, if intervals were not fixed by simple ratios, classic ratio theory lost its paradigmatic application, and evaporated, leaving algebra or logarithms.[83] Third, Aristoxenus, when taken up by figures such as Galileo's father Vincenzo, suggested alternative models for how experience should figure in mathematical accounts of nature.[84] Lefèvre shows how such a synthesis was already shaped within the university curriculum. The *Elementa musicalia* of 1496 so thoroughly synthesized Pythagoras and Aristoxenus that it was twice dismissed by Gioseffo Zarlino, the influential Venetian music theorist and teacher of Vincenzo Galileo. In his *Institutioni Harmoniche* (1571) Zarlino regarded Lefèvre as a simple follower of Boethius and the Pythagoreans; by 1588 he had come to see Lefèvre as a defender of Aristoxenus.[85] In fact, both charges were partly true: Lefèvre self-consciously offered a middle way between Pythagoras and Aristoxenus that took account of both mathematics and the senses.

In the early 1490s there were several potential sources for such a synthesis. One was musical practitioners. In fact, it is very likely musicians did adopt something like Aristoxenus' approach to tuning. We know that by the later Middle Ages musicians were adopting various forms of 'equal temperament', apparently by trial and error. A second source was medieval music theorists, who were generally persuaded by the Pythagorean use of mathematics to systematize intervals, and wary of simply handing theory over to the sensory judgement of practitioners. The Pythagorean approach was, from late antiquity on, based on Boethius' *Institutio musica*, supplemented by paragraphs in encyclopaedic works such as Martianus Capella and Isidore of Seville and in commentaries on Plato's *Timaeus*. Boethius gave two strategies for dividing the tone. The first was the Pythagorean, which he aimed at Aristoxenus, who 'attached little value to reason but yielded to aural judgement' by

[81] Aristoxenus' *Harmonica* is in Andrew Barker (ed.), *Greek Musical Writings*, Vol. II: *Harmonic and Acoustic Theory* (Cambridge: Cambridge University Press, 1990), 119–84.

[82] Benjamin Wardhaugh, *Music, Experiment and Mathematics in England, 1653–1705* (Farnham: Ashgate, 2008), 29–58.

[83] Benjamin Wardhaugh, 'Musical Logarithms in the Seventeenth Century: Descartes, Mercator, Newton', *Historia Mathematica* 35, no. 1 (2008): 19–36.

[84] Michael Fend, 'The Changing Functions of Senso and Ragione in Italian Music Theory of the Late Sixteenth Century', in *The Second Sense: Studies in Hearing and Musical Judgment from Antiquity to the Seventeenth Century*, ed. Charles Burnett, Michael Fend, and Penelope Gouk (London: Warburg Institute, 1991), 199–221; Claude V. Palisca, 'Scientific Empiricism in Musical Thought', in *Studies in the History of Italian Music and Music Theory* (Oxford: Clarendon Press, 1994), 200–35.

[85] Gioseffo Zarlino, *Le institutione harmoniche*, (2nd edn, Venice: De Franceschi, 1573), 139. Then see Gioseffo Zarlino, *Sopplimenti musicali* (Venice, 1588), 176: 'E per finire dico, che mi par vedere Aristosseno essere stato anche cosi ben difeso dal Fab[r]o . . . nella Divisione del Tuono contra Tolomeo.'

asserting that a tone can be divided in half.[86] Boethius countered with the classic Pythagorean argument: there is no way to divide a ratio of 9:8 equally in two without creating an irrational ratio.[87] Therefore an octave required five tones and two 'minor' semitones (i.e. 256:243)—not quite the six tones of Aristoxenus. The second strategy Boethius reported from Philolaus, which made a tone out of nine 'commas'. Boethius himself showed that these commas could not be equally divided.[88] Not all medieval theorists followed Boethius. In the early fourteenth century, Marchetto of Padua famously suggested a synthesis based on dividing the tone into five equal parts, where two parts closely approximate the Pythagorean semitone. As we shall see, Lefèvre never pursued these approximations, preferring instead to incorporate a geometrical technique for evenly dividing the tone.

A third source for rethinking these strategies was the Renaissance recovery of Aristoxenus' Greek tradition, which later justified the judgement of the ears through classical authority. But this recovery took a long time to register. The one treatise we have by Aristoxenus himself was not printed in Latin until 1562.[89] Similarly, the two other great authors on the topic, Ptolemy and Euclid, both gave firmly Pythagorean divisions of intervals, Ptolemy explicitly critiquing Aristoxenus.[90] A number of smaller handbooks transmitted some of Aristoxenus, including Cleonides, Aristides Quintilianus, and Alypius.[91] But these were first available in Valla's translation of Cleonides in 1497; I have found no traces of manuscript circulation beyond Italy.[92] Lefèvre himself only ever cited a few sources directly, including Boethius, Martianus Capella, and Euclid's *Elements of Geometry*; his references to Aristoxenus and other authorities seem to be filtered entirely through authors available to medieval writers.

The novelty of Lefèvre's *Elementa musicalia* was therefore strictly within the framework of late medieval efforts to rethink the mathematics in light of sensory experience. The very title advertised it as an effort to dress up Boethian music theory in the Euclidean demonstrative style explored in the last chapter. Recall that Lefèvre and his students rehabilitated Boethius' two extant paraphrases of ancient Greek mathematical works: the *Institutio arithmeticae* and the *Institutio musicae*. And just as Lefèvre found in Jordanus' *Arithmetica* an axiomatized version of the

[86] *De institutione musica*, V.13 (translated by Calvin M. Bower as Boethius, *Fundamentals of Music*, ed. Claude V. Palisca (New Haven, CT: Yale University Press, 1989), 173); cf. III.1 (Bower, trans., 88).

[87] *De institutione musica*, III.1 (Bower, trans., 88). At III.3 Boethius offers another argument.

[88] *De institutione musica*, III.8 (Bower, trans., 97). See the very helpful discussion of Wardhaugh, *Music, Experiment and Mathematics*, 37.

[89] Antonio Gogava (ed.), *Aristoxeni Harmonicorum Elementorum libri III. Cl. Ptolemæi Harmonicorum, seu de Musica lib. III. Aristotelis de Obiecto Auditus Fragmentum ex Porphyrij commentariis* (Venice, 1562). The Greek *editio princeps* was published later, edited by Johannes Meursius (Leiden, 1616).

[90] The *Harmonica* of Ptolemy was first published with Aristoxenus; Euclid's *Sectio canonis* was first translated by Giorgio Valla, *Cleonidae Harmonicum introductorium [incl. Euclidis sectio canonis] interprete Georgio Valla Placentino. L. Vitruvii Pollionis de Architectura libri decem. Sexti Frontini de Aquaeductibus liber unus. Angelic Policiani opusculum, quod Panepistemon inscribitur. Angeli Policiani in priora analytica praelectio, cui titulus est Lamia* (Venice: Simone Bevilacqua, 1497).

[91] See Palisca, *Humanism*, 49–50.

[92] Palisca, *Humanism*, 67–87, shows Valla's use of these texts in the *De expetendis et fugiendis rebus* (Venice: Aldus Manutius, 1501). Cf. J. L. Heiberg, *Beiträge zur Geschichte Georg Vallas und seiner Bibliothek* (Leipzig: Harrossowitz, 1896), 113, 123.

Boethian *Arithmetica*, the *Elementa musicalia* axiomatized Boethius' *Musica*. Even Lefèvre's style of commentary drew on the long tradition of Boethian arithmetic. The *Elementa musicalia* began with common notions and postulates, and then built a systematic series of deductively interlinked enunciations. A methodological note at the beginning highlights the position of music as a *scientia subalternata* which depends on the principles of arithmetic. The remainder of the book constantly reminds the reader of that dependency, since the book's proofs are studded with references back to the arithmetical treatise with which it was published. Early readers already pointed out that the *Elementa musicalia* was difficult to understand without first reading the *Elementa arithmetica*.[93] The book did not wear its mathematical apparatus lightly.

But the book was no less an effort to take account of the senses. In fact, musical practice seems to have motivated Lefèvre's music theory. In a remarkable introduction, he thanked two teachers for their long instruction, musicians whom we only know by their names, Iacobus Labinius and Iacobus Turbelinus.[94] To woo his readers, he wove together instances of music's wonderful effects, from Pythagoras calming drunken youths to its reputation in antiquity for healing fevers and wounds, as well as grief.[95] To some extent, this is a tapestry of topoi—mostly gathered from the standard encyclopaedic work of Martianus Capella—to adorn a bookish treatise. But elsewhere Lefèvre hinted that he expected such topoi to move from books into practice; in his commentary in the *Nicomachean Ethics* Lefèvre referred to the practical power of song to heal melancholy with reference to antiquity and especially his friend Giovanni Pico della Mirandola, who would sing and play the lute (*cythara*) in order to overcome fatigue.[96]

The *Elementa musicalia* is therefore delicately balanced around the problem of retaining full mathematical control while also doing justice to the sensory experience of music. The first book builds up the Pythagorean system out of five ratios of 9:8 and two minor semi-tones of 256:243. Lefèvre explores the arithmetical qualities of these intervals, how they are added, subtracted, or multiplied. These operations set the limits of the system. Citing Boethius and his sources Archytas and Nicomachus, he shows how such ratios intervals are superparticular, and therefore cannot be divided into two equal smaller ratios (I.5).[97] It is clear that Lefèvre has

[93] Lefèvre recognised this himself in the *Epitome Boetii* that he put directly after the work. Pedro Ciruelo also made the point in his own edition of the *Elementa musicalia*, in *Cursus quatuor mathematicarum artium liberalium* (1528), a2r.

[94] Lefèvre, *Elementa musicalia*, f1r (PE, 29–33).

[95] Lefèvre, *Elementa musicalia* (PE, 30).

[96] See Chapter 2, note 52. Lefèvre also raised the point in his commentary on the Psalms, regarding King David.

[97] Dividing a tone is actually a matter of dividing a ratio into two smaller ratios, by finding a suitable mean between the two numbers that are the ratio's terms. For example, what is half of the tone, if the tone (9:8) is defined as the difference between a fourth (4:3) and a fifth (3:2)? The answer is not an arithmetical mean (8 1/2), which divides the tone into two unequal ratios (18:17 and 17:16). The harmonic mean, in which the differences between the terms are in the same ratio (8 8/17, with the property that 9-minus-this and this-minus-8 are themselves in the ratio 9:8) suffers from the same problem. The geometrical mean $\sqrt{(9x8)}$ divides the tone into two equal ratios, but unfortunately they are irrational ratios. Therefore they are unsuitable for use in music, which was traditionally constrained (because of its basis in arithmetic) to employ only ratios of numbers, not ratios of magnitudes which, as in this case, might be incapable of expression as (whole) numbers.

Aristoxenus in view here, because book one concludes with a series of arguments that show the impossibility of fitting six whole intervals of 9:8 into an octave.

The arithmetical foundation set, book two of *Elementa musicalia* brings in sound to set up the classic problem of dividing the tone. The first sentence seizes on the world of sensation by defining a consonance: it 'comes to the ears sweetly and uniformly, by means of a multiple or superparticular perfect ratio', while dissonance mixes different sounds as 'a harsh and unpleasant blow reaching the ear'.[98] But quickly the world of sound is regulated mathematically. The monochord is a geometrical given: 'Let there be a string a b c on which we may set any tone to be located...'.[99] The monochord is assumed to be a real instrument, genuinely to test the senses: 'in this way you may set to the judgement of the senses as many tones as you like to play, discerning their mixtures both sweet and lacking harmony (when heard, these make one bristle or flee, just like a scrape)'.[100] Yet the effect is to bind the Pythagorean ratios to the harmonies of the physical world, so the reader is not surprised to find the remainder of the book explicitly aimed at the errors of Aristoxenus, who used aural judgement to divide the tone evenly.[101] If the ratios match experience, then it makes no sense to cast ratios aside at this point! The verdict on Aristoxenus is unambiguous: 'In this way, those who follow the coarse judgement of the senses take leave of their intellect, and easily find themselves cast away from the gates of learning.'[102] Even the Pythagorean Philolaus comes off poorly, for he had suggested that perhaps dividing the semitone into smaller parts, schismata and diaschismata, might provide the minimal parts for assembling an even half tone.[103] Book two builds towards the theorem that 'the ratios of a schisma and diaschisma are unknown and irrational'.[104] At the end of the second book he explains why he had followed Boethius so closely: 'Greek curiosity is not deterred by the labour of calculation from investigating how many commas are in a diesis, how many are in an apotome, and even how many are in a tone. For unless I had understood this project through what went before, even though it is more work than is helpful or useful in musical modulations (it seems to me), I would have dismissed it.'[105] Lefèvre here explains why he calculated a table of ratios for the tiniest components of a tone. Exhausting in its detail—his

[98] *Elementa musicalia*, f6r: 'Consonantia est soni gravis acutique mixtura, suaviter uniformiterque auribus incidens, ex multiplici aut superparticulari ratione perfecta. Dissonantia est duorum sonorum non se natura suaviter miscentium, ad aurem perveniens aspera iniocundaque percussio.'

[99] *Elementa musicalia*, f6r: 'Sit a b c chorda quecumque supra quam iubeamur tonum collocare'.

[100] *Elementa musicalia*, f6r: 'et hoc pacto sensuum iudiciis quodquot voles tonos deprehendendos committeres, et eorum mixturas tum suaves tum inconcinnas (quas auditus tanquam offensus horret refugique) decernendas.'

[101] *Elementa musicalia*, f6v–f7r: 'Aristoxenus musicus auriam iudicio cunta committens perparum esse probandus'. Lefèvre observes that Martianus Capella had even taken the error of Aristoxenus further.

[102] *Elementa musicalia*, f7r: 'Sic enim qui stolidum sensus iudicium sequentes intellectum relinquunt, facile ex disciplinarum aditis se explosos sentiunt.'

[103] Boethius, *Musica*, III.8 is the only witness to Philolaus' system. Boethius presents it in outline, but tacitly critiques Philolaus by defining the schismata and diaschismata differently.

[104] *Elementa musicalia*, f7r: 'Rationes schismatis atque diaschismatis sunt ignotae atque irrationales.'

[105] *Elementa musicalia*, g1v: 'Non est greca curiositas calculi labore deterrita quo minus quot commata in diesi quot in apotome quot denique in tono sint, pervestigaret. Quod nisi a prioribus

calculations extend to thirty-three figures—the table helps to prove the impossibility of calculating an arithmetical division of the tone using minimal parts. This leaves the Pythagorean semitone where Boethius left it: slightly smaller than half a tone, in the ratio 256:243.

So much, so Boethian. But then Lefèvre changes his angle of approach. He moves from arithmetic to geometry, led more firmly by the ear. I have already noted Lefèvre's regular allusion to the sensation of sound in the *Elementa musicalia*, albeit carefully matched to Pythagorean arithmetic. The third book begins in earnest with musical sensation, opening with an account of the limitations of Pythagorean theory in tuning real instruments beyond a couple of octaves. 'The Pythagoreans do not go higher because voices that go higher seem to make a certain buzzing, and because nature puts limits and measure on each and every voice at the consonance of two octaves.'[106] Here Lefèvre brought in the great problem of experience in Pythagorean music theory, which is, as he observes, that a buzzing or dissonance is 'easily experienced' above two octaves. In the Pythagorean tuning the fifths and octaves are 'pure' or consonant. Over the course of several octaves, however, other intervals (especially thirds) grow 'impure', becoming more and more dissonant the further they are stretched over the course of the scale.[107] Music theorists had normally ignored this problem, as Lefèvre also notes, 'thinking themselves contained by the limits' that Pythagoras had taught. He claims to follow this same path: 'this is what I will try to imitate.'

And indeed the third book of the *Elementa musicalia* begins with the unproblematic consonant intervals of the octave, fourth and fifth, suiting Pythagorean intuitions. But quickly Lefèvre finds a way to bridge from these consonant intervals to a more geometrical and material account of sound. He adduces titbits on the physics of sound from (pseudo-)Aristotle's *Problemata*. While giving a method for reconstructing an octave on a string, he assures readers that 'Aristotle testifies that this is the most elegant and beautiful of the consonances.'[108] Aristotle also agrees with the Pythagoreans, he shows, that simple consonances of the octave, fourth, and fifth

tentatum cognivissem, cum id quoque plus laboris quam (ut michi visum est) in musicis modulationibus usus utilitatisque afferat, missum fecissem.'

[106] *Elementa musicalia*, g2v: 'Nec altius ascendunt Pythagorici quod altius ascendentibus voces quoque pacto illis stridule vise sunt, et quod unicuique ferme sue vocis modum limitesque ad consonantiam bis diapason natura fecerit. Quoque habita contemplatione musica adusque consonantiam bis diapason, reliquam ut habeatur quamfacillimam putaverunt ut qui ad ter atque quater diapason musicos modulos aptare voluerint. Et hac quoque de causa musici ferme omnes in definienda determinandaque atque tradenda disciplina musica limites Pythagore non transcendunt, putantes eius limitatibus contenti, et priscam veteremque autoritatem secuti sufficienter determinasse, quod et nos in hoc opere tentabimus imitari.'

[107] The specifics of this problem would require far more space than I have here; for further information, see Thomas J. Mathiesen, *Apollo's Lyre: Greek Music and Music Theory in Antiquity and the Middle Ages* (Lincoln, NE: University of Nebraska Press, 1999); Andrew Barker, *The Science of Harmonics in Classical Greece* (Cambridge: Cambridge University Press, 2007). Useful overviews are Murray Barbour, *Tuning and Temperament: A Historical Survey* (1951; Mineola, NY: Dover, 2004), and Wardhaugh, *Music, Experiment and Mathematics*, 9–20.

[108] *Elementa musicalia* III.18, g4r: 'Hec consonantiarum ut in libro probleumatum testatus est Aristoteles elegantissima pulcherrimaque est.'

are all superparticular intervals.[109] As such, it is impossible to divide them into equal intervals expressible as 'a certain and definite number', Lefèvre recalls from the previous book, confirming the central problem of Pythagorean tuning.

The *experience* of sense justifies the first step towards geometry. This is made explicit in a constructive proposition, which describes 'how to regulate all consonances in order on a given string, and how to experience them perceptibly, by sense'.[110] Take a compass, Lefèvre tells his reader, and divide a string into thirds, quarters, and halves. Then he teaches us to test these consonances by ear:

> So place a musical hemisphere [i.e. a bridge], step by step, on the individual notes of the string. Compare these sounds to the sound of the whole string and you will take note of the proposed consonances in sequence. You will experience this more quickly if you find the equal tone or unison on string a–b [i.e. by dividing at half]; you will quickly notice this tone when you have plucked each section of a–b.[111]

It is this experience that guides the next ten enunciations of the book, as Lefèvre uses the sensation of these consonances to cut a path between Aristoxenus and Pythagoras (and the Boethian tradition that Lefèvre claims to 'imitate'). On the one hand, as we saw with book two, the experience justifies the intuition of Pythagoras that 'simple consonances' conform to ratios of small whole numbers.[112] On the other hand Lefèvre now has the structure for building tones from experience, the way the musicians do. The whole third book of the treatise therefore builds to its final enunciation III.35, in which Lefèvre uses a geometrical diagram to accomplish in material form what numbers could not: to equally divide the tone (in a harmonic mean), or indeed 'any consonance you like in the musical art'.

At proposition III.35, geometry thus breaks the rules of arithmetic, for 'the foregoing [propositions] contend that this cannot be done by a certain, definite, rational ratio; but it *can* be done geometrically, without a certain or definite ratio of numbers.'[113] The technique is from Euclid's *Elements* book 6, where proposition 9 shows that for any perpendicular drawn from a line to a semicircle constructed on that line, the perpendicular is the harmonic mean between the segments it divides (see Figure 5.8).[114] Lefèvre's proof works because the perpendicular is the diagonal of a square equal in area to a rectangle with sides equal to the interval ratio. That is, it may well be in irrational ratio. But, expressed as lines, irrational ratios are no

[109] *Elementa musicalia* III.24, g5r: 'Omnis enim consonantia aut in superparticulari aut in multiplici genere ex diffinition consistit. Et in hoc facile cognosci potest ex novo Probleumatum libro quod Pythagoricis consensit Aristoteles.'

[110] *Elementa musicalia* III.28, g5r: 'Omnes consonantias in data chorda suo ordine subiungere et eas sensu perceptibiliter experiri.'

[111] *Elementa musicalia* III.28, g5r: 'Suppone igitur musicale hemispherium sensim singulis chorde notis, et sonos ad totius chorde sonum diligenter attende, et suo ordine propositas concinentias annotabis, quod promptius experiri valebis si chorde a b chordam equisonam unisonamque etiam collocaveris, cuius sonum cum singulis sectionum a b percussionibus non segniter attenderis.'

[112] This is confirmed by at *Elementa musicalia* III.34.

[113] *Elementa musicalia* III.35, G6v: 'Nam precedentes condendunt id effici non posse Arithmetice certo constitutoque numero atque rationali habitudine; hec vero id effici posse geometrice sine numeri certa constantique ratione.'

[114] This proposition is VI.9 in Campanus; in the modern edition it is VI.13.

M

Cõſonãtie ĩ Arithmetica medietate	A primavnitate	A ſcđa vnitate	A tertiavnitate
3 | 6 | 9 | 12	1 | 2 | 3 | 4	10 | 20 | 30 | 40	100 | 200 | 300 | 400
duplus diapaſon | ſeſquiſti⁹ diateſ.	duplarẽ | epitrit⁹	diapaſ. | diateſ.	diapaſon | diateſſa.
ſeſqualt diapẽte	hemiol.	diapẽ.	diapente
Triplus. diapaſon diapẽ.	Triplaris	diapa. diap	diapa. diapẽte
Quadruplus. bis diapaſon	Quadruplaris	Bis diapaſon	Bis diapaſon

¶ Intelligit ternaria ꝓgreſſiõe ad ſe addi numerus qñ additur p̃mo ad ſe ſemel/ deinde bis/ deinde ter. dico ergo quẽcunq; numerũ ad ſe hoc pacto additum: oẽm cõſtituere conſonãtiaʒ muſicam. Nam ſemel ſibi adiectus: duplex ad ſe efficit/ ⁊ diapaſon conſonãtiam. et bis ſibi additus: ad primã additionem ſeſqualteꝝ cõſtituit ⁊ diapente: et ad ſe triplũ atq; diapaſon ac diapente. nam in p̃ma additiõe bis/ ⁊ in ſecũda ter cõtinetur. ter v̓o ſibi additus ad ſcđaʒ additiõem facit ſeſquitertiũ ⁊ diateſſaron: nam primus nũerus hic quater/ illic v̓o ter cõtinetur. et ad ſe quadruplũ facit et bis diapaſon. in hac igitur ꝓgreſſione oĩs cõtinetur cõſonãtia muſica. et qꝛ cõtinue numerorũ ex eius additiõe ſurgentiũ ipſe numerus additus eſt differẽtia: ergo reperte conſonãtie in Arithmetica medietate continẽtur. et ex hac cognoſcitur cur in oĩm vnitatum quaternaria ꝓgreſſione que vbiq; denarũ Pythagorici plenitudinem implet: oẽs conſonantie muſice et in Arithmetica quidem medietate reperiãtur.

¶ Tonum et oẽm conſonantiam ſimplicem: in duo equa partiri/ veraq; medietatum puncta in chorda: geometrice monſtrare. 35

¶ Precedẽtes ſeptima ſcđi/ ⁊ viceſimatertia tertij pretẽdũt tonũ/ diateſſaron/ diapẽte ac diapaſon: in duo equa diuidi non poſſe. hec v̓o mõſtrat quo pacto ea omnia poſſint in duo equa partiri. nec hoc quidẽ repugnat. Nã precedẽtes contẽdunt id effici non poſſe Arithmetice certo/ cõſtitutoq; numero/ atq; ratiõali habitudine: hec v̓o id effici poſſe geometrice ſine numeri certa/ cõſtantiq; ratione.

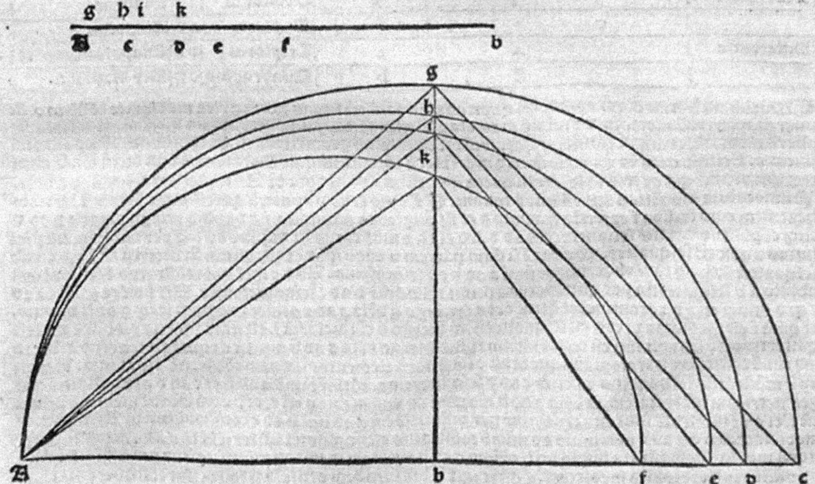

¶ Sit ergo data chorda a b ſuperior in qua iubeamur integrum ſemitonium/ et ꝯſonãtiarũ diateſſaron/ diapente/ ac diapaſon vera media reperire. facio a b ⁊ c b tonum. a b et d b diateſſaron. a b ⁊ e b diapente. a b et f b diapaſon eo qui in precedẽtibus mõſtratus eſt modo. deinde in inferiori linea a c indefinite quãtitatis capio a b equalẽ linee ſuperiori a b et b c ꝯtinue equalem linee b c ſuperiori. et a puncto b verſus c: capio lineam b d equalem chorde b d et b e equalẽ chorde b e et b f chorde b f. et intelligo q̃ttuor dimidios circulos a c/ a d/ a e/ a f. et a puncto b educo perpẽdicularẽ linee a c ad circũferẽtias ſemicirculorũ a c/ a d/ a e/ a f. et pũcta vbi eos cõtingit linea ſint g/ h/ i/ k: ad que pũcta educo a g/ c g/ a h/ d h/ a i/ e i/ a k/ f k. at per nonã ſexti geometrie a b ad b g vt b g ad b c. facio igit in chorda ſuperiore a b lineam b g equalẽ inferiori linee b g et cũ ꝓportio a b ad g b vt g b ad c b vt preoſtẽſum eſt: ſeqtur tonum a b ⁊ c b eſſe in duo equa diuiſum: ⁊ pũctũ g eſſe medium veri ſemitonij ſignũ. et per eãdem que ꝓportio a b ad b h ea ſit b h ad b d. eſto igitur h b in chorda a b equalis linee b h per idem vt prius: ea erit ꝓportio a b ad h b que h b ad d b. quare a b et d b diateſſaron in duo equa partita eſt. et eodem pacto facta chorda i b ſuperiori equali linee b i. et chorda k b equali linee b k: mõſtrabis cõ-

Figure 5.8. Lefèvre shows how to divide any interval equally, ‘without a certain or fixed ratio of number’. From his *Elementa arithmetica, musicalia, etc.* (1496), G6v, proposition III.35. By kind permission of the Syndics of Cambridge University Library, Shelfmark: Inc.3.D.1.21.

longer a problem. What the mind cannot calculate, the eyes can see and physically measure—and the ears can hear.

Lefèvre clearly knew he was breaking with accepted norms. But his justification came from within the standard Boethian classification of music as a mixed, subaltern science. At the beginning of the *Elementa musicalia*, Lefèvre explained that he would stray from the norm:

> A subalternate science, such as music is to arithmetic, uses the principles and enunciations (*demonstrata*) of the prior science. However, I have tried to have this happen as rarely as possible in this discipline, but, when it seems relevant, I use another method of demonstration than is usual in arithmetic, a method by which individual things seemed to be done more appropriately, even if the same might be done more quickly and easily by arithmetical judgement and by what is shown in arithmetic. However, anyone can very easily see when this is done and when it is not done from any of following demonstrations.[115]

By saying that it will be obvious where he has departed from the arithmetical mode of explanation, Lefèvre seems not to recognize the most obvious feature of his 'elements' of music: that they are structured according to the Euclidean *mos geometricus*. Unlike Boethius, he begins with tables of *dignitates* and *petitiones*; these support the web that underlies the enunciations. Yet this does not make the book geometrical. As seen in the last chapter, the style of elements is not uniquely geometrical for Lefèvre, but is simply the mathematical mode of commentary.

Instead, we find geometry here linked with the visual, sensible stuff of measurement. Throughout, Lefèvre invokes diagrams and tables to support intuitions, 'to get it faster' (*ut promptior habeatur*). The notion of intuition is also linked with geometrical practice: Lefèvre must have been practising geometry for some time, taking his new technique from book VI, one of the later, less read books of Euclid's *Elements*. The very novelty of the technique he imported suggests that by the early 1490s Lefèvre had already made a close study of Euclid,[116] and in his geometrical construction he refers to the *exercitationes geometricae*, the embodied practices through which one gains an 'immediate' understanding of the argument.[117] Geometry overlaps with the senses also in his order of exposition, as I have argued. To prepare for geometry, he gets his reader to focus on the 'sensible, physical experience' of the sound. Geometry enters already at this earlier stage, when measuring

[115] Lefèvre, *Elementa arithmetica, musicalia etc*, f2r: 'Scientia subalternata qualis ad Arithmeticam Musica est, principiis et demonstratis scientie prioris subalternantisque utitur, at studuimus ut quam fieri potest rarissime id in hac disciplina fiat, verum plerumque ubi oportunum videbitur alio quam in arithmeticis factum est utemur demonstrandi modo, quo singula magis ex propriis facta videantur quamvis arithmetico suffragio atque iis que in arithmeticis monstrata sunt eadem promptius faciliusque fierent. Attamen quando id fiet et quando non sequentibus demonstrationibus cuilibet perquam perspicuum evadere poterit.'

[116] It is possible, of course, that he learned of this technique from some other music theorist rather than from Euclid; but that seems unlikely since all later musical authors (Zarlino, Glarean, etc.) cite Lefèvre as its author.

[117] He suggests that his techniques 'will be immediately understood from a little time in geometrical practice' (*hec statim aliquantulum in geometricis exercitatis nota esse possunt*). *Elementa musicalia* III.35, G7r.

out string lengths with a compass (III.28). The reader is thus prepared with the tools of geometry before arriving at the surprising culmination of the work in the geometrical division of the tone at *Elementa musicalia* III.35. What makes this enunciation geometrical is the material practice of measuring and drawing lines with compass and straightedge—eye and hand work together with ears.

Lefèvre's own view of his novelty can be measured in the final fourth book of *Elementa musicalia*, which returns to a classic Boethian division of the monochord. Lefèvre regularly pushes back against medieval practice, setting the 'ancient authority' of the Pythagoreans against 'moderns' such as Gregory and medieval musical modes. Perhaps this ambivalence betrays an anxiety over novelty, in an effort to disguise it. But another way is to see it in light of Lefèvre's own aims. In the final lines of the treatise, Lefèvre offered a series of commonplaces on music: the music of the spheres ('the magi discuss this more fully'), and the power of music to excite and calm the soul. Pythagoras returns as a voice for ethical moderation—music should not be too fast or too slow, and minds ruled by it will find happiness. 'Happy will be those who seek this end in music and in all earthly philosophy.'[118] One seeks novelty in music for the sake of, not despite, such experience.

In sum, the immediate result of this dense reworking of Boethius' harmonic theory is an irrational ratio, a small enough matter on its own. But it represents the incursion of one field, geometry, into music, a field ruled by arithmetic according to the entire force of the Pythagorean music-theoretical tradition. Lefèvre's own methodological throat-clearing alerts us to the effort it took to make this step. And its implications were recognized by a string of commentators including, besides Gaffurio and Zarlino, Erasmus of Höritz, Heinrich Schreiber, Lodovico Fogliano, Michael Stifel, and Juan Bermudo.[119] The one geometrical construction in Lefèvre's *Elementa musicalia* deployed the Boethian tradition against itself, eroding the ancient barrier between arithmetic and geometry, between numbers and lines. It is precisely at this site of erosion where early modern mathematics developed new practices of algebra, logarithms, methods of indivisibles, and so on.[120] Indeed, Lefèvre's geometrical construction ultimately relies on visual inspection to divide the tone—which shares much with Descartes' use of the proportional compass to

[118] *Elementa musicalia*, IIII.30, h6v: 'felices ii erunt qui hoc fine et musicen et omnem mundanam philosophiam quesierint'.

[119] Erasmus of Höritz, *Musica* (1506), Heinrich Schreiber (1518), Lodovico Fogliano, *Musica theorica* (1529), 36r; on these see Claude V. Palisca, *Music and Ideas in the Sixteenth and Seventeenth Centuries* (Urbana, IL: University of Illinois Press, 2006), 145. On Michael Stifel, *Arithmetica integra* (1544), see Pesic, 'Hearing the Irrational', 506–10. On Bermudo, see Abdounur, 'Ratios and Music in the Late Middle Ages', 6.

[120] e.g. Wardhaugh, 'Musical Logarithms'. The classic study is Jacob Klein, *Greek Mathematical Thought and the Origin of Algebra*, trans. Eva Brann (1934; New York: Dover, 1992). On the early modern breakdown of the distinction, besides Pesic, see Michael S. Mahoney, 'The Beginnings of Algebraic Thought in the Seventeenth Century', in *Descartes: Philosophy, Mathematics and Physics*, ed. Stephen Gaukroger (Brighton: Harvester Press, 1980), 141–55, Henk J. M. Bos, *Redefining Geometrical Exactness: Descartes' Transformation of the Early Modern Concept of Construction* (New York: Springer, 2001), 119–32; Katherine Neal, *From Discrete to Continuous: The Broadening of Number Concepts in Early Modern England* (Berlin: Springer, 2002), Antoni Malet, 'Renaissance Notions of Number and Magnitude', *Historia Mathematica* 33 (2006): 63–81.

classify curves.[121] We can see the formation of these intuitions in slow motion, grain by grain, if we follow a reader trying to make sense of these books.

A READER

The visual and material composition of these books aimed at practice, at the technical skills of actually using a compass and pens to divide a musical interval, or using tables to calculate the longitudes and latitudes of earthly cities and planets on the horizon—techniques of mathematical craft. So did readers become technicians? At stake is the question of how print contributed to making the mathematical culture assumed by the later new philosophies. One does not need to assume a rigid technical determinism, where print somehow *must* cause innovation. But there is space here for a 'soft' determinism: once they exist, new technical cosmographies shift the range of possible readings and technical practices, for example. Objects such as the Fabrist *Textus de sphera* encode practices such as calculating the heavenly coordinates of Aries, and such objects presented new possibilities to new readers. Did these books then render these readers practitioners?

Certainly readers did not miss the more practical bent of the *Textus de sphera*. Within three years of its publication, the Paris theologian Pedro Ciruelo published a similar commentary and edition of the *Sphere*.[122] He underscored the material explanatory power of Sacrobosco's definitions: 'it is clear that the movement of this proposed sphere is not quidditative or formal, but causal. For the fact that the sphere is caused by the imagined revolution of the semicircle holds for the senses, and for this reason the text adds that the sphere is this sort of round solid, etc.'[123] Reading this passage, Beatus Rhenanus noted that Ciruelo had not gone far enough, writing that 'this is Euclid's definition of a sphere; instead one should give the definition of the workman's method of making a sphere which our Lefèvre gives'.[124] A page later, Ciruelo repeated nearly word for word the lathe example that Lefèvre had given, where Beatus Rhenanus commented 'Lefèvre is of this opinion'.[125] Ciruelo himself was not convinced the example was right, for he noted that 'although this may seem beautiful and ingenious, it was not what Euclid had in mind, but rather something

[121] William R. Shea, *The Magic of Numbers and Motion: The Scientific Career of René Descartes* (Canton, MA: Science History Publications, 1991), chapter 3.

[122] Pedro Ciruelo, *Uberrimum sphere mundi commentum intersertis etiam questionibus domini Petri de Aliaco* (Paris: Jehan Petit, 1498).

[123] Ciruelo, *Uberrimum sphere*, a8v: 'Et sic patet quod ista proposit[a] sphera est transitus non est quiditativa aut formalis sed causalis, est enim sensus quod sphera ex transitu semicirculi ymaginario causatur ideo subiungitur in textu id est spera est tale solidum rotundum etc.'

[124] Ciruelo, *Uberrimum sphere*, BHS K 950, a8v, *in margine*: 'Hec spherae diffinitio ab Euclide Megarensi assignata; magis fabricandae spherae modum industriam quam Faber noster potius descriptio dicenda est.'

[125] Ciruelo, *Uberrimum sphere*, BHS K 950, b1v, *in margine*: 'Huius opinionis est Faber Stapulensis'. See my account of Lefèvre's language about the sphere above (note 60), and compare Ciruelo's wording: 'Quare dicunt hanc mire efficacie descriptionem que (et si latenter philosofico more) spherarum conficiendarum quantum sensibilis materia recipere videlicet artificium et instrumentum insinatur. Si enim ex levi calibe aut ferro semicirculus (excavata area dimissaque circumferentia acuta cum duabus extremalibus dyametri portionibus) accipiatur instrumentum esset tornandis spheris aptissimum.'

we add to him'.[126] Whatever the disagreement, colleagues and students alike were alert to the creative implications of Fabrist mathematics.

Surviving copies turn up readers in later generations who made these books sites for practice. Some clearly found these books useful for drawing connections within their broader humanist learning. For example, readers of the *Textus de sphera* were especially quick to mark cities and climes that came up in their readings of classical antiquity. In a copy at the Huntington Library, one reader copied sentences from Cicero's legal oration *Contra Rullum*, section 95, where Cicero expanded upon the agricultural qualities of land relevant to his case.[127] On the advice of Sir Philip Sidney, the Tudor reader Gabriel Harvey turned to the *Textus de sphera* as the point of entry into mathematical learning, but he used the margins of his copy to proclaim Lefèvre, Clichtove, and Bovelles as *polytechni*, practitioners of the practically oriented learning that he hoped would fortify England.[128] In some hands, therefore, the book could become part of a broader set of mathematical practices. A copy now at Houghton Library, Harvard, is bound together with more advanced books on geometrical and instrumental calculation of days, distances, and planetary locations, Jean Fernel's *Monalosphaerium* and *Cosmotheoria* (both Paris, 1527).[129] This copy places the *Textus de sphera* in the context of practice, so the pages of all three books are filled with extended calculations, most concerning the periods of planets in order to find planetary conjunctions. Likewise, in other copies we find genitures for horoscopes, drawn on blank sheets or in the margins.[130] Lefèvre and his students had enriched the school text with enough technical apparatus that some readers, at least, could use it as a stepping stone to practice.

In the remainder of this chapter I will focus on a reader we have already met. Beatus Rhenanus did *not* become a mathematical practitioner, but this makes him all the more useful for the historian. Rather than representing a small specialist readership, he stands in for the generally educated reader, the Renaissance university student who had to endure a few mathematical lectures en route to the more glorious study of letters, law, or theology. His tangential touching of mathematical learning apprenticed him in this broader mathematical culture.

Let us scrutinize the books Beatus read on mathematics at Paris, bound together in one volume, *Bibliothèque humaniste de Sélestat* K 1046. The volume comprises a set of three printed books, all by Lefèvre, Clichtove, and Bovelles:

a. *Epitome compendiosaque introductio in libros arithmeticos divi Severini Boetii [Fabri], adiecto familiari [Clichtovei] commentario dilucidata. Praxis numerandi*

[126] Ciruelo, *Uberrimum sphere*, b1v: 'Sed quamvis hec pulchra et ingeniosa videantur, hanc tamen non fuisse Euclydis mentem sed quam supra dedimus.'

[127] Joannes Sacrobosco, *Sphera cum commentis* (Venice: Ioannes Rubeus et Bernardino Vercellensis, 1508), The Huntington Library, Burndy 751765, fol. 61r.

[128] Jacques Lefèvre d'Étaples, *Textus de Sphaera* (Paris: Simon de Colines, 1527), BL 533.k.1. See especially notes on 1r, 3r, 4r, 5r.

[129] Lefèvre, *Textus de sphaera* (Paris: Colines, 1527), Houghton Library f EC.Sa147s.1527.

[130] e.g. BnF res.v209 (Paris: Colines, 1521), 13v; Huntington, Burndy 751765 (Venice, 1508), 55v; BL 8562.f.34 (Paris: Estienne, 1534), 29r.

certis quibusdam regulis (auctore Clichtoveo). Introductio in geometriam Caroli Bovilli. Astronomicon Stapulensis (Paris: Wolfgang Hopyl and Henri Étienne, 1503).

b. Lefèvre, *Elementa arithmetica; Elementa musicalia; Epitome in libros arithmeticos divi Severini Boetii; Rithmimachie ludus que et pugna numerorum appellatur* (Paris: Johannes Higman and Wolfgang Hopyl, 1496).

c. Lefèvre, *Textus de Sphera Johannis de Sacrobosco, Cum Additione (quantum necessarium est) adiecta: Nouo commentario nuper edito ad utilitatem studentium Philosophice Parisiensis Academie: illustratus. Cum compositione Anuli Astronomici Boni Latensis. Et Geometria Euclidis Megarensis* (Paris: Henri Estienne, 1500).

The volume's binding indicates that these fascicles were bound together early, before Beatus used the book. On the top of first guard-page's verso is his habitual *ex libris*: 'Est Beati Rhynaw Schletstattini Anno Verginei Partus 1.5.0.3. Decem et octo duodenis emptus. Parrhisiis.' The name 'Rhynaw' instead of 'Rhenanus' confirms that the *ex libris* was indeed written in the year the book was purchased; not until the following year did Beatus begin to use the Latin spelling.[131] This is significant because that flyleaf, the first of several sheets added to the beginning and end of the binding, was originally pasted down to the inner side of the front binding: the glue- and ink-marks that bled from the original binding vellum are evident around the edges of this folio (see Figure 5.12). After the initial binding, the folded-over vellum would have obscured the original *ex libris*, so Beatus repeated it on the title page of the first fascicle.[132] The repeated spelling of 'Rhynaw' indicates Beatus had the book bound very quickly after purchase, in the short period before he latinized his name. Furthermore, the *ex libris* is not repeated on the other two fascicles, allowing several probable judgements about how Beatus first bought the book. First, the fact that Beatus placed his ownership mark on the inside of the first guard-page and then only on the first title page is evidence that Beatus bought the books together—either they were sold as a block together, or he had the bookseller tie them together immediately. This observation is strengthened by the presence of the extra several flyleaves at the front and back of the text block. These suggest that someone, likely Beatus or the bookseller, meant these books to be together, and furthermore expected that they would be annotated. Beatus must have had these books bound with 'softcover' vellum before he used the volume, since his notes on the first flyleaf fit within the outline of the folded-over vellum (Figure 5.12). Certainly it was he who had them bound, and he had done so in or before 1503. It seems all the more likely that booksellers marketed these books together, before they were bound by students, since there exist several editions of the Fabrist mathematical works from the same period bound together in the same way, in

[131] The transition can be observed in the *ex libris* copied by Knod, *Aus der Bibliothek*, 62.

[132] 'Est Beati Rhynaw Schletstattini Anno 1.5.0.3. Parrhisiis. Ma[nus]. Pro[pria].'

plain vellum covers.[133] Perhaps teachers at Cardinal Lemoine arranged such course packs with the booksellers, perhaps Henri Estienne himself as he sold them out of his shop on rue St.-Jacques. Whatever the case, this codicological line of evidence places the book within the context we explored in Chapters 2 and 3.

As a student of mathematics, Beatus showed, however, rather selective assiduity: he carefully annotated the books on arithmetic (what we might call number theory), basic geometry, a little music, and the astronomy of Sacrobosco's *Sphere*—but he ignored Bovelles's short treatises on the squaring of the circle, cubing of the sphere, and perspective. That is to say, Beatus strayed little beyond the basic quadrivium reflected in university statutes.

Nevertheless, Beatus did learn the basics of computation. This is especially evident in a series of notes in his Sacrobosco, which include an extensive gloss on the rules Lefèvre gave for calculating in sexagesimal arithmetic (Figure 5.6 above), explaining precisely how to transfer remainders, a task more difficult than in decimal arithmetic. On another page, Beatus works through the calculations that yield the different thicknesses of the various spheres in the cosmos. As discussed in Chapter 3, the neat repetition, without erasures—including errors that do not affect the outcome—suggests that Beatus copied these operations as exercises in class (Figure 5.9). On the opposite page (Figure 5.10), Beatus improvised a reference table of the proportions between various units of measurement, from the inch to the mile, and he then supplied the means for converting between units.

Measurement, in the case of astronomy, required tools, and although there is no indication in his notes that Beatus actually used any such instruments, he did mark Lefèvre's explanation of 'by what clever tool philosophers take down the path' of a star, giving directions for 'measurers of the world' to take observations with an astrolabe.[134] Lefèvre's instructions at least accurately describe the instrument, and the book gestures towards the value of such objects by including De Lattes's treatise on the astronomical ring at the end of the edition Beatus owned.

For a more comprehensive sense of what mathematics meant to Beatus, we must move far beyond mere calculation. Annotations are spread throughout the Sammelband's couple of hundred folios. For us, the endpages are a rich point of entry into the notes that trace the movements of Beatus' mind. Additional pages were added during binding, where bits of text, diagrams, and calculations fill seven folio sides. On those end pages, Beatus copied many schemata and definitions, aimed at helping him recall basic terms of the quadrivium, such as a handy taxonomy of geometrical shapes from rhombus to equilateral triangle. Another synoptic memory tool is a series of hammers and harps (Figure 5.11).

[133] Examples of *Sammelbände* of these and related books from Paris that can be found still bound together in their original vellum, or with annotations that indicate that these books were read together: BnF Tolbiac, Rés-M-R-55; BnF Tolbiac RES-M-R-58; BL IB.40135 (bound with BL IB.40128). The last volume, in particular, bears the same style of red and blue rubrication that is found in some of Beatus' nicer volumes, purchased, one might suppose, directly from Henri Estienne's shop.

[134] BHS K 1046c, a6v–a7r: 'Sed quo ingenio philosophi terre ambitum deprehenderint... mundi mensores stelle notate altitudinem notarunt.'

The loops draw out harmonic relations of the Pythagorean intervals that fit within the tuning of the Dorian heptachord, a reminder that even though Beatus did not mark up his copy of the *Elementa musicalia*, he did engage music.[135] The diagram functions more as a reminder than a tool of calculation. But other figures in these endpages do heavier conceptual work, extending beyond the book's text—it is here that we can see tensions within the quadrivium, leading to fissures and fusions across disciplines. Consider a figure which we met first in Chapter 2, the 'Lullian pyramid' or cone on the very first leaf, which appears nowhere in the printed text (Figure 5.12).

The cone appears to ascend through various verbs: *Esse*, *Vivere*, *Sentire*, *Phantasiari*, *Ratiocinari*, *Intelligere*, *Mentificare*. A label links these terms to Ramon Lull, who experimented with a dynamic philosophy constructed around activity rather than Aristotelian substances, thus layering reality as verbs instead of the usual nouns.[136]

Layering the universe from matter to God was a common motif. Nicholas of Cusa used such a *conus* or circular pyramid in his *De coniecturis* as a way to structure a vision of the world as unfolding divine unity into the multiplicity of matter.[137] This was a particularly number-driven hierarchy inspired partly by Dionysius the Areopagite's categorization of the world as divine intelligible light, emanating into angels, principles, all the way down to *sensibilia*. Marsilio Ficino loosely organized the first book of his *Theologia Platonica* around this Dionysian schema; in his commentary on the Areopagite's *Hierarchia ecclesiastica*, Lefèvre himself visualized this relation of God to world as a 'pyramid, at whose point light lives inaccessible'.[138] Lefèvre even suggested it was possible to explain Aristotle's theory of substances with such figures—a pyramid, he said enigmatically, is like an 'ensouled body'.[139] The pyramid was a rich figure to think with.

The particulars of the cone in Beatus' mathematical book likely come from Charles de Bovelles. In a pair of later publications, Bovelles used the triangle as a way to describe how cognitive modes align with objects in the world. In *De sensu* Bovelles set out five cognitive modes: *sensus*, *imaginatio*, *ratio*, *intellectus*, and *mens*. These align with *mundus*, *corpus*, *anima*, *angelus*, and finally *deus*, as Bovelles clarifies in a table (Figure 5.13, left). The triangular motif emerges two pages later, making the point that just as the all the world, bodies, souls, and angels are traceable back to God—indeed ultimately *within* God—so all the powers of the senses

[135] Cf. Lefèvre, *Elementa musicalia*, book IV. For details see Barbour, *Tuning and Temperament*, chapter 1.

[136] A good introduction to this metaphysics is Charles H. Lohr, 'Mathematics and the Divine: Ramon Lull', in *Mathematics and the Divine: A Historical Study*, ed. T. Koetsier and L. Bergmans (Amsterdam: Elsevier, 2005), 211–29.

[137] Cusanus, *De coniecturis*, IV.94.

[138] Lefèvre, *Theologia vivificans*, 19r: 'ad anagogen et divinam assurrectionem divinus hic pater ignis assignat proprietates. Sed ignis ad sensibilia dumtaxat conferatur, deus autem ad universa. De igne intelligantur sensibiliter, de deo autem divine et intelligibiliter. Et he ex adiecta pyramide (cuius mucronem lux habitat inaccessa, et eius basem tenebre) deprehendi possunt'.

[139] Jacques Lefèvre d'Étaples, *Libri logicorum ad archteypos recogniti cum novis ad litteram commentariis* (Paris: Hopyl and Estienne, 1503), 23r: 'liceret ad dimensiones et figuras comparare... Corpus animatum, pyramidi'.

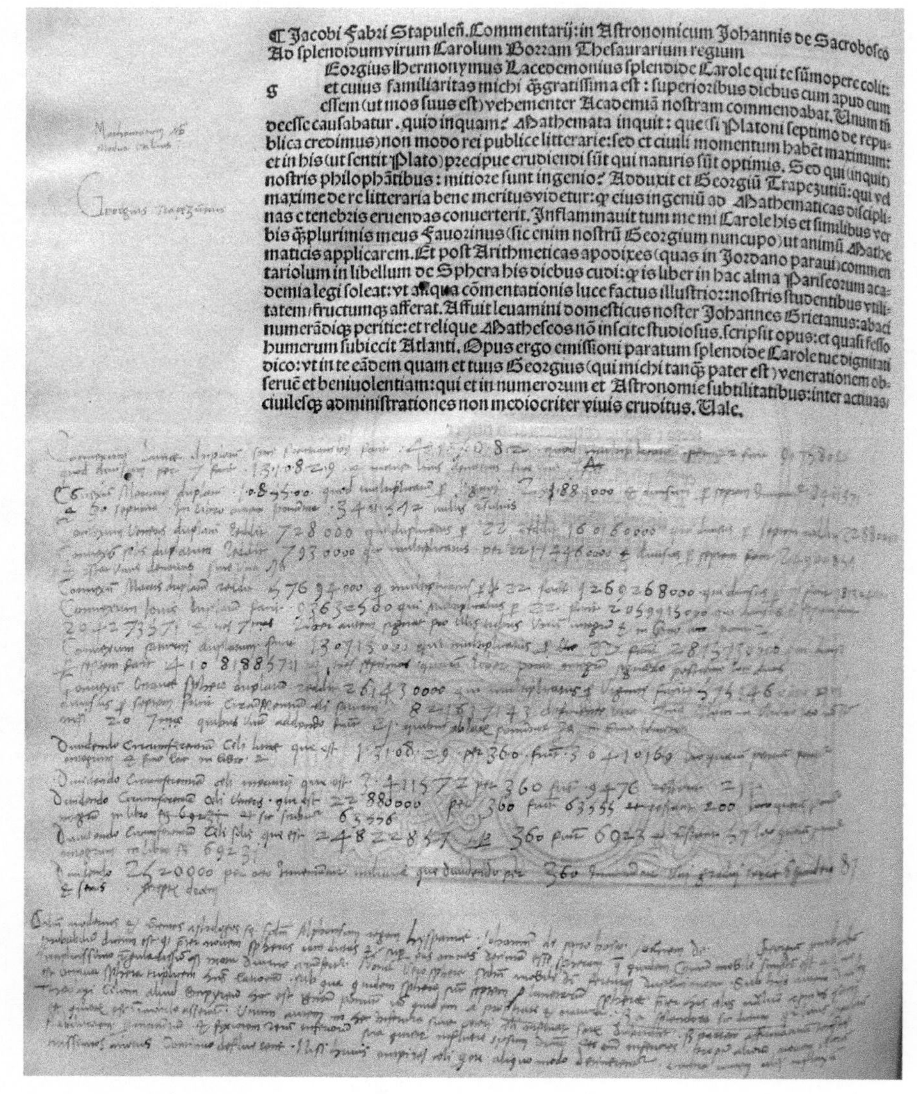

Iacobi Fabri Stapulen. Commentarij: in Astronomicum Iohannis de Sacrobosco Ad splendidum virum Carolum Borram Thesaurarium regium

Georgius Hermonymus Lacedemonius splendide Carole qui te summopere colit: et cuius familiaritas michi quam gratissima est: superioribus diebus cum apud eum essem (ut mos suus est) vehementer Academiam nostram commendabat. Unum tamen deesse causabatur. quid inquam? Mathemata inquit: que (si Platoni septimo de republica credimus) non modo rei publice litterarie: sed et civili momentum habent maximum: et in his (ut sentit Plato) precipue erudiendi sunt qui naturis sunt optimis. Sed qui (inquit) nostris philophantibus: mitiore sunt ingenio? Adduxit et Georgium Trapezuntium: qui vel maxime de re litteraria bene meritus videtur: quod eius ingenium ad Mathematicas disciplinas e tenebris eruendas convertit. Inflammavit tum me mi Carole his et similibus verbis quam plurimis meus Favorinus (sic enim nostrum Georgium nuncupo) ut animum Mathematicis applicarem. Et post Arithmeticas apodixes (quas in Iordano paravi) commentariolum in libellum de Sphera his diebus cudi: quod is liber in hac alma Parisiorum academia legi soleat: ut aliqua commentationis luce factus illustrior: nostris studentibus utilitatem fructumque afferat. Affuit levamini domesticus noster Iohannes Grietanus: abaci numerandique peritie: et relique Mathesos non inscite studiosus. scripsit opus: et quasi fesso humerum subiecit Atlanti. Opus ergo emissioni paratum splendide Carole tue dignitati dico: ut in te eadem quam et tuus Georgius (qui michi tanquam pater est) venerationem observet et benivolentiam: qui et in numerorum et Astronomie subtilitatibus: inter activas civilesque administrationes non mediocriter vivis eruditus. Vale.

Figure 5.9. Beatus Rhenanus records the performance of calculations of the distances between the various planetary spheres: on the bottom of the page, he records various 'modern' opinions on the number of these heavenly spheres—the omission of what is evidently 'Regiomontanus' in the first line suggests he missed the name during dictation and intended to fill it in later. Detail from Beatus Rhenanus' copy of Lefèvre, *Textus de sphera* (1500), BHS K 1046c, a1v.

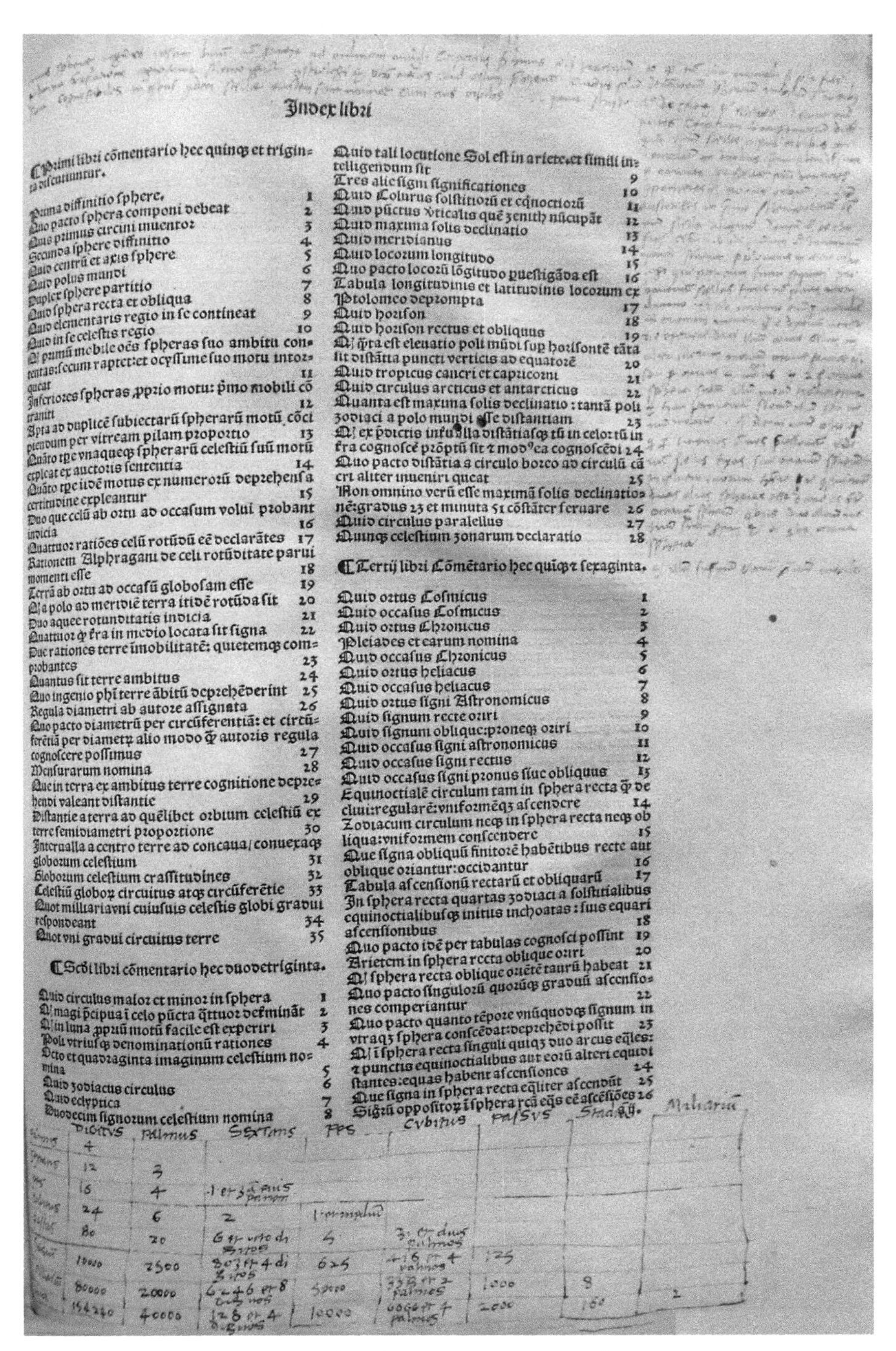

Index libri

Primi libri cōmentario hec quinqȝ et trigin-ta discutiuntur.

Secdi libri cōmentario hec duodetriginta.

Tertij libri Cōmētario hec quinqȝ et sexaginta.

	Digitus	Palmus	Sextans	Pes	Cubitus	Passus	Stadium	Miliarium
[illegible]	4							
[illegible]	12	3						
[illegible]	16	4	1 et 3a [illegible]					
[illegible]	24	6	2	1 [illegible]				
[illegible]	80	20	6 et [illegible]	5	3 et duos palmos			
[illegible]	10000	2500	803 et 4 di[illegible]	625	416 et 4 palmos	125		
[illegible]	80000	20000	6246 et 8 [illegible]	5000	3333 et 2 palmos	1000	8	
[illegible]	154240	40000	128 et 4 [illegible]	10000	6066 et 4 palmos	2000	16	2

Figure 5.10. Beatus draws a table for converting units of measurement. Detail from Beatus Rhenanus' copy of Lefèvre, *Textus de sphera* (1500), BHS K 1046c, a2r.

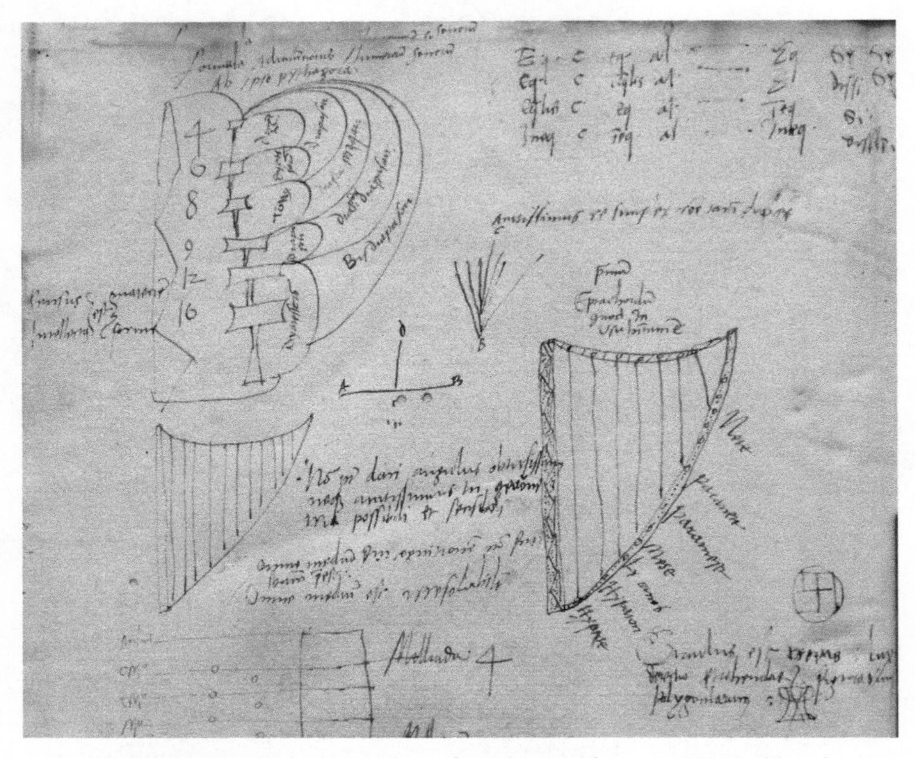

Figure 5.11. Beatus recalls the terminology of Greek music theory: BHS K 1046, end papers.

exist ultimately within the mind.[140] The figure that sums up this line of thought (Figure 5.13, right) suggests the same layers found in Beatus' notes, where the cone unites all things in God or, in its cognitive analogue, *mens*. In an introduction to natural philosophy, Bovelles dedicated a final book to mathematical analogies, in which he addressed how surfaces meet to produce a triangular pyramid, and so model fire and other physical bodies.[141]

In his mathematics textbook, the notes Beatus clustered around the pyramid offer some ways to use this 'cone coinciding in God'. First, near a triad of dots: 'a triangle is characterized by the outflow of accidents of all things, corruptible and incorruptible, from that divine monad which is God.' Below the pyramid: 'many things are philosophized from the base of a pyramid, which is a sort of arrangement and beginning of things.' More generally, 'with a pyramid we can philosophize

[140] Bovelles, *Liber de intellectu etc.*, 29v: 'Omnia sunt in mundo: omnia in corpore, omnia in anima, omnia in angelo, omnia in deo; et de horum...Omnia rursum in sensu sunt, omnia in imaginatione, omnia in ratione, omnia in intellectu, omnia in mente'. As Klinger-Dollé points out, this formulation is strongly reminiscient of Cusanus as cited above: Klinger-Dollé, *Le* De sensu *de Bovelles*, 467 n. 92. A striking omission, however, is the concept of folding that Cusanus uses to express the movement from unity to multiplicity.

[141] Charles de Bovelles, *Libri physicorum elementorum* (Paris: Josse Bade, 1512), 41r–42r.

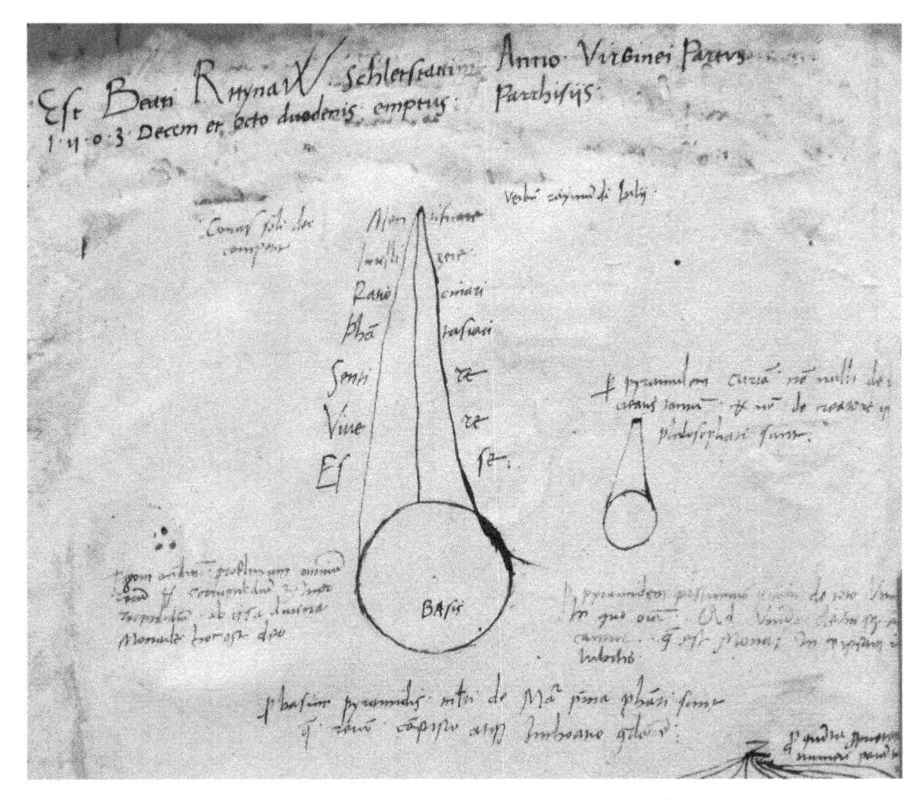

Figure 5.12. Beatus and the 'Lullian pyramid': BHS K 1046, end papers.

about the whole one on which everything is predicated—about the one, God, the monad which is divisible by itself.' A final note accompanies a smaller pyramid drawn at the right, with its tip broken off, as if to underscore that this unity is separable from everything that flows to and from it: 'through the short pyramid [i.e. one without its point] many things can be philosophized about the creation alone, and not about the creator himself.'[142] These thoughts embed a series of assumptions, held together with the idea that mathematics somehow supplies figures that help one to move between different domains of study.

But how does this work? This remarkable Lullian pyramid ranges across the apparently disparate domains of mind, nature, and the divine, and I suggest that tracing this movement illumines how Lefèvre could bring geometry into music,

[142] BHS K 1046, end paper: '[1] Trigoni accidentium profluxum omnium rerum et corruptibilium et incorruptibilium, ab ipsa divina monade hoc est deo. [2] Per basim pyramidis multi de materia prima philosophati sunt que rerum [compositio?] atque inchoatio quedam est. [3] Per pyramidem possumus philosophari de toto unum in quo omnia, ad unum, deum scilicet [predi]cantur, que est monas in se ipsum resolubilis. [4] Per pyramidem curtam non nulli de creatis tantum et non de creatore ipse philosophati sunt.' For the definition of the *curta pyramis* see 6v.

L.

mur: Deum/ Angelum/ Animam/ Corpus/ Mundum.

1	Deus
2	Angelus
3	Anima
4	Corpus
5	Mundus

¶Nam eternus/ increatus/ immensusq; deus cunctis haud immerito eminens creaturis: primū obtinet locum. Angelus vt simplicissimus/ primusq; rerum omniū productus a deo actus: secundo statuēdus est loco. Est enim & ipsi deo attiguus & proximꝰ. Porro anima licet omnium creaturarū sit vltima: tamen(vt in libro de intellectu docuimus)supremū sensibilium/ & angelis proximum sortita est locū. Est eni anima: sensibilium omniū actus & finis. Deinde anime vltimū & proximū est corpꝰ: quod illi famulatur subditurq;. quodue eādē in sublunarē hanc regionē: pegrinatum ductitat ac vehit. Extremo autem atq; imo loco/ positus est mundus: vt humani corporis circūstantia ac natiuum eius domicilium.

¶Vnde fit vt secundum totidem obiecta: distribuende sint anime cognitiones/ in singulas singulis aptatas ac proprias. 5

¶Anima internoscitat mūdū/ proprium corpus/ seipsam/ angelū ac deum. Hec autem sūt diuersissima obiecta. varie igitur sunt/ ac singulis proprie anime vires. Qua enim vi intuetur mundum/ sensitiuā nūcupamus/ seu exteriorē sensum. Qua corpus inspectat proprium: imaginatiuū/ interiorēue sensum. Qua sese conspicatur/ sibiq; fit manifesta: vocitanda est ratio. vt que anime propria est: & a qua peculiarius ipsa rationalis censetur. tanq̄ huiusmodi appellatione/ illi ab equali cognitione indita: qua neq; inferiora despicit/ neq; superiora inspectat: sed in sua sede cōtinet/ sibiq; coheret. Qua vero angelus illi aperitur/ reuelatur/ fitq; notus: vis intellectiua vocetur. Intellectus eni angelis proprius est/ vt & anime ratio. Qua deniq; deū ipsum/ quouis modo cōtēplari & speculari lubet: illi mentis portio adesse dicenda. Est enim mens ratione/ atq; intellectu superior: & consūmata dei cognitio.

Cognitiōes aie	Mens	Intellectus	Ratio	Imaginatio	Sensus
Obiecta	Deus	Angelus	Anima	Corpus	Mundus

Fit iterum vt anima sit numerus/ mēsura et equalitas oim: sitq; cūctis plens.

¶Nam quotquot sunt substantiarū genera: tot illi cognitiuas potestates inesse mostraui= 6

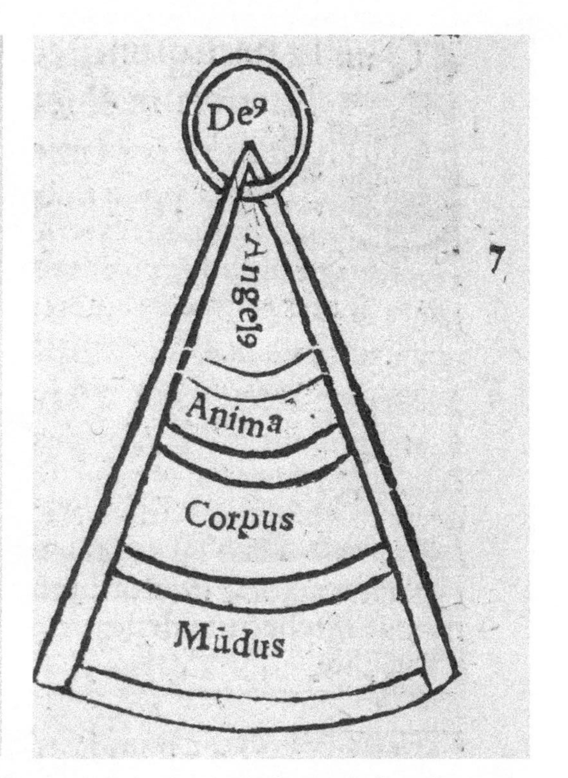

Figure 5.13. Bovelles links cognitive modes with objects in the world. Bovelles, *De intellectu, etc.* (1511), 28v. A page later, Bovelles offers a cone of the senses that matches the cone of reality, which contracts into God. Bovelles, *De intellectu, etc.* (1511), 29v. By kind permission of the Syndics of Cambridge University Library, Shelfmark: Acton.b.sel.32.

and what work such mathematical images were doing in Beatus' classroom. To explain what I think is going on, let me identify some of the techniques Beatus tries out in the margins of this volume.

These tools help Beatus to navigate a conundrum, which I should summarize first. Recall a basic commitment of these Boethian texts from our earlier section on music, the commitment to a distinction of discrete and continuous magnitudes. In this tradition, mathematics is an ontological enterprise of classification: it classifies magnitudes into certain kinds of numbers (odd/even, perfect, figurate, etc.). This ontology absolutely distinguishes discrete number from continuous magnitudes, which is the basis for distinguishing arithmetic from geometry. Thus Clichtove explained the central insight of Pythagorean arithmetic in his gloss on Lefèvre's *Introductio Boetii* that all numbers are aggregates of unities or monads. The monad is an entirely different species of number than the dyad; it is neither even nor odd, since it cannot be divided. Furthermore it is endlessly fertile; other numbers emerge when unities are set in ratio, forming the diad, the triad, and so on. The distinctions in this ontology were reinforced by techniques, for example, of ratio theory—ratios were compounded differently in music than they were in geometry, emphasizing distinctions of discrete and continuous magnitudes.[143] The first pages of Beatus' book are crammed with scribbled observations on this Pythagorean tradition, invoking Pythagoras' name,[144] and even citing Aristotle on Pythagorean number theory.[145]

This ontology, with its clear distinction of number and line, presents the conundrum, for Beatus no less than for the entire tradition of Renaissance mathematics. On the one hand, Beatus learns to make careful distinctions between disciplines on the basis of these distinctions between objects. His notes reinforce the distinctions between arithmetic, geometry, and the subaltern disciplines of music and geometry. On the other hand, however, his notes refer to authorities who give mathematics larger roles beyond the mathematical disciplines themselves. The most important authority here is Boethius himself, who suggested in the proemium to the *Arithmetica* that mathematics was important for theology. Beatus also cites Nicholas of Cusa on the 'utility of the mathematical disciplines' to buttress Clichtove's comments on how important mathematics is for studying logic, natural philosophy, moral philosophy, and particularly theology. 'Unity, as Nicholas of Cusa says, is the example of the whole exemplified multitude.'[146]

[143] In music they are added 'continuously', end to end; in geometry they can be multiplied by 'denomination'. An overview is given by Edith Dudley Sylla, 'The Origin and Fate of Thomas Bradwardine's *De Proportionibus Velocitatum in Motibus* in Relation to the History of Mathematics', in *Mechanics and Natural Philosophy Before the Scientific Revolution*, ed. Roy Laird and Sophie Roux (Dortrecht: Springer, 2007), 67–119.

[144] BHS K 1046a, 3r, inter alia: 'Pythagoras numeros impares masculos appelavit pares vero foeminos' and 'Numeri nichil aliud sunt quam explicationes unitatis.'

[145] BHS K 1046a, 3r: 'Inter numeros primos ternaris omnium primus est. Hinc ait Aristoteles in li. de anima. Ternarius est numerus utrobique primus.' BHS K 1046a end page: 'Acutissimus re simplex ratione tamen duplex. Non [inter dari] angulus obtusissimus neque acutissimus in geometria possibili et sensibili.'

[146] BHS K 1046a, 3r: 'Utilitas disciplinarum mathematicarum: Unitae ut inquit Nicolaus de Cusa est omnis exemplate multitudinis exemplar.' Cf. Cusanus, *Dialogus de ludo globi*, §64.

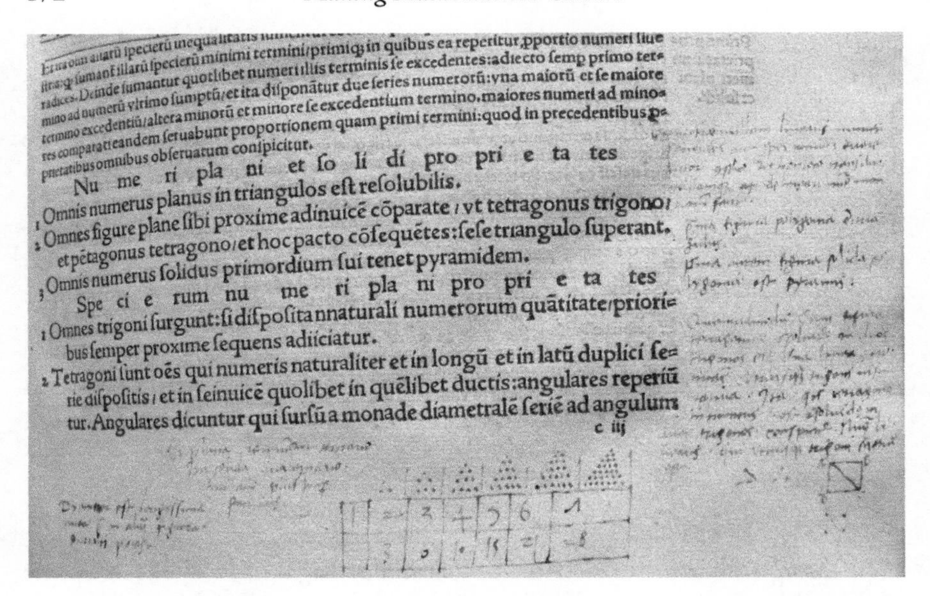

...ſpecierū inequalitatis ... in quibus ea reperitur, ꝓportio numeri ſiue
...ſumantur illarū ſpecierū minimi termini/primiq; in quibus ...
...Deinde ſumantur quotlibet numeri illis terminis ſe excedentes: adiecto ſemp primo ter...
...numerū vltimo ſumptū/et ita diſponātur due ſeries numerorū: vna maiorū et ſe maiore
...termino excedentiū/altera minorū et minore ſe excedentium termino, maiores numeri ad mino...
...comparati eandem ſeruabunt proportionem quam primi termini: quod in precedentibus p...
...omnibus obſeruatum conſpicitur.

Nu me ri pla ni et ſo li di pro pri e ta tes

1 Omnis numerus planus in triangulos eſt reſolubilis.
2 Omnes figure plane ſibi proxime adinuicē cōparate/vt tetragonus trigono/ et pētagonus tetragono/et hoc pacto cōſequētes: ſeſe triangulo ſuperant.
3 Omnis numerus ſolidus primordium ſui tenet pyramidem.

Spe ci e rum nu me ri pla ni pro pri e ta tes

1 Omnes trigoni ſurgunt: ſi diſpoſita naturali numerorum quātitate/prioribus ſemper proxime ſequens adiiciatur.
2 Tetragoni ſunt oēs qui numeris naturaliter et in longū et in latū duplici ſerie diſpoſitis/ et in ſeinuicē quolibet in quēlibet ductis: angulares reperiūtur. Angulares dicuntur qui ſurſū a monade diametralē ſeriē ad angulum

c iij

Figure 5.14. Beatus ruminating on a series of triangular numbers. BHS K 1046a, 20r.

These two kinds of notes show Beatus and his teachers balanced between policing boundaries and crossing them.

The conundrum has a visual solution. Beatus learns three techniques for crossing such boundaries, first crossing from arithmetical reasoning to geometry, and on to natural philosophy and theology. The first of these techniques involves figurate numbers. One of the major classes of numbers in Pythagorean number theory are 'figurate' numbers, numbers that can be arranged into a regular shape. Three pebbles or dots, for example, make a triangle, so three is a 'triangular' number. By adding three dots to one side, we can make 6, the next triangular number. On one page Beatus explores the series of triangular numbers: 3, 6, 10, 15, 21, 28 (etc.) (Figure 5.14).

Such series can have fascinating properties. We could, with Beatus, add up the differences between each member of the series, which produce a second series of integers: 1, 2, 3, 4, 5, 6, 7 (3 is 2 more than 1, 6 is 3 more than 3, 10 is 4 more than 6, and so on). Such properties strike the eye when lined up in tables, which Beatus does elsewhere in the book. Next to the Lullian pyramid he sets out two series: first the odd integers, then below them their doubles (Figure 5.15). The table tests one of Lefèvre's claims about the class of 'evenly odd' (i.e. odd numbers multiplied by evens): 'Of all evenly odd numbers arranged in a continuous series, when two intermediate numbers are joined from some neighbouring number on each side, and those numbers are likewise collected into one [i.e. brought to unity], then necessarily the collected numbers equal the same number.'[147] This contorted

[147] *Epitome Boetii*, 12v: 'Omnium numerorum pariter imparium pari continuaque serie dispositorum, duas medietates simul iunctas suis proxime altrinsecis numeris et illis qui super illos sunt usque ad unitatem, simul in unumque collectis equari necesse est.'

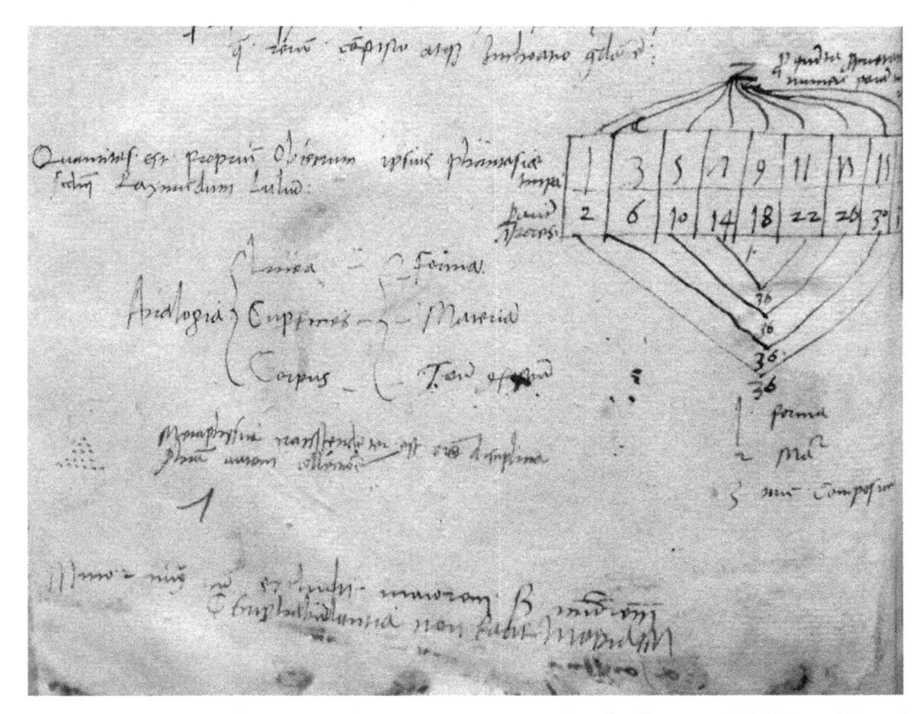

Figure 5.15. Beatus brings together series to test a rule of inference. BHS K 1046, end page, detail. (This is the bottom of the page shown above in Figure 5.12.)

proposition culminates a sequence of propositions about the properties of evenly odd numbers, so it is somewhat easier to understand in context. But visualizing was even easier.

In this series, Beatus experiments in a way that does not quite correspond to any figure in the text. The series of odd numbers (1, 3, 5, etc.), multiplied by 2, produces an evenly odd series (2, 6, 10, 14, 18, 22, etc.). Then he picks a central reference point (in this case 18, which doubles to 36), and adds the numbers equidistant on either side. Thus 14 and 22 equal 36. And so do the next numbers: 10 and 26 is 36. The striking insight is that the same equivalence holds true no matter where in the series one starts the procedure—Clichtove's commentary makes this clear with other such visual tables of series.

Such visual techniques are what make these texts persuasive, precisely because of a feature described in Chapter 4: none of them have proofs—not Lefèvre's *Introductio Boetii*, not Clichtove's commentary, and not Bovelles's various introductions to geometry. Lefèvre simply narrates the propositions; Clichtove explains them in expanded prose and with multiple examples, introducing some tables; Bovelles supplies only enunciations and diagrams. Therefore the power of these propositions to persuade is either through authority—Lefèvre indexes these propositions to sources in Jordanus and Boethius, which Clichtove reinforces with occasional citations—or through the reader's own experience. The propositions use the language of universality and necessity, but without proof; their demonstration is left to the reader's actual experience. How do we find rules for constructing perfect numbers, the class of numbers equal to the sum of their divisors? We construct a series of the ones we know and see if we can infer new properties.[148] Beatus infers the general truth of the propositions from mathematical operations on series whose truth he must *do* and *see* to understand.

This reliance on physical inspection—literally, intuition—gets to a problem that has been at the heart of mathematics since the ancient Greek philosophers. To what extent do mathematical forms relate to the messy matter of nature? Beatus' notes show him wavering on this point, committed to the distinction of physics from mathematics, yet drawn to the idea that mathematics could still model reality in some sense.

The second technique I describe here is haptic, allowing a leap from matter to mathematics and back to matter: drawing a pen's inky point across paper to make a line. In the text on geometry, Charles de Bovelles describes how magnitude is produced by a point's movement: '[a point] does not produce (*gignere*) any magnitude out of its essence or from that which it is in itself, but only by its proper power or its motion.'[149] In the margin, Beatus supplied a gloss: 'according to the intellectual philosophy of the Pythagoreans, this dictum is one in which a line is imagined from the flow of a fertile point. However, a physicist will never allow this.'[150] So by adducing the terminology of a 'flowing point' Beatus explicitly recognizes that he is pushing into areas forbidden in Aristotelian natural philosophy, yet the bulk of his notes commit him to this view of the 'Pythagoreans'. Who are they? The terminology of the flowing point is striking, and it is tempting to find an echo of Proclus, who supplied the un-Euclidean definition of a line as flowing point in his commentary on the first book of Euclid's *Elements*. There is a small possibility that Beatus, Bovelles, or some other teacher at Lemoine encountered Proclus in Giorgio Valla's *De expetendis et fugiendis rebus* (Venice, 1501), which contained large sections of the commentary on Euclid.[151] But usually Lefèvre and his circle identified 'Pythagoreans' with Boethius and his followers, and Beatus' notes often set Nicholas of Cusa as a representative of this Pythagorean tradition. In fact, Cusanus is clearly behind other notes in the same section of Bovelles's text, where the point is fertile as the centre and so the origin of the circle in a process of unfolding. 'According to the intellectual method of doing philosophy, a centre is a

[148] This is the method Bovelles uses in his study of perfect numbers in *Liber de intellectu etc.*, as I discuss in Oosterhoff, 'Connaissance visuelle comme méthode élémentaire chez Charles de Bovelles', in *Bovelles philosophe et pédagogue*, ed. Anne-Hélène Klinger-Dollé, Emmanuel Faye, and Jocelyne Sfez (Paris: Éditions Beauchesne, forthcoming).

[149] BHS K 1046a, 55r–v. This is one in a list of the properties of a point: '5. Ipsum non sui ipsius essentia sive eo ipso quod est, sed propria solum virtute suove motu, omnem de se magnitudinem gignit.'

[150] BHS K 1046a, 55r: 'Secundum intellectualem pythagoricorum philosophiam hoc dictum est qui ex fluxu puncti lineam procreanti imaginantur. Non tamen hoc physicus unquam admittit.'

[151] Axworthy, for instance, has shown how Oronce Fine adapted sections of Valla's translation of Proclus: Angela Axworthy, *Le Mathématicien renaissant et son savoir: le statut des mathématiques selon Oronce Fine* (Paris: Classiques Garnier, 2016), 58–69.

circle enfolded; and a circle is a centre *un*folded.'[152] In this way, the point is the origin and even the equivalent of the circle—a Cusan equivalence or coincidence of opposites. The theme becomes even more explicit in the endpapers—though still attributed to Pythagoras for instance when he writes 'The Pythagorean definition of God: God is a homocentric sphere whose centre is everywhere and the circumference is nowhere'—a statement famously advanced by Cusanus.[153]

The theological lessons of the line as a flowing point assume a specimen performance, in which the student performs the technique with his own hand. The starting point was Beatus' experience of construction in the matter of the page. As Beatus wrote in the margin of Bovelles's introduction to geometry: 'you will place one foot of the compass on some point and put the other on the diameter, working it around from beginning to end. You will see the point in the centre.'[154] Within the precisely penned circle, an indentation in the paper marks the centre; this circle has been drawn with a compass, unlike several other circles Beatus drew in the margins—such as a hastily scrawled, squished circle a few lines above—as if Beatus were just learning to use the instrument. The direct verbal construction ('you will place one foot of the compass...') recalls not Euclid's geometry, or even the Latin introduction of Beatus' textbook, which loftily ignored the instruments needed to enflesh the conceptual structures: 'let there be a circle o...'. Instead, such language echoes the rules for material construction found in medieval practical geometries.[155]

In the endpapers, the theological lessons continue to prod Beatus to practice. Next to circles that still show the deep prick of the compass foot at the centre are theological maxims: 'Felicity is a return to one's origin'; 'The figure of a circle is the most proper...definition of God'; 'Only man returns on himself. Only man apprehends external things.'[156] The experience of actually constructing a circle lends new force to the ancient account of the liberal arts as an *encyclos paedeia*, a

[152] K 1046a, 56r: 'Centrum, secundum intellectualem philosophandi modum, circulus est complicatus; circulus autem centrum explicatum.'

[153] BHS K 1046, back endpaper: 'Diffinitio dei pythagorica: Deus est sphera \[homocentris?]/ cuius centrum est ubique et circumferentia nusquam.' Cf. Cusanus, *De docta ignorantia* I.23, picking up an the topos *deus est sphaera infinita cuius centrum est ubique, circumferentia nusquam*, first found in the *Liber XXIV philosophorum*, now associated with the fourth-century Christian Marius Victorinus, but credited to Hermes Trismegistus throughout the medieval period. See Françoise Hudry (ed.), *Le livre des XXIV philosophes: résurgence d'un texte du IVe siècle* (Paris: Vrin, 2009), 7–8, 158. It circulated diffusely through Platonic thinkers, notably Alan of Lille, Vincent de Beauvais, and Meister Eckhart.

[154] BHS K 1046a, 56r: 'Super quantumque punctum [efficantem] initiantem aut finemque, unum circini pedem fixum posueris, et aliam fac in diametrum. Viduri punctum in centrum.'

[155] See Stephen K. Victor, *Practical Geometry in the High Middle Ages, Artis Cuiuslibet Consummatio and the Pratike de Geometrie* (Philadelphia, PA: American Philosophical Society, 1979). Bovelles adopts this language in his practical geometry: Charles de Bovelles, *Géométrie en françoys. Cy commence le Livre de l'art et Science de Geometrie : avecques les figures sur chascune rigle au long declarees par lesquelles on peut entendre et facillement comprendre ledit art et science de Geometrie* (Paris: Henri Estienne, 1511).

[156] BHS K 1046, endpaper: 'Felicitas regressus est uniusquoque suam principiam. [*circle*] Figura circularis dignissima est. [*circle, in which is written:*] deus diffinitio...Solus homo redit in seipsum [*image of circle, with four points, one at each quarter*] solum homo apprehendit externitates'.

circle of learning: as Beatus noted, humane disciplines are the circle which in their flow and return always refer to the one point at the centre—divine beatitude.

These theological lessons enfold assumptions about mathematics and physics too. On the same endpaper, Beatus repeated and expanded the point about the flowing line: 'Those who philosophized intellectually say that the point relates to the line in the same way as the unit to number, so that they say a line is known from a flowing point (and a surface is from a flowing line). A physicist should deny this absolutely.'[157] This note is partly explained by a schema below it, which visually correlates motion, time, and magnitude (*motus*, *tempus*, *magnitudo*) of natural philosophy with the change, instant, and point (*mutatum esse*, *nunc*, *punctus*) of the drawn line. Statements of this kind are careful. The claim is not that a point *is* a flowing line, or that the series of numbers is *equivalent* to a line. Naming the *physicus* shows that natural philosophy is in mind here, and Beatus is quite right to say that an orthodox Aristotelian natural philosopher will find this nonsense. Yet the overall point is that there is *some* meaningful similarity between arithmetic and geometry, and somehow, natural philosophers notwithstanding, the flowing mathematical point produces surfaces, a point made all the more persuasive in Beatus' manipulation of pen and compass on the surface of the page.

The third technique I find in Beatus' margins aims to help the mind's eye identify this meaningful similarity between different domains: *analogiae* arranged in brackets. Such brackets pervade Beatus' notes, their sheer number begging for explanation. In Chapter 3, we saw that Beatus explicitly presented analogy as an art of arts rooted in number.[158] We encountered the analogy Beatus drew between grammar and the 'true political philosophy' in Aristotle's *Politica* (Figure 3.5) so that political philosophy bears 'an analogy, that is a ratio and similitude, to grammar'.[159] Here, in his mathematical book, we see efforts to apply this in different ways, such as on folio 81v, where he uses long sweeping brackets to set up relations between different mathematical objects (Figure 5.16).

Such brackets were traditional ways of visualizing certain kinds of ratios—harmonic proportions in music, as we saw above.[160] But this diagram attempts to do more, dissecting more finely the kinds of sameness and difference between different objects: Beatus highlights the 'dissimilar' relations between lines, surfaces, and bodies, while establishing 'similar proportions' between some constitutuent parts, such as straight lines, planes, and angular bodies (all have straight edges). As a conceptual technique, the analogy functions like the tables and series discussed above; it sets related objects next to one another, allowing inferences from one class into another.

These three techniques—conceptualizing analogy, drawing points that flow into lines, and visualizing series—all help explain why the Lullian pyramid was such a powerful figure for Beatus and his teachers. A particularly telling *analogia* is on the

[157] BHS K 1046, end page. See Chapter 3, note 106. [158] pp. 79–85.

[159] BHS K 1046, end page: 'per Analogiam quoque id est proportionem ac similitudinem in Grammatica.'

[160] On this form of analogy in the twelfth century, see Somfai, 'Calcidius' Commentary'; for its use in music, see e.g. Boethius, *Institutio musica*, II.25 (trans. Bower, 80).

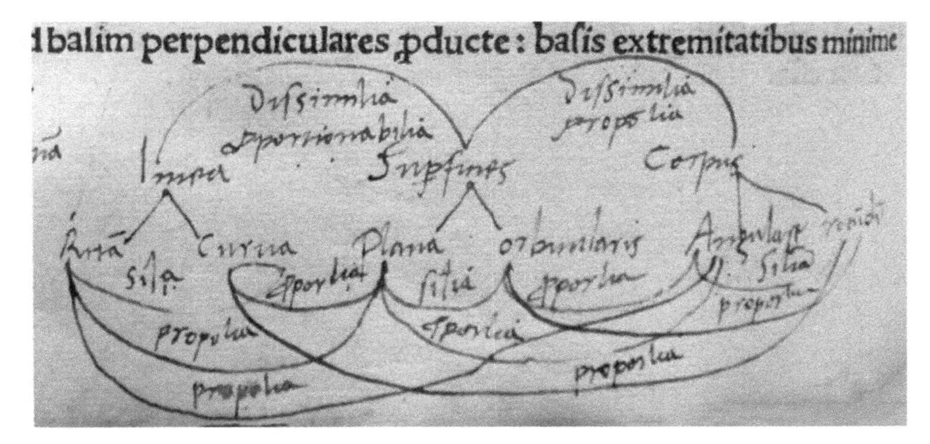

Figure 5.16. Beatus classifies forms of 'proportionality' between lines, surfaces, and bodies. Lefèvre, Clichtove, Bovelles, *Epitome Boetii, etc.* (1503), BHS K 1046a, 81v.

endpaper, just below the Lullian pyramid and to the left of the series of triangular numbers (Figure 5.15). On the left side, brackets line up a series of objects: line, surface, body. On the right side, the analogy is: form, matter, and the *totum compositum*. When we turn to the text, we find Beatus also drawing various pyramids on folio 6v. There too is an analogy of line, surface, and body (Figure 5.17, top left margin). But now the basis is a set of numbers: linear, planar, and solid numbers. In this portion of the text, Lefèvre has set out the various types of figurate numbers. In the margins Beatus draws out, as number theorists had done since antiquity, this Pythagorean taxonomy in dots arranged in patterns. Thus the first linear number is the dyad, since a line between two dots is the simplest, minimal line. The first planar number is the triad; lines between three dots make a triangular surface. But the next category of 'solid' numbers is founded on the pyramid. A dot above that triangle delineates the most fundamental solid. With these principles, though, one can add dots in various patterns. But note that so far as number theory is concerned, this is all arithmetical. Dots are not geometrical points, but represent arithmetical units: the monad, the dyad, the triad, and so on. To shift into the geometry of lines, surfaces, and bodies requires an imaginative move, a move primed by asking one to think of the dyad as a 'linear' number.

In the text, then, Beatus applies this pattern of thought to the form of the pyramid (Figure 5.12). A pyramid is constructed of layers of surfaces, stacked up. By making the same mental leap that allows points to flow into lines, and a series to support a general rule, these grids of dots become bodies. The pyramid therefore captures within itself all the various structures, from the tip of the cone (monad), including the surfaces stacked on the base, to the composite whole.

In Beatus' endpages, we see him toying with this imaginative move, as his pen hovers around the Lullian image of the pyramid. Below it, the *analogia* moves from geometrical forms to material objects; the geometrical body is analogous to the composite whole. Beside it, Beatus examines the series of evenly odd numbers.

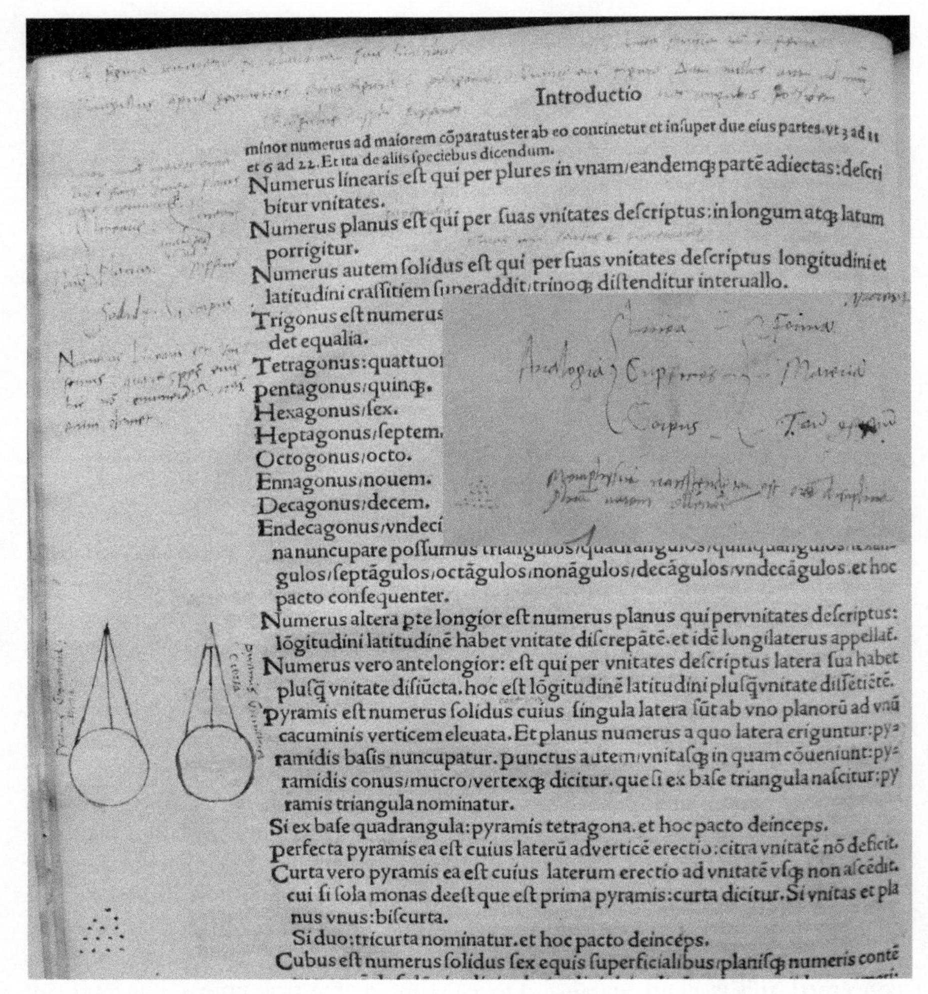
Introductio

minor numerus ad maiorem cõparatus ter ab eo continetur et insuper due eius partes.vt 3 ad 11 et 6 ad 22.Et ita de aliis speciebus dicendum.

Numerus linearis est qui per plures in vnam/eandemq3 partẽ adiectas:descri bitur vnitates.

Numerus planus est qui per suas vnitates descriptus:in longum atq3 latum porrigitur.

Numerus autem solidus est qui per suas vnitates descriptus longitudini et latitudini crassitiem superaddit/trinoq3 distenditur interuallo.

Trigonus est numerus

det equalia.

Tetragonus:quattuo

pentagonus/quinq3.

Hexagonus/sex.

Heptagonus/septem.

Octogonus/octo.

Ennagonus/nouem.

Decagonus/decem.

Endecagonus/vndeci

na nuncupare possumus triangulos/quadrangulos/quinquangulos/hexan gulos/septãgulos/octãgulos/nonãgulos/decãgulos/vndecãgulos.et hoc pacto consequenter.

Numerus altera pte longior est numerus planus qui per vnitates descriptus: lõgitudini latitudinẽ habet vnitate discrepãtẽ.et idẽ longilaterus appellat.

Numerus vero antelongior: est qui per vnitates descriptus latera sua habet plusq̃ vnitate disiũcta.hoc est lõgitudinẽ latitudini plusq̃ vnitate differẽtẽ.

Pyramis est numerus solidus cuius singula latera sũt ab vno planorũ ad vnũ cacuminis verticem eleuata.Et planus numerus a quo latera eriguntur:py ramidis basis nuncupatur.punctus autem/vnitasq3 in quam cõueniunt:py ramidis conus/mucro/vertexq3 dicitur.que si ex base triangula nascitur:py ramis triangula nominatur.

Si ex base quadrangula:pyramis tetragona.et hoc pacto deinceps.

perfecta pyramis ea est cuius laterũ ad verticẽ erectio:citra vnitatẽ nõ deficit.

Curta vero pyramis ea est cuius laterum erectio ad vnitatẽ vsq3 non ascẽdit. cui si sola monas deest que est prima pyramis:curta dicitur.Si vnitas et pla nus vnus:biscurta.

Si duo:tricurta nominatur.et hoc pacto deinceps.

Cubus est numerus solidus sex equis superficialibus/planisq3 numeris contẽ

Figure 5.17. The point of analogy in Beatus' notes is to justify inferences from one domain to another. BHS K 1046a, 6v, detail of a schema between figurate numbers and geometrical forms. *Inset*: BHS K 1046, end page, detail of the schema for an analogy between geometrical forms and material objects.

And the pyramid itself becomes the emblem for moving between different modes of cognition, both within the mind (the various cognitive verbs) and without (as different cognitive powers mirror different layers of the world). In this pyramid, analogies, flowing points, and series of numbers thus become techniques for crossing disciplinary boundaries, transforming mathematics into a total picture of reality.

* * *

A cosmographer sits at the centre of a city with five gates. Through those gates messengers arrive from all the corners of the world, delivering accounts of their experience. One brings colour, another taste, and the others bring sounds, images, and the feeling of things. Nicholas of Cusa offered this analogy for the cooperation of the mind and sense in knowing the world. As the *homo cosmographus* jots down what is brought to him, he begins to build up a *mappa*. If one of the five gates were blocked, Nicholas observes, the cosmographer's description would be deficient, and therefore 'he strives as hard as he can to keep all the gates open, to continually listen for the news of new messengers, always making his description truer'.[161] Having completed his map, the cosmographer then turns to the *artifex et causa* of the sensible world, God. Beatus Rhenanus, perhaps reading this passage in Paris, observed here the 'human inclination to its own perfection'.[162] As I have argued in this chapter, Lefèvre cultivated mathematics for other-worldly ends, but by this-worldly means—the Fabrists explored the pedagogical possibilities of print because their scholarly vocation urged them to pause over the idea of tracing the invisible things of God in the visible world. They aimed at the intellectual vision of God, attained through the complete cycle of knowledge rooted in the senses.

This other-worldly empiricism heightened a sensitivity to the necessity and limits of physical experience in manipulating mathematical objects. The visual programmes of Fabrist books were inventive, recombining old elements to make new wholes—a material performance of the practices of introduction I explored in the last chapter. Indeed, these practices of recombining, reintroducing, teaching, and commenting on the medieval quadrivium opened up possibilities of learning new skills and crossing old conceptual boundaries. In the *Textus de sphera*, a qualitative astronomy took on the tasks of a calculating cosmography; a medieval astronomy handbook transformed into a primer in the mathematical techniques required to map the earth. In the *Elementa musicalia*, a traditional study of music theory demanded the reader to attend to the sensory problems of practice, using that experience to justify a disciplinary shuffle: what started out a branch of arithmetic now began to include geometry.

All of these cues were not in vain, as Beatus Rhenanus' reading shows. An education such as this did not make mathematical practitioners—but it *did* cultivate the vision of a new culture of learning in which mathematics held the promise of untold wisdom and practical power. Beatus exercised habits of reading that put pressure on mathematical domains, contributing to the long-term reshaping of mathematical disciplines, such as the distinction of number and line, or the practice of compounding ratios. Such reading could also reconfigure how mathematics related to other disciplines. The next chapter turns to this question in natural philosophy.

[161] Nicholas of Cusa, *Opuscula theologica et mathematica* (Strassburg: Martin Flasch, 1488), v3r: 'studet igitur omni conatu omnes portas habere apertas et continue audire novorum semper nunciorum relationes, et descriptionem suam semper veriorem facere.'

[162] Nicholas of Cusa, *Opuscula*, BHS K 951, *in margine*: 'inclinatio humana ad sui perfectionem'.

6

The Mathematical Principles of Natural Philosophy

The biggest shift in early modern philosophy was arguably the collapse of a view of nature as a teeming realm of active Aristotelian substances. Where Aristotelian hylomorphism had once reigned, a range of alternatives thrived, from various versions of atomism to quasi-teleological frameworks of seeds and inborn principles.[1] But from Descartes to Laplace such alternatives increasingly involved mathematical and mechanical explanations. Where Aristotle had limited mathematics to 'formal' causes, the new philosophies made extension an efficient cause too. Aristotelian hylomorphism collapsed into various versions of the intuition that substances are simply extended mathematical objects whose best analogy is machines.

How do we account for this shift? How do we get from the rich substance metaphysics of medieval school philosophy to the taste for mathematical proof underlying Descartes's spare metaphysics of extension, or to Galileo's optimism that machines and mathematical models are interchangeable? A once popular answer was that *novatores* such as Descartes and Galileo simply preferred Platonic idealization to Aristotelian abstraction.[2] Newly discovered Platonic sources such as Proclus' *Commentary on the First Book of Euclid* offered a powerful view of the imagination and its ability to divine a *mathesis universalis* in the world. Yet this account of the 'mathematization of the world' has largely fallen out of favour because we have changed our understanding of the textual and material culture in which mathematical disciplines were practised. First, we no longer see Aristotelianism as a monolith, but detect early modern natural philosophers working out all manner of innovative physics under the broad banner of Aristotle.[3] Under scrutiny, Aristotelianism now

[1] Christoph Lüthy, 'The Fourfold Democritus on the Stage of Early Modern Science', *Isis* 91, no. 3 (2000): 443–79; Hiro Hirai, *Le Concept de semence dans les théories de la matière à la Renaissance: De Marsile Ficin à Pierre Gassendi* (Turnhout: Brepols, 2005); Dmitri Levitin, *Ancient Wisdom in the Age of the New Science: Histories of Philosophy in England, c. 1640–1700* (Cambridge: Cambridge University Press, 2015), chapter 5.

[2] Classic studies include Edwin Arthur Burtt, *The Metaphysical Foundations of Modern Science* (1924; 2nd edn, London: Routledge & Kegan Paul, 1932); Ernst Cassirer, 'The Relationship between Metaphysics and Scientific Method', in *Galileo, Man of Science*, ed. Ernan McMullin, trans. Edward W. Strong (New York: Basic Books, 1967), 338–51; Alexandre Koyré, *Metaphysics and Measurement: Essays in Scientific Revolution* (Cambridge, MA: Harvard University Press, 1968). For an alternative account of this historiography, see Sophie Roux, 'Forms of Mathematization (14th–17th Centuries)', *Early Science and Medicine* 15 (2010): 319–37.

[3] Pioneering studies include Charles B. Schmitt, *Aristotle and the Renaissance* (Cambridge, MA: Harvard University Press, 1983); F. Edward Cranz and Charles B. Schmitt, *A Bibliography of Aristotle Editions 1501–1600* (2nd edn, Baden-Baden: Koerner, 1984).

looks capacious and no longer supports the hoary distinction of (Platonist) humanism and (Aristotelian) scholasticism. When we actually examine particular works of early modern natural philosophy, we find that Jesuits and Lutherans fostered rich mathematical cultures within broadly Aristotelian frameworks.[4] Second, we see more clearly how mathematical learning shot through the ever richer material culture of early modern science. Objects and tools such as magnets and telescopes spurred new kinds of thinking and making, and the rise of calculating merchants and measuring artisans made space for the middling printers, artists, architects, and engineers who shared a culture as mathematical practitioners.[5] The star case here is often Galileo, whose intellectual credit grew with his ability to fabricate instruments and to think with objects.[6] Like other practitioners, he approached making books as an extension of the manual crafts. Indeed, heirs to Lefèvre's project such as Oronce Fine, Peter Ramus, and their students perfectly represent a culture of mathematics as practical craft as much as bookish discipline.[7]

Lefèvre, Bovelles, and their circle illuminate both accounts, since they wrote on Aristotelian natural philosophy as well as on mathematics and mathematical instruments. In this chapter I will suggest a convergence between these two approaches to knowing nature, the philosophical and the practical. On the one hand, these scholars reread Aristotle in a way that put special emphasis on the 'principles' of number in the world and in the mind. On the other hand, their reinterpretation gave new attention to practices of making, describing the mind's operations as creative forms of measuring. Together, these positions made it possible to see mathematical objects as operating causally, tied to real change in the natural world.

FRIENDSHIP AND PHYSICS

Lefèvre's most sophisticated physics occurs in dialogues often taken to support his reputation as a humanist heroically taking on the barbarous scholastics. He wrote four dialogues on metaphysics and another two on physics.[8] He was associated with the genre later in the sixteenth century, when humanists wrote dialogues

[4] See Chapter 1, notes 6, 14, 22.

[5] The classic formulation here is Edgar Zilsel, 'The Sociological Roots of Science', *American Journal of Sociology* 47 (1942): 544–62. See scholarship cited by Henrique Leitão and Antonio Sánchez, 'Zilsel's Thesis, Maritime Culture, and Iberian Science in Early Modern Europe', *Journal of the History of Ideas* 78, no. 2 (2017): 191–210. A fertile statement of the question remains Thomas S. Kuhn, 'Mathematical vs. Experimental Traditions in the Development of Physical Science', *Journal of Interdisciplinary History* 7, no. 1 (1976): 1–31.

[6] Mario Biagioli, *Galileo, Courtier: The Practice of Science in the Culture of Absolutism* (Chicago: University of Chicago Press, 1993); Mario Biagioli, *Galileo's Instruments of Credit: Telescopes, Images, Secrecy* (Chicago: University of Chicago Press, 2006); Domenico Bertoloni Meli, *Thinking with Objects. The Transformation of Mechanics in the Seventeenth Century* (Baltimore, MD: Johns Hopkins University Press, 2006); Matteo Valleriani, *Galileo Engineer* (Dordrecht: Springer, 2010).

[7] See literature in the epilogue, Chapter 7, at notes 31, 32.

[8] The dialogues on physics are appended to Lefèvre, *Totius Aristotelis philosophiae naturalis paraphrases* (1492); those on metaphysics were written earlier (see Chapter 4) but printed first in Lefèvre, *Introductio in metaphysica* (1494). Both are published with later editions of the natural philosophical

between Lefèvre, Clichtove, and a student, discussing moral philosophy.[9] It seems puzzling, then, that Lefèvre wrote his most advanced dialogue on one of the schoolmen's most abstract topics: the quantification of qualities, also known as the latitude of forms. The dialogue offers no novel insights, and only at the end makes a single sideways reference to the mean-speed theorem that is the most famous inheritance of this tradition.[10] Nevertheless, the dialogue insistently explores how mathematics applies to physical change.

The latitude of forms had been central to the fourteenth-century philosophical culture that Jean Gerson and others university reformers had warned against. Edith Sylla has stressed that the doctrines of the fourteenth-century 'Oxford Calculators' crystallized within undergraduate practices of disputation known as *sophismata*.[11] Students were expected to unknot logical puzzles, thinking sharply, quickly, on their feet. They found mathematical techniques a useful supplement to arguments in logic, as a way to use proportions to reason about physical change. In turn, this mathematical mode of reasoning was applied in other domains: certainly in logic, but also in theology and natural philosophy or physics.[12] By the fifteenth century, calculatory techniques were widely used in European universities for thinking about 'degrees and latitudes' of physical change. As a genre of analysis, the latitude of forms was explicitly theorized as a mixed or 'middle science' of mathematics, of the sort discussed in the previous chapter.[13]

introductions, beginning with Jacques Lefèvre d'Étaples and Josse Clichtove, *Totius philosophiae naturalis paraphrases, adiecto commentario* (Paris: Henri Estienne, 1502).

[9] The first dialogue to include Lefèvre himself is Champier, *Ianua logice et physice* (1498). Lefèvre's student Alain de Varennes wrote such a dialogue in his *De amore dialogus, de luce dialogi, etc.* (Paris: Henri Estienne, 1512). So did Giuli Landi, decades later in Italy: Luca Bianchi, 'From Lefèvre d'Etaples to Giulio Landi: Uses of the Dialogue in Renaissance Aristotelianism', in *Humanism and Early Modern Philosophy*, ed. Jill Kraye and M. W. F. Stone (London: Routledge, 2000), 41–58. See also E. Garin, 'Echi italiani de Erasmo e Lefèvre d'Étaples', *Rivista critica di storia della filosofia* 24 (1971): 88–90 and F. Lenzi, 'I dialoghi morali e religosi di Giulio Landi, Lefèvre d'Étaples ed Erasmo', *Memorie domenicane* 4 (1973): 196–216.

[10] Literature on the mean speed theorem is large. A helpful entry is John E. Murdoch and Edith Dudley Sylla, 'The Science of Motion', in *Science in the Middle Ages*, ed. David C. Lindberg (Chicago: University of Chicago Press, 1978), 206–66.

[11] Edith Dudley Sylla, 'The Oxford Calculators', in *The Cambridge History of Later Medieval Philosophy*, ed. Norman Kretzmann, Anthony Kenny, and Jan Pinborg (New York: Cambridge University Press, 1982), 542–7; Edith Dudley Sylla, 'The Oxford Calculators in Context', *Science in Context* 1 (1987): 257–79.

[12] On this broader trend, see John E. Murdoch, '*Mathesis in Philosophiam Scholasticam Introducta*: The Rise and Development of the Application of Mathematics in Fourteenth Century Philosophy and Theology', in *Arts libéraux et philosophie au Moyen Âge: Actes du quatrième congrès international de philosophie médiévale* (Paris: Vrin, 1969), 215–54.

[13] For the German-speaking lands, see now Daniel A. Di Liscia, *Zwischen Geometrie und Naturphilosophie: die Entwicklung der Formlatitudenlehre im deutschen Sprachraum* (Munich: Universitätsbibliothek, 2002), chapter 2. In this chapter, I draw extensively on the work of Di Liscia and Edith Sylla, as well as Curtis Wilson, *William Heytesbury: Medieval Logic and the Rise of Mathematical Physics* (Madison, WI: University of Wisconsin Press, 1956). For a valuable survey of the historiography, see John E. Murdoch, 'The Medieval and Renaissance Tradition of Minima Naturalia', in *Late Medieval and Early Modern Corpuscular Matter Theories*, ed. Christoph H. Lüthy, John E. Murdoch, and William Newman (Leiden: Brill, 2001), 91–132.

University critics deplored the habits students learned by disputing sophisms, as we saw in Chapter 2.[14] Lefèvre compared practitioners of such disputes to Euthydemus and Dionysius, the sophists that Plato had lampooned in the *Euthydemus*.[15] Such perverse philosophizing drew from the mild-mannered Picard the strong language that would become common among early sixteenth-century humanists such as Erasmus, Vives, and Rabelais: 'To weave useless speeches and sophistic machinations against such high and sublime teachings seems more laughable than the fact that the earth once upon a time produced a ridiculous little ape in contempt of the gods.'[16] Lefèvre, who freely credited friends but rarely called out enemies by name, later aimed his strongest vituperation against the *calculatores*:

> What about this tramp who, like some sort of trickster or fraud, long ago was rightly exiled from Italy and found shelter among us—who calls herself 'calculation' while perverting all rational calculating? First off, remove the too rough and unlearned absurdity of speech, which confounds men and angels with asses and mingles gods above with realms below. The absurdity propagates every day in the poisonous manner of weeds; indeed they ensnare and ruin tender wits, spoiling the bread of all teaching.[17]

The calculating tramp personifies Richard Swineshead's *Liber calculationum*, one of the founding texts of the *calculatores*.[18] Note the link between improper practices of reasoning and moral care; Lefèvre criticized logic-chopping as a failure to habituate wits in true learning. Therefore it was an ulcer which, once ruptured, should discharge and 'be completely vomited out from our school, as motherly as it is famous'.[19] The subtle mastery of certain logical genres—sophismata in particular—emblematized the obsession with the disputatious practices that Fabrists believed plagued the university.

[14] See pp. 104–6, and cf. Chapter 3.

[15] Lefèvre, *Totius philosophiae naturalis paraphrases*, b1v (PE, 6): 'Praeterea indignum putant et a philosophiae dignitate quam plurimum alienum in sophisticam expositionem incidere et amica sophistarum syncathegoremata sequi vimque in ipsis ullam facere.' Lefèvre raised similar criticisms in his commentary on Aristotle's logic, where he pointed out that Aristotle generally agreed with Plato's criticism of sophistry in the *Meno*. Lefèvre, *Libri logicorum* (1503), 187v–79r. The comparison of scholastic disputation to sophistry was by no means new, and practitioners of different scholastic schools did the same. See the section 'Mathematics in Paris' in Chapter 2.

[16] Lefèvre, *Totius philosophiae naturalis paraphrases*, b1v (PE, 6): 'Et contra haec alta et sublimia dogmata captiosas orationes texere machinamentaque sophistica ridiculius esse videtur quam terram olim ridiculam simiolam in contemptum deorum peperisse.'

[17] Lefèvre (ed.), *Ricardi [Sancti Victoris] De Trinitate* (1510), a2r (PE, 226): 'Quid quod inculcatoria quae pridem tamquam circulatrix quaedam et subdola iure ab Italis exulat apud nostros invenit asylum et se calculatoriam nominat quae omnem rationis pervertit calculum? Tolle insuper rudem nimis et indoctam sermonis absurditatem quae homines et angelos cum asinis confundit et superos Acherontaque miscet, loliorum more sese in dies noxie propagantem et tenella implicantem immo perdentem ingenia, omnis doctrinae panem inficientem.'

[18] John E. Murdoch and Edith D. Sylla, 'Swineshead (Swyneshed, Suicet, etc.), Richard', in *Dictionary of Scientific Biography*, vol. 13 (New York: Scribner, 1976), 184–213.

[19] Lefèvre (ed.), *Ricardi De Trinitate*, a2r (PE, 225): 'et rupto ulcere totum hoc fetulentum et noxium virus suppuretur et a nostro tam insigni quam almo studio evomatur penitus.'

The related criticism of the *calculatores* was that they mixed the methods of the disciplines. Lefèvre decried 'Peripatetic opinions' as a 'great and noxious plague of the mind'.[20] The operative word here is *opiniones*, to be contrasted with the certitude of *scientia*. Uncertainty was symptomatic of a basic failure of method, a failure to obey Aristotle's prohibition against using the products of one discipline as the basis for demonstration in another.[21] Each discipline proceeds according to its own method: 'physical matters are understood physically; metaphysical affairs divinely; logic logically.'[22] Scholars failed to do this when they did not search out the fundamental axioms proper to each discipline, which are 'understood publicly and are the lights of the disciplines, and which are never learned from the lights belonging to other principles'.[23]

Notwithstanding all this strong language against them, Lefèvre took up the topics and methods of these despised *calculatores* in his most distinctive work of natural philosophy, the *Introductory Dialogue on More Difficult Physics*. The work addressed the latitude of qualities (also known as the intension and remission of forms), approached through a sustained mathematical analysis. Why contribute to the very discipline that was to blame for the university's deepest problems?

Part of the answer lay in dialogue and the habits it inculcated. The choice of dialogue was a statement of pedagogical goals. Lefèvre explained this choice. Guillaume Gontier, a student who accompanied him to Rome, had suggested dialogues because 'if you do, you will advise those who are learning how they should ask questions and how they should answer them; at the same time you will usefully counsel both student and teacher'.[24] Lefèvre later recalled that when visiting Rome he had marvelled at a pair of youths pleasantly disputing in a form of dialogue they evidently had learned from George of Trebizond's rhetoric.[25] The dialogues on physics therefore present a model of how discussion *ought* to happen. The sorites and sophisms of adversarial disputation was to be contrasted with a sociable search for the truth. The prefatory note performs the sociability that Lefèvre hopes to foster: 'let outsiders marvel at how much goodwill there is among those who cultivate the liberal arts here in Paris study (where we know how it is)!'[26] He decries squabbling modes of philosophizing, for 'what is philosophy but love of wisdom? What is a philosopher but a lover of truth? It rightly behoves them to be friends of

[20] Lefèvre, *Totius philosophiae naturalis paraphrases*, b1r (PE, 5).

[21] See Chapter 5, note 6.While Aristotle stressed the uniqueness of a discipline's starting points in the introductory comments of most works, he offered a focused discussion in the *Analytica posteriora*, 1.7, 75a38–75b20. He made some exceptions for the 'mixed sciences' in 1.9, 75b37–76a30. He again distinguished the principles of disciplines at *Metaphysica* 13.3, 1077b31–1078a6.

[22] Lefèvre, *Totius philosophiae naturalis paraphrases*, b1r (PE, 5).

[23] Lefèvre, *Totius philosophiae naturalis paraphrases*, b1r (PE, 5): 'Quae publica intellectus et disciplinarum sunt lumina, et ex aliorum propriorum principiorum luminibus nequaquam dinoscenda.'

[24] Lefèvre, *Introductio in metaphysica*, b1v (PE, 22): 'Si ita feceris, [Gontier] inquit, admonebis qui docturi erunt quo pacto interrogare debeant, interrogataque docere, et simul utiliter discipulo consules et docenti.'

[25] Lefèvre (ed.), *Trapezontii dialectica*, a1v–a2r (PE, 190–1).

[26] Lefèvre, *Totius philosophiae naturalis paraphrases*, j8r–v (PE, 15): 'quanta sit animorum benevolentia inter liberalium artium cultores in hoc nostro Parisio studio (ubi res cognita esset) exteri mirarentur.'

one another (as they rightly suppose).'[27] Philosophy is rooted in love and must be conducted within the sociable bonds of Aristotelian friendship (*amicitia*) and goodwill (*benevolentia*).[28]

The two dialogues on natural philosophy exemplify this ethos of philosophical friendship. The interlocutors bear Greek names that match their function, listed at the outset:

Hermeneus	Interpres
Oneropolus	Coniector
Polypragmon	Negociator
Noerus	Intellectualis
Epiponus	Laboriosus
Enantius	Contrarius
Homophron	Concordans
Neanias	Adolescens

Thus the first dialogue is led by the teachers Hermeneus and Oneropolus, who respectively interpret the subject and offer conjectures to reason about it. Enantius offers contrasting arguments, while Homophron suggests ways to harmonize them: the *personae* perform a kind of concordance of opposites.

The two dialogues take different approaches. In the first dialogue, the father Polypragmon places his son Epiponus in the care of several teachers. Lefèvre translates the father's name as *negociator*; we might see him as a 'busybody', regretting that his own business ties leave him unable to pursue the greater contemplative wisdom he wishes for his 'intellectual' son.[29] As the teachers talk with Epiponus, they set him the task of reading a book of 'introductory physics' much like Lefèvre's own: 'he should read it three or four times over again, and set it in his memory. Meanwhile, we will go for a walk as we wait' until he finishes.[30]

On his return, the teachers point to the circles—the ones introducing the very introduction to physics in which Lefèvre publishes this dialogue: 'Do you see this figure placed at the beginning of our introduction?'[31] And so the lesson begins, just as we saw Beatus Rhenanus record in his lectures on natural philosophy, by

[27] Lefèvre, *Totius philosophiae naturalis paraphrases*, j8r–v (PE, 15–16): 'Quid enim philosophia nisi sapientiae amor? Quid philosophus nisi verus eiusdem amator? Iure decet itaque (ut recte sentiunt) ipsos esse amicos.'

[28] I explore this language of friendship more fully in Richard J. Oosterhoff, 'Lovers in Paratexts: Oronce Fine's Republic of Mathematics', *Nuncius* 31, no. 3 (2016): 549–83.

[29] The term *polypragmon* has an uncomplimentary history: Matthew Leigh, *From Polypragmon to Curiosus: Ancient Concepts of Curious and Meddlesome* (Oxford: Oxford University Press, 2013). Lefèvre may have been aware of Angelo Poliziano's play on the term: Denis J.-J. Robichaud, 'Angelo Poliziano's Lamia: Neoplatonic Commentaries and the Plotinian Dichotomy between the Philologist and the Philosopher', in *Angelo Poliziano's 'Lamia': Text, Translation, and Introductory Studies*, ed. Christopher S. Celenza (Leiden: Brill, 2010), 131–89.

[30] Lefèvre and Clichtove, *Totius philosophiae naturalis paraphrases*, 119v: 'Noere, huic coaequali tuo Physicam introductionem procura. Quam ter quarterve repetitis vicibus legat, memoriaeque mandet. Nos obiter expectantes deambulabimus.'

[31] Lefèvre and Clichtove, *Totius philosophiae naturalis paraphrases*, 119v: 'Figura igitur huic nostrae introductioni praepositam vides?'

drawing out of the student's memory definitions of NaCaMILUT: *Natura*, *causa*, *motus*, *infinitum*, *locus*, *vacuum*, and *tempus*.[32] Each term presents a basic topic in Aristotle's physics. Nature is composed of matter and form. Matter is the subject of change underlying generation, corruption, augmentation, diminution, alteration, and change of place. Form is what generates the change, just as fire is the form that alters wood through burning. There are four causes. And so on. The teachers praise the youth when he answers well, comparing his intellectual wealth with that of Croesus (studies give him more than he would gain with his father...).[33] Sometimes he prods his teachers for glimpses of future delights. Oneropolus conjectures about how the union of form and matter can be compared to a line and a surface, which fuse together to make an object. Hermeneus swiftly intervenes:

> 'Don't tease the boy with analogies!'
>
> Epiponus: 'Do you mean then that all natural things arise whole and composite from lines and surfaces, since they are fused from matter and form?'
>
> Hermeneus: 'See whether or not I'm right to think you're teasing the boy.'
>
> Oneropolus: 'I speak, boy, but I do not speak as you suppose, for now you cannot understand. But keep on and you will understand eventually.'[34]

Meandering through the definitions Epiponus has memorized, the dialogue is punctuated by brief asides on the value of knowledge, the order of nature towards the *summum bonum*, and the aims of philosophy.

At various points in this first dialogue, mathematics threatens to enter but is successfully rebuffed until the end, where the discussion turns to time. For Aristotle, time was an example of continuous magnitude—it can be properly considered as a continuous series of instants, just as a line is a series of points.[35] Together, Oneropolus and Epiponus labour through a series of analogies. Just as extension characterizes change in the quantity of objects, so qualities such as whiteness change by *intensio* or intensification; so also, the succession of time is *latio*. Oneropolus teaches his student how to visualize the change of time on a line from A (past) to E (future) (Figure 6.1).

The second dialogue is framed instead as a 'difficult' one, beginning with the title: *Dialogus difficilium physicalium introductorius*. The student is no longer the *puer* Epiponus, but the *adolescens* Neanias, who takes on a more active role. While the first dialogue is more basic, an exercise in memorizing and recalling basic definitions, reining in exploratory asides, the second turns to a deeper exploration of truth. Enantius sets the scene by recalling an earlier conversation with Neanias, when the young man had declared his commitment 'to the study of beautiful

[32] Cf. Figures 2.1 and 3.4.

[33] Lefèvre and Clichtove, *Totius philosophiae naturalis paraphrases*, 120v.

[34] Lefèvre and Clichtove, *Totius philosophiae naturalis paraphrases*, 121v: 'Her. Noli o Oneropole puerum analogiis ludere. Epi. Visne ergo rem naturalem omnem ex lineis et superficiebus consurgere perfectam, compositamque esse, quandoquidem ex materia et forma conflata sit? Her. Vide an ne non recte sentirem te puerum ludere. One. Dico fili, et non dico ut concipis, nunc autem intelligere non possis. Sed tu aliquando intellecturus reserva.'

[35] e.g. *Physica* IV, 222a28–b7.

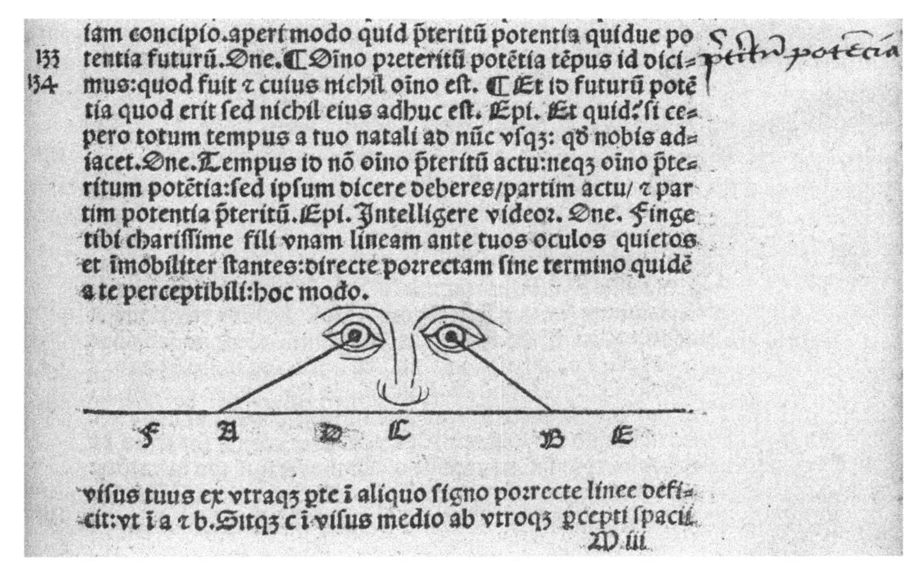

iam concipio. apert modo quid p̄teritū potentia quidue po
133 tentia futurū. Dne. ¶ Dĩno preteritū potētia tēpus id dici=
134 mus: quod fuit ⁊ cuius nichil oĩno est. ¶ Et id futurū potē
tia quod erit sed nichil eius adhuc est. Epi. Et quid: si ce=
pero totum tempus a tuo natali ad nūc vsq3: qd̄ nobis ad=
iacet. Dne. Tempus id nō oĩno p̄teritū actu: neq3 oĩno p̄te=
ritum potētia: sed ipsum dicere deberes/partim actu/ ⁊ par
tim potentia p̄teritū. Epi. Intelligere videor. Dne. Finge
tibi charissime fili vnam lineam ante tuos oculos quietos
et ĩmobiliter stantes: directe porrectam sine termino quidē
a te perceptibili: hoc modo.

visus tuus ex vtraq3 p̄te ĩ aliquo signo porrecte linee defi=
cit: vt ĩ a ⁊ b. Sitq3 c ĩ visus medio ab vtroq3 p̄cepti spaciū.
M iii

Figure 6.1. How to visualize past, present, and future along a continuum. Lefèvre, *Totius philosophiae naturalis paraphrases* (1492), M3r. By permission of the British Library, Shelfmark IA.40121.

letters', but not to the grammar, rhetoric, and poetics that he had once enjoyed, for he now had moved on to focus on philosophy.[36] The boy's wise recognition of the trivium's limits prompts Enantius to reflect on age: 'It seems wrong to me for an old man to marvel at a boy, since wonder arises from ignorance, even of those things boys seem to understand. But it seems to me worse to preserve one's ignorance to the grave.'[37] The young man embodies the philosophical wonder that sparks the beginning of more advanced studies, which the other interlocutors then take up.

The difficulty of the dialogue's advanced topic is thematized in a pair of tantalizing *notae*—which also gloss the significance of the dialogue genre. First, Proteus is presented as an ancient sage who would only willingly foretell the truth if overcome and bound. Second, Milo of Croton is the Greek wrestler who 'killed a bull with a barehanded blow in an Olympic contest, carried it for a hundred yards, and then ate the whole thing that very day'.[38] Both stories would become powerful

[36] Lefèvre and Clichtove, *Totius philosophiae naturalis paraphrases*, 127r, where Enantius asks: 'quaeso si pulchrarum litterarum ludo indulgeas, si ipsum colis. Colo inquit... Sed quam disciplinarum inquam profiteris? An Grammaticam, Rhetoricam, Poeticam? Quid? Quadam benignitate respondit. In his o Enanti non amplius versor, aliquando sum versatus, meum siquidem nunc studium circa philosophiam intentum est.'

[37] Lefèvre and Clichtove, *Totius philosophiae naturalis paraphrases*, 127r: 'Turpe michi videtur senem virum, puerum admirari. Siquidem admiratio ex ignorantia nascitur, eorum etiam que pueri michi cognoscere videntur. Sed turpius suam ignorantiam ad sua sepulcra servare michi visum est. Puerum tamen fateor sum admiratus.'

[38] Lefèvre and Clichtove, *Totius philosophiae naturalis paraphrases*, 127r: 'Proteus filius Oceani et Tethyos, vates maximus, nonnisi coactus, victusque volens ora veridica soluere. Milo Crotonensis

metaphors for the difficulty of pursuing natural knowledge. Over the sixteenth century, the metaphor of binding Proteus would become a telling image of nature itself, to be wrestled into revealing its secrets, or else of the investigator, chasing down barehanded experience.[39] Here in Lefèvre's dialogue, however, Proteus is a friendly conversationalist. Near the end of the dialogue Oneropolus claims that he has finished the subject, but Enantius and Homophron together plead with him to continue. 'Just as Aristeus once bound up the varied and multiformed truth-telling Proteus with arms and chains, and so compelled him to tell the truth, so Homophron and I bind you in our arms, and we compel you to teach us what we asked.'[40] Oneropolus acquiesces: 'you're a friend, Enantius, and finely you compel me—fearing the worthy arms of Milo—to assent to your requests.' The toil of knowing, the wrestling of the natural philosopher is not in the savage cut and parry of disputatious one-upmanship, but the encouraging conversation of friends who share the goal of knowledge.[41]

Indeed, the dialogue effectively reframes the latitude of qualities as a matter of friendship. As we shall see in the next section, as a whole the work does examine how mathematics can be applied to natural change. Nevertheless, these opening *notae* suggest that Lefèvre also views this as a project of using analogy to harmonize philosophical dissent: 'Analogy is a certain ratio of one thing to another, which tests uncertain things by means of certain ones. In a wonderful way, it was started by Plato, perfected by Aristotle; it is unknown to many, and hidden from the majority of philosophers.'[42] The continuity between Plato and Aristotle suggests this method is so universal that it enfolds all lovers of truth, while the claim of hiddenness tacitly reproaches the lack of consensus among contemporaries—a kind of disagreement rooted in wrongheaded disputation. At one point, Homophron rebukes Enantius when he evades a question: 'you answered badly.' Enantius relents and accepts Homophron's rebuke, saying 'I ought to have answered openly. For we do not compete like School Sophists [i.e. university disputants], vainly fighting and holding our position whether right or wrong. We have achieved something

atheleta fortissimus, qui nude manus ictu in certamine olympico taurum interfecit, stadio uno spiritu retento portavit, quem totum die illo comedit.'

[39] Proteus appeared in the *Odyssey* 4.382–569, Plato *Euthydemus* 288b, Virgil *Georgics* 4.387–529, Ovid *Metamorphoses* 8.731, 11.221–56, Diodorus Siculus *Antiquities of Egypt* 1.62. The identification of Proteus as a *vates maximus* suggests that Lefèvre had Diodorus at least partly in mind, while a later reference to Aristaeus reveals Virgil as another source. For the early modern period see, inter alia, Edgar Wind, *Pagan Mysteries in the Renaissance* (London: Faber and Faber, 1958), 158–75, William E. Burns, '"A Proverb of Versatile Mutability": Proteus and Natural Knowledge in Early Modern Britain', *Sixteenth Century Journal* 32, no. 4 (2001): 969–80, Peter Pesic, 'Shapes of Proteus in Renaissance Art', *Huntington Library Quarterly* 73, no. 1 (2010): 57–82.

[40] Lefèvre and Clichtove, *Totius philosophiae naturalis paraphrases*, 148v–149r: 'ut Aristaeus olim varium, multiformemque veridicum tamen Protea brachiis et vinculis implicuit, compulitque verum fateri, ita quoque ego et Homophron brachiis te implicabimus, cogemusque quod quaerimus te nos docere. One. Comis es o Enanti, et pulchre me cogis emerita Milonis brachia timentem, vestris petitionibus acquiescere.'

[41] Elsewhere, Lefèvre, Clichtove, and Alain de Varennes expand on the Aristotelian view that true friendship involves mutual goodwill around shared aims: Oosterhoff, 'Lovers in Paratexts', 565–6.

[42] Lefèvre and Clichtove, *Totius philosophiae naturalis paraphrases*, 127r: 'Analogia, certa rei ad rem proportio, quae incerta certis probat, a Platone mirifice inchoata, ab Aristotele completa, multis ignota, et a plerisque philosophorum occultata.'

more beautiful, something suitable to our great age. Indeed, we are not conquered ourselves, but having conquered our own ignorance we stand victors.'[43]

Throughout the dialogue, friendship is not used to eliminate struggle, but rather friendship makes the struggle for truth fruitful. Thus Enantius (*Contrarius*) and Homophron (*Concordans*) together draw out Oneropolus (*Coniector*)—their names surely are intended to reference obliquely the conjectural harmony of opposites that characterizes the philosophy of Cusanus, in which it is precisely the tension of opposites, the dynamic movement from one to the other, that enables understanding. The work concludes with an avowal of Socratic unknowing and a rebuke of sophistry, couched within a performance of friendship. Enantius praises and thanks Oneropolus, who demurs:

> 'O Enantius, I know nothing on my own, but I learn freely from everyone...I freely impart to everyone, as long as they are not sophists who deride the sacred things of philosophy, and as long as they are not pertinacious attached to some master and wrapped up in pertinacious opinions.'[44]
>
> Enantius: 'And rightly so. For those sorts labour in philosophy while deathly ill. Once there is good opportunity, Homophron and I will visit you again, and we will deal with several other things we want to learn thoroughly from you.'
>
> Oneropolus: 'I hope I can! Whenever you like, my heart is all yours.'
>
> Enantius: 'We desire what lies within [your heart]. To you, here, farewell, our Oneropolus!'
>
> Oneropolus: 'Farewell indeed, O Enantius and Homophron, my friends.'[45]

The last word goes to friends—those whose mutual goodwill is buoyed up by shared love for intellectual goods. *The More Difficult Dialogue on Physics* underlines the deep meaning of friendship within Lefèvre's circle during this period, to which I alluded in Chapter 1. Indeed, it reveals friendship as a methodological corrective to pathological versions of school disputation. By writing mathematical physics as a dialogue that exemplifies the virtues of friendship, Lefèvre calls the domain he finds most susceptible to philosophical vices (mathematical physics) back to intellectual fruitfulness.

[43] Lefèvre and Clichtove, *Totius philosophiae naturalis paraphrases*, 151r: 'Enantius: Haud habeo quid dicam. Homophron: Immo vero. Enan: Quidnam? Ho: Te perperam respondisse. Enan: Probe ais O Homophron: ita plane respondere debui. Non enim certamus ut Gymnastici Sophiste, vane altercantes, et nostram positionem sive iure, sive iniuria observantes. Hoc profecto nostram grandevam etatem dedeceret, et nobis pulchrius vinci; immo nos ipsi non vincimur, sed nostra ignorantia victa manemus victores.'

[44] *Pertinax* applies to those who insist in wrong views even after being shown their falsity. Medieval canon law made this the litmus test for heresy: simply having heterodox views is not heresy unless one pertinaciously holds to those views after being shown their error.

[45] Lefèvre and Clichtove, *Totius philosophiae naturalis paraphrases*, 151v: 'Oneropolus: O Enanti ex me nulla novi, verum ab omnibus libens disco...ego omnibus libens effundo, modo sophiste non sint, qui philosophie sacramenta rident, et modo alicuius magistri secte pertinaciter additi non sint, et opinionibus pertinacius obvoluti. Enantius: Nec iniuria quidem. Nam hi in philosophia extremo morbo laborant. Ego et Homophron te, cum primum tibi et nobis bona facultas aderit, revisemus, tractabimusque nonnulla alia que summopere abs te perdiscere cupimus. Oneropolus: Utinam id possim. Cum voletis, hoc pectus totum vestrum est. Enantius. Volumus quod intra latet. Et tu in hoc vale noster Oneropole. Oneropolus. Bene valete o Enanti et Homophron amici.'

MATHEMATICAL PHYSICS

The *More Difficult Physics* circles around the role of mathematical reasoning in natural philosophy, revealing points of harmony and tension within the Aristotelian tradition. In doing so, it draws on practices I have outlined in previous chapters, so it will be worth our time to go beyond the work's rhetorical frame to trace its arguments.

This second dialogue picks up where the first left off, with the contrast of continuous and discrete quantities. The opening claim is that, as in the case of flowing time, change in the natural world is defined by continuous quantities. In the background here hangs a world of assumptions set by Aristotle. *Physica* II.2 had fixed the assumption that mathematics is of limited use for understanding physical change.[46] There Aristotle had distinguished the proper tasks of physicists and mathematicians around the notion of abstraction. As Lefèvre paraphrased it, 'the physicist considers magnitudes and figures as the boundaries of a natural thing, as conjoined to matter, and he intends to refer everything to motion; the mathematician however abstracts those things from the natural object, away from matter and motion, by means of thought and understanding', looking only to lines and shapes.[47] For this reason, the physicist does not demonstrate things using mathematical principles.[48] Observe that what keeps maths out of physics is the *philosophy of mathematics*, or the status of mathematical objects—these are a hidden version, Lefèvre volunteers in his paraphrase, of Plato's ideas.[49] But the more systematic account of mathematical objects was set out in Aristotle's *Categories*.[50] I will turn to this account more fully in the next section; here I only note that Aristotle, apart from repeating orthodox ancient distinctions of discrete and continuous magnitude, argued for two claims. First, that it is nonsense to speak of 'contrary' magnitudes. It seemed obvious to Aristotle that 'three cubits long' has no contrary; and 'great' is a relative term to 'small', not a contrary. Second, that it is nonsense to speak of a magnitude having varying degrees. For example, 'fiveness' does not admit of degrees: how can one handful of five fingers be more or less than another? Both claims come under strain in the *More Difficult Physics*.

[46] A second *locus classicus* is *Metaphysica* I.7–9. On the relevance of these passages for late medieval natural philosophy, see Di Liscia, *Zwischen Geometrie und Naturphilosophie*, 18–23, and the remainder of chapter 1.

[47] Lefèvre and Clichtove, *Totius philosophiae naturalis paraphrases*, 23v: 'Nam physicus magnitudines et figuras considerat ut termini sunt rei naturalis et coniuncta materiae, et intentio eius omnia ad motum referre; Mathematicus autem ea a re naturali materia et motu cogitatione et intellectu abstrahit.'

[48] Lefèvre, like Clichtove's commentary, outlined Aristotelian theory of subalternation, with music, perspective, and astronomy as the classic examples. In the edition of 1502, a printed marginal note observes that 'these mathematical sciences come closer to physics' (Lefèvre and Clichtove, *Totius philosophiae naturalis paraphrases*, 23v: 'Que scientie Mathematice ad physicam propius accedunt').

[49] Lefèvre and Clichtove, *Totius philosophiae naturalis paraphrases*, 23v: 'et haec latuerunt Platonem ideas introducentem'.

[50] This is not to say his views are straightforward. A collation of Aristotle's myriad statements about mathematics is Giuseppe Biancani, *Aristotelis loca mathematica* (Bologna: Bartholomaeus Cochius, 1615). A useful effort to grapple with them is John J. Cleary, *Aristotle and Mathematics: Aporetic Method in Cosmology and Metaphysics* (Leiden: Brill, 1995).

The discussants begin with the difference between form and quantity. The doctrine of forms in late medieval philosophy is vast and convoluted, but everyone agreed that forms could overlap in the same object. A plant has a material form as well as vegetal, and its structure exhibits qualities of both. Oneropolus asks us to imagine two suns in the sky. We would experience light and heat from both; such qualities can, like forms, occupy the same space. But unlike qualities, quantities do not overlap, for by definition they occupy space. One chunk of steel can be fused to another, but it cannot occupy the same space as the other. Dimensions cannot overlap.

Enantius points out another contrast between soul and dimensions, taking the example of his own body. If he cuts himself up into, say, four pieces, will each have a different soul? It seems not, since four body parts does not imply each has a different soul. Rather, 'our whole soul exists throughout our whole body', while taking one bit of the body away doesn't mean that a certain part of the soul is in that bit. This results in a conundrum: the soul is throughout the body, but found in no part of it. The conundrum is resolved with an analogy, which is linked to Parmenides, Pythagoras, and Zeno—the idea that the soul fills a body just as a point is the essence that defines a sphere. Clichtove clarified the source: 'God is a point filling the universe', everywhere and nowhere at the same time.[51] The analogy introduces a crack into the account, for a mathematical object (the point) here takes on characteristics of soul and essence, occupying various places at once.

The discussants share these anxieties, which undermine the theory of abstraction. A big buzzing fly spoils the ointment: how do mathematical objects exist in the material world? A point, for example, is indeed 'conjoined to matter' (*materiae coniuncta*). Experience sets the problem: 'By experience we understand that sensible things, once divided, are again constituted continuously as one, as when water is joined to water and one flame to another.'[52] These examples make it difficult to consider things as extensions, as continuous lines. Experience suggests that the lines can be cut, and rejoined. But what happens at the point where the line is cut in two? A single cut is placed at a single point. Yet that one point serves as the end for two new lines. So it appears that a major assumption was wrong: two points *can* exist in the same place. Enantius first resolves the conundrum as a linguistic problem, suggesting that a *punctum* is spoken of in two ways: 'the first is the way by which we name a continuing potential point; the second way we speak of a point potentially is that which is not yet the terminus of the line, but is so potentially, which we can name the not-yet continuing potential point.'[53] The explanation is linguistic, but it implies that the physical world is full of potential points which

[51] Lefèvre and Clichtove, *Totius philosophiae naturalis paraphrases*, 130r (Clichtove's commentary at 131v): 'Et deus est punctum [*sic*] replens universum.' On this topos in the Platonic tradition, from Victorinus to Cusanus, see Chapter 5, note 153. Lefèvre may have associated Parmenides with Pythagoras thanks to Diogenes Laertius, who identified him as the son of a Pythagorean.

[52] Lefèvre and Clichtove, *Totius philosophiae naturalis paraphrases*, 134r: 'et experientia cognitum habemus res sensibiles divisas, iterum continuari et uniri natas esse, ut aquam aquae et flamam flamae.'

[53] Lefèvre and Clichtove, *Totius philosophiae naturalis paraphrases*, 132v: 'Attamen bifariam dicemus punctum in potentia: primo eo modo quod dictus est, quod punctum potentia et continuans nominamus; secundo punctum potentia dicemus quod nondum est linee terminus, sed esse potest, quod punctum potentia et nondum continuans appellare possumus.'

can somehow be activated. Experience again supports this notion, for when we strike something or light a lamp, the sound and light immediately fill the space. The reason seems to be that sound and light extend from their punctiliar source, acting like a radiating series of lines, in turn bounded by the surfaces and bodies around them.[54] There is no mechanism here; rather, Lefèvre draws on the active ability of Aristotelian nature to generate change, so that 'sensible things naturally have a desire to do this', a certain 'inborn drive of nature' by which sparks rise up and so on.[55] This desiring activity is the orthodox foundation of Aristotelian physics. What is remarkable here is that nature's activity is used to explain how points, lines, and surfaces emerge in physical objects. Quantitative change in natural objects—size, shape, and place—starts to look like any other change.

This elision of qualities and quantities works in several ways. Even at this early stage in the discussion, visual tools emphasize the similarity of nature and mathematics. A point's behaviour in cutting a single line, then becoming two—terminating both segments—and rejoining again upon reassembling the line, is accompanied as a labelled line. The line accompanies the reader as the most concrete example of the phenomenon of physical movement. Thus a running series of labelled lines (Figure 6.2) leaves the reader chiefly seeing physical change as a problem of points and lines.

The line is the main visual tool for modelling the kind of physical change that this dialogue ultimately will address: change of qualities such as hot and cold, black and white, and so on. In fact, the discussants bridge to the topic of qualitative change with a line representing an object ten feet long, from *a* through to *k*, set alongside a one-foot object *l* (Figure 6.3). This diagram is intended to illustrate a topic that seems unproblematically one of quantity: density and rarity. If the same quantity were spread through *a–k*, it would be rare; while if the same quantity were forced into *l*, it would be dense. Indeed, example of a sponge raises a line of reasoning which opens the possibility of an atomistic or mechanical account of density, in which dense objects have particles pressed more closely together. A sponge turns out to be an unsatisfying analogy, for the dilation of cellules suggests a vacuum, not quite what is intended by 'rarity'. So the discussion hastens past this 'too difficult' topic (atoms or particles) to the 'quamfacillimam' topic of heat and cold, black and white (qualities).

The line model of density *assumes* that density is a quantity, since it is 'what possesses much matter within a small quantity'.[56] But such a notion of density put pressure on Aristotle's effort to fence quantities off from qualities. First, the divided line hinted that density could admit degrees—it could be more or less. Second, even though density and rarity are relative in some cases (air is rare relative to

[54] Lefèvre and Clichtove, *Totius philosophiae naturalis paraphrases*, 134r: 'Statim igitur (cum punctum, linea, et superficies separata non sint) rei sensibilis actus, punctum, lineam, aut superficiem suae molis terminum efficiet, ut res ipsa sit, quemadmodum ipsa esse nata est.'

[55] Lefèvre and Clichtove, *Totius philosophiae naturalis paraphrases*, 134r: 'res sensibiles ad hoc naturale desiderium habent... ingenuum conatum naturae'.

[56] Lefèvre and Clichtove, *Totius philosophiae naturalis paraphrases*, 136v: 'Quod sub multa materia quantitatem habeat parvam.'

℃Arbitror hos duos modos aut eoſdē eſſe/aut perparū diſtare. Nā fere ter= 25
minus p id ipſum cuius eſt terminus ſubiecto adeſt. One. imo plurimū & qua
ſi ex oppoſito diſtāt. Enan. Quo pacto iſtud? One. hoc pacto. Nam accidēs
ſeparari/ſeorſumq; a ſuo ſubiecto ſaluari nequit. Enan. nequit indubie. One.
At linee a b nondū diuiſe c terminus eſt linee c b. Enan. eſt. One. diuidat igit
a b in puncto c maneatq; c verſus a. Enan. placet. One. Nam id poſſibile. verū
tamen c prius erat in linea c b vt terminus. Enan. Erat. One. et nō amplius eſt/
et ſeorſum manet. Enan. Hec vera ſunt. et intelligo quorſum hec. One. & acci
dens de ſubiecto cui aſſit: nunq̄ trāſit/emigratq; in alterū. Enan. Nunq̄. One.
Terminus autē vt terminus de vno in alterū tranſit. Enan. Quo pacto? One.
d terminus eſt linee d b/& nullo pacto eſt linee a c. Enan. nullo pacto. One. in=
telligo ergo lineas a c et d b cōtinuari/deleriq; c. Enan. Eſto ita. One. Facta eſt
linea a d b: quā rurſum intelligo diuidi in d: ipo quidē d ad partē a manēte. nul
la eni potior ratio ipſum poſſe manere ad partē b q̄ a. Enā. Nulla prſus. One.
Trāſiuit igitur de linea in lineā/vt de linea b in lineā a. Enan. perſpicuū euaſit.
One. Quo fit vt pmiſcue punctū de linea in lineā/et linea de ſuperficie in ſup=
ficiem/et ſupficies de corpore in corpus tranſire queat. vt verbi cauſa punctū
d tranſire poteſt in lineā b/& poſtea e in lineā a. Iungo eni e b linee f g: abeatq;
g manēte e. Deinde ſeco lineam f b in puncto e/ipſo quidē e manente ad partē
f: reparato puncto h. et iungo lineā a d et h b/manente d et ipſo h deleto: quā
quidē lineā a b rurſum diuido in puncto d/ipſo quidē manente in parte b: iam
de a tranſijt in b. Enan. Tranſijt. One. Copulo a i et e f: abeunte i/fitq; linea
a f: quā diuido in puncto e/ipſo quidē manente in parte a. Nonne cōſtat e po
ſtea trāſiſſe in partē a/et ppoſitum? Enan. Conſtat. One. Rurſus accidēs in di
uerſis ſubiectis nō cōmunicātibus eſſe nō poteſt. Enā. nō. One. Et idē pūctū
vt e prepoſite linee a f/diuerſarū linearū a e et e f nō cōmunicantiū terminus

Figure 6.2. The behaviour of mathematical objects in physics. Sequence of diagrams illustrating the behaviour of points in dividing lines, lines in dividing surfaces, and surfaces in dividing bodies. Detail from Lefèvre and Clichtove, *Totius philosophae naturalis paraphrases* (1502), 134v. FC5 L5216 502t. Houghton Library, Harvard University.

water, but dense relative to fire), the interlocutors suggested that there could be absolute extremes. Therefore, the boundaries of this quantitative inquiry are not set by logical possibility, but by physical characteristics: 'Indeed we do not address what *can be* said about quantity, for no doubt they do not have logical contraries; but here we seek an examination of physical quantities, and plainly we admit that they do have contraries.'[57] Quantities, when *physical*, do admit of degrees and have absolute extremes (i.e. contraries) like qualities—contradictions of Aristotle.

Oneropolus points to the overlap between quantitative and qualitative modes of inquiry: 'In fact, to philosophize about rarity and density is *just like* philosophizing about white, black, and the intermediate colours.'[58] Homophron cries out in wonder, for he has long tried to sort the problem out; he listens to Enantius and Oneropolus push the overlap between quantities and qualities here with the example

[57] Lefèvre and Clichtove, *Totius philosophiae naturalis paraphrases*, 137r: 'Sane quantitatis dicibilia hic non agimus, que indubie Logica contraria non habent, sed physicarum quantitatum hic contemplationes exquirimus, quarum plane fatemur esse physica contraria.'

[58] Lefèvre and Clichtove, *Totius philosophiae naturalis paraphrases*, 137v: 'Et revera de raro et denso philosophari, perinde est atque de albo, nigro et mediis coloribus philosophari.'

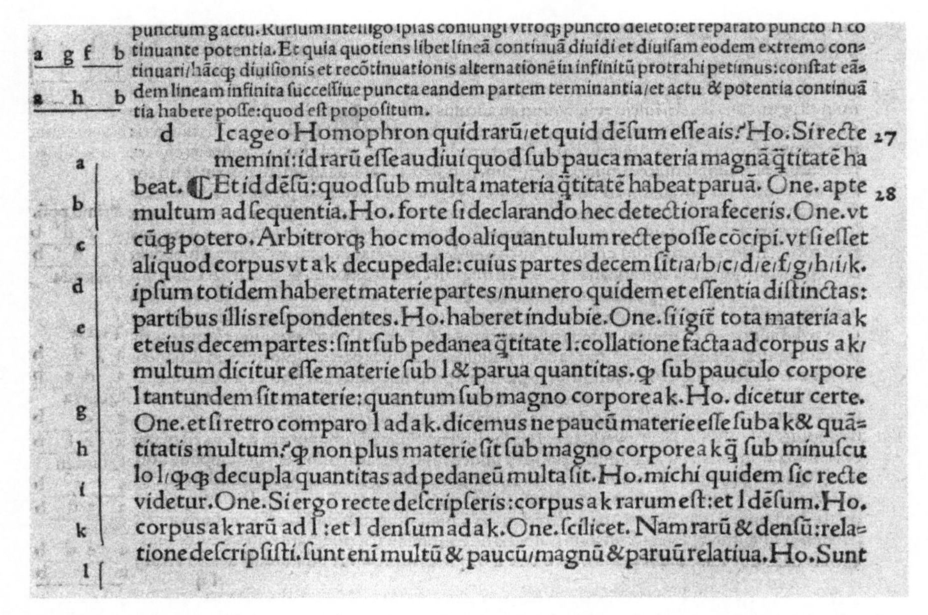

punctum g actu. Rursum intelligo ipsas coniungi vtroq; puncto deleto: et reparato puncto h co-
tinuante potentia. Et quia quotiens libet lineā continuā diuidi et diuisam eodem extremo con-
tinuari/hācq; diuisionis et recōtinuationis alternationē in infinitū protrahi petimus: constat eā-
dem lineam infinita successiue puncta eandem partem terminantia/et actu & potentia continuā-
tia habere posse: quod est propositum.

d Icage o Homophron quid rarū/et quid dēsum esse ais? Ho. Si recte 27
memini: id rarū esse audiui quod sub pauca materia magnā q̄titatē habeat. ¶ Et id dēsū: quod sub multa materia q̄titatē habeat paruā. One. apte 28
multum ad sequentia. Ho. forte si declarando hec detectiora feceris. One. vt
cūq; potero. Arbitrorq; hoc modo aliquantulum recte posse cōcipi. vt si esset
aliquod corpus vt a k decupedale: cuius partes decem sit/a/b/c/d/e/f/g/h/i/k.
ipsum totidem haberet materie partes/numero quidem et essentia distinctas:
partibus illis respondentes. Ho. haberet indubie. One. si igit̄ tota materia a k
et eius decem partes: sint sub pedanea q̄titate l: collatione facta ad corpus a k/
multum dicitur esse materie sub l & parua quantitas. q̄ sub pauculo corpore
l tantundem sit materie: quantum sub magno corpore a k. Ho. dicetur certe.
One. et si retro comparo l ad a k. dicemus ne paucū materie esse sub a k & quā-
titatis multum? q̄ non plus materie sit sub magno corpore a k q̄ sub minuscu-
lo l/q̄q; decupla quantitas ad pedaneū multa sit. Ho. michi quidem sic recte
videtur. One. Si ergo recte descripseris: corpus a k rarum est: et l dēsum. Ho.
corpus a k rarū ad l: et l densum ad a k. One. scilicet. Nam rarū & densū: rela-
tione descripsisti. sunt enī multū & paucū/magnū & paruū relatiua. Ho. Sunt

Figure 6.3. Seeing density. In the margin, a line *a–k* helps visualize a ten foot object, then compressed into line *l*, to illustrate the concept of density. Detail from Lefèvre and Clichtove, *Totius philosophae naturalis paraphrases* (1502), 136v. FC5 L5216 502t. Houghton Library, Harvard University.

of breath, which cools as it is more distant from one's mouth. Warm breath is rare, and so quickly rises—but as it rises, it disperses, cools, and loses its relative rarity. The qualities of hot and cold appear linked to rarity and density, a linkage which is developed over several pages as a justification for visualizing rarefaction and condensation, and qualities such as hot and cold as extremes. As extremes, such physical contraries can be set on a line.

In the series of arguments we have just surveyed, the relationship between qualities and quantities is limited to analogy. But how strong is the analogy? Consider the dialogue's culminating example, which claims something even stronger than analogy at the outset: 'In fact, to philosophize about the intension and remission of qualities and their contraries *is* to philosophize about greatness and smallness and the contrariety of quantities.'[59] The example revolves around a line. Is the line a mere convenient convention? If so, perhaps the analogy is not very close. Oneroplus posits a line of qualities divided into ten degrees *a–k* (Figure 6.3). This measure of heat is matched with an inverse line of cold *l–u*. The basic idea is that the minimal amount of heat corresponds to the maximal amount of cold. By agreeing on what the line represents, the interlocutors perform the conventional

[59] Lefèvre and Clichtove, *Totius philosophiae naturalis paraphrases*, 144v: 'Et revera philosophari de qualitatum intensione et remissione et earum contrarietate est philosophari de magnitudine et parvitate et quantitatum contrarietate.'

aspects of using mathematics in physics: they agree that the total *a–k* will represent the maximum heat of fire, 'for what is hotter than fire, what is rarer than fire?... Let us establish this between us.'[60] The interlocutors agree that the total of ten degrees is likewise a matter of convention: they are not like astronomical degrees, which are absolute, each being 1/360th of a circle's circumference; and the number ten is a convenient reflection of the Pythagorean preference for ten, but in the end only 'follow our will and decision'.[61] So the particular details of the line are mental constructs.

So far, it appears the analogy of between quantity and quality, mathematics and physics, is quite loose. But the argument also foregrounds mathematics. The line represents a range that is equally mathematical and physical. This is what Enantius challenges when he asks about mixing qualities, for he wonders why—since a given degree of heat matches its opposite degree of cold—other degrees of heat might not be substituted for that degree of cold.[62] Oneropolus leads Enantius through a *reductio ad absurdum*. Imagine if one were to add a bit more heat (x) to *a–k*. Then the total heat of the object would be *a–x*—and the original proposition that *a–k* was maximal would be false. A mathematical argument, then, can say something about what is physically possible.

In this example, Lefèvre joins a late medieval tradition that eroded Aristotle's claims that quantities do not have degrees or contrariety. First, degrees. To reason about qualitative changes, medieval philosophers postulated a line that is divided into degrees representing the amount of an object's qualities. The *Physica difficilia* chose the perfectly orthodox case of a line with ten degrees, from minimum to maximum heat. Such a line exploited two assumptions already standard among the fourteenth-century calculators. One assumption was that qualities can be arranged on a continuum that is divided mathematically.[63] As just mentioned, earlier parts of the dialogue legitimized this assumption around the analogy of quantitative dimensions to qualitative intensions as an example of a continuum. Considered this way, quanties do have degrees.

Second, this account undercuts Aristotle's claim that quantities do not have contraries. A second assumption in the tradition of calculators concerned the maximum and minimum of physical qualities. Aristotle had gestured towards this notion in *De caelo* I.11, which fourteenth-century philosophers had developed a sophisticated genre *De maximo et minimo*.[64] Lefèvre's dialogue legitimized this assumption on the same basis: the analogy of extension, succession, and intension. Thus, just as changing a section of plank would require removing the old plank to make space for a new one, so changing one quality with its opposite involves

[60] Lefèvre and Clichtove, *Totius philosophiae naturalis paraphrases*, 146v–147r: 'Quid calidius igne, et quid igne rarius?... Sic esse inter nos statuamus.'

[61] Lefèvre and Clichtove, *Totius philosophiae naturalis paraphrases*, 147r: 'voluntatem arbitriumque nostrum sequi'.

[62] Lefèvre and Clichtove, *Totius philosophiae naturalis paraphrases*, 146v.

[63] For the sake of brevity, I move past a series of arguments about the continuum, rooted in *Physica* VI. See Wilson, *Medieval Logic*, chapter 2.

[64] This genre concerned the limits of physical powers of any sort (i.e. not just qualities). See Wilson, *Medieval Logic*, chapter 3. See also *Physica* V.2.226b, which describes change as occuring between two extremes.

D II 147

exiſtimo . One. Sic eſſe inter nos ſtatuamus. Erit ergo ignis natura calidiſſi=
mus. Enan. Erit ſane. One. Et vbicunq; vnum contrariū eſt: alterum aduenire
natum eſt: tum intenſione/ tum extenſione. Alioquin ſe ab eodem ſubiecto
non expellant. Enan. Ex ſuperiore ſermone cognitum eſt. vt neq; dimenſio
dimēſionem expelleret: niſi ſimul ſecundum extenſionem coaduenirent. One.
Et ignis a nobis conceſſus eſt calidiſſimus. Enan. Conceſſus. One. Maximam
igit̄ ignis caliditatē in decē partes intēſione equas a/b/c/d/e/f/g/h/i/k/ cogita=
tiōe diuidamus: quarū quālibet: gradū vocemus. hoc abs te requiro: an vnus
frigiditatis gradus vt l: ſiml̄ eſſe poſſit cum maxima caliditate a k? Enā. Quid
ꝓhibet? One. Nō ſimul intēſione cū a erit: nā duo cōtraria ſimul eſſent. pro=
bata ſunt autē prius eſſe cōtraria. Enan. Fateor. One. Neq; eadē rōne cū gra=
du b: aut aliquo aliorū decem nominatorū. Enan. non certe. ſed nichil ꝓhibet
illum gradum extra illos decē aduenire: & eidē ſubiecto & eidē ſubiecti parti
adeſſe. hoc eni prius intelligebam: cum admiſi l ſimul cū maxima caliditate a k
eſſe poſſe. One. idipſum querebam. & tibi ſic eſſe videt̄: michi vero ſecus. Eſto
attamē vt vis ſi poſſibile eſt: adueniatq; gradus frigiditatis l eidem ſubiecto cū
maxima caliditate a k: intenſione tamē extra/a/b/c/d/ & reliquos gradus. Enā.
Admitto. One. vbicūq; vnum contrariū nature potentia eſſe valet: illic poteſt
& reliquū. Enan. poteſt. One. Alioquin non expelleret ipſum: niſi ipſi adueni
ret ſecundū ipſius extenſionis aut intenſionis ſitum. Enā. nō expelleret. One.
At l frigiditatis gradus nature potētia aduenit ſubiecto a k extra decem calidi
tatis gradus a/b/c/d/ & reliquos. Enan. conceſſum eſt. One. illic igitur et ſecū
dum eius intenſionis ſitū aduenire poteſt caliditas: que ipſum expellat. Enā.
Sic oportet. One. Adueniat igitur: quandoquidē id poſſibile eſt/ vt x. Enan.
michi admittendum eſt. One. vides igitur a x totam caliditatē: & a k eius ptē?
Enan. video plane. One. non erit igitur a k maxima natura caliditas: aut dabis
maius maximo. Enan. id certe impoſſibile eſt: & me pulchre ad id quod deſide
52 rabam perduxiſti. Et gaudeo me fuiſſe concluſum. One. ¶ O Enanti quo pa
cto diffinit Aſtronomus circuli gradum? Enan. Aliquādo tenui: michi forte
nūc e memoria lapſum eſt. Credo ipſos circuli gradum vocare eius tricenteſi=
mam atq; ſexageſimā partem. One. haud lapſum eſt. At vero hic nō ita nomie
gradus vtimur. Quia tamen nomen poſitum/ aptūq; non habemus: noſtrum
triū mei/ tui/ & Homophronis inſtructionis gratia: gradū diffiniamus. Enan.
Quo pacto? One. Sine controuerſia in re & nomine gradus: concordemus/
conſentiamuſq; gradum qualitatis eſſe maxime natura qualitatis in decē in=
tenſione equas partes diuiſe: quālibet decimam aut decime equalem. Enan.
maximā natura qualitatē intelligis (vt arbitror) maxime nature poteſtate intē
ſam. One. intelligo. Enan. proſequere vt voles: ipſe tecum concordo. Ho. et
ego: & eo maxime ꝙ numerum denarium pythagorici maximū crediderunt:
& ſeriem numerorum haud vltra procedere. Maximam itaq; qualitatem mi=
chi iuſte videris in denarium graduum numerū ſoluiſſe. One. Tametſi o Ho=
mophrō prime rerū nomibus īpoſitiones volūtatē/ arbitriūq; noſtrū ſequi vi=
tv

a b c d e f g h k x | m n o p q r s t u

Quid gradus circuli ſcdm aſtronomos.

Pythagorici.

Figure 6.4. The mathematically absurd and the physically impossible. The argument *ad absurdum* visualizes why qualities can only be mixed within an absolute maximum. Note the references to *gradus* and the Pythagorean *denarium* in the margin. Lefèvre and Clichtove, *Totius philosophiae naturalis paraphrases* (1502), 147r. FC5 L5216 502t. Houghton Library, Harvard University.

expelling the old.[65] For example, when a cold iron is set next to a hot iron, the hot iron *gains* a quality of coolness that *expels* a certain quality of hotness.[66] If qualities possess a finite range defined by extremes (maximum and minimum), then opposite qualities cannot simply be added together; adding one pushes out another, so that the sum remains constant. In a word, such extremes behave as *contraries*.

Thus underlying the two lines *a–k* and *l–u* and the transfer of qualities from one to the other is a commitment to physical contraries. Within the framework of physical contraries, any given quality *must* correspond with its opposite quality. It is impossible, therefore, to conceive of cold as simply an absence of heat. The existence of one entails the existence of the other; a little heat entails the existence of a lot of cold, so that diminishing the one proportionally augments the other. As Enantius summarizes, physical qualities must exist in three configurations of contraries: (1) maximal and simpliciter (no mixing): (2) secundum quid (mixed so that the sum of both equal 10), (3) neutra/mixta (in mixed proportion).[67] In this account of mixtures, physical qualities begin to put pressure on the assumption that mathematical objects cannot have contraries.

This overlap between mathematics and physics should not be pushed too far. As we have found at various other points in this book, the concept of analogy was useful as a mathematically loaded term for relating disciplines, but it remained a fuzzy concept. Analogy therefore retained a distance between quality and quantity, a long step away from the Cartesian reduction of one to the other.

Still, the repetition of this analogy constantly justified applying reasoning from one to the other. Near the end of the dialogue, this sort of reasoning achieves a new level of generality, as the discussants try to isolate what is shared between these different but analogous sorts of change, between intension/remission of qualities, increase and decrease of quantities—and, they add, speed and slowness of movement. The analogy lies not in the qualities or powers of nature themselves, but in quantifying their *changes*. Generalizing in this way, it becomes possible to quantify not only the changes themselves, but also the *rate of change*: 'the intension of a latitude is speed, and its remission is slowness. Therefore it also happens (if we do wish to transfer intensity to latitudes) that the fastest motion is also the most intense, and if it were naturally the slowest motion, it would be also the most remitted.'[68] Here we approach something like the famous mean speed theorem, by applying the measurement of change to itself.

[65] Lefèvre and Clichtove, *Totius philosophiae naturalis paraphrases*, 144v–5r.

[66] It seems that Lefèvre adopts a version of the 'admixture' theory of qualitative change. For the range of options see Stefan Kirschner, 'Oresme on Intension and Remission of Qualities in His Commentary on Aristotle's Physics', *Vivarium* 38, no. 2 (2000): 255–74, Elzbieta Jung, 'Intension and Remission of Forms', in *Encyclopaedia of Medieval Philosophy: Philosophy Between 500 and 1500*, ed. Henrik Lagerlund (Dortrecht: Springer, 2011), 551–5.

[67] Lefèvre and Clichtove, *Totius philosophiae naturalis paraphrases*, 148v.

[68] Lefèvre and Clichtove, *Totius philosophiae naturalis paraphrases*, 151r: 'Haud aliter atque ego concipis. Et lationis intensio est velocitas, et eius remissio, tarditas. Quo fit iterum (si intensum ad lationes transfere velimus) ut motus velocissimus, sit et intensissimus. et si esset natura motus tardissimus, illae esset et remississimus. Et motum caeli circa aequatorum velocissimum, et circa axem tardissimum; diversitate non discrepare, neque ambulationem et cursum.'

Around 1492 (when the first edition of the *Paraphrases totius philosophiae naturalis* appeared), this form of mathematical physics is difficult to set in a discursive or visual context: in part, as often is the case, because Lefèvre cites no sources; in part because there is a lack of scholarship on fifteenth-century manuscript traditions relating to these genres.[69] Certainly I have found no printed precedents for the dependence on working from diagrams. One candidate might be the medical commentaries on Galen's *Ars medica*, which discussed the 'latitude of health', and several major examples were printed in the late 1480s.[70] But none of these early printed versions use comparable lines of degrees, and Lefèvre never makes an example of the latitude of health. A second candidate might be the revival of the calculating tradition among Lefèvre's colleagues in Paris. But here the timing would make Lefèvre a source, not epigone. Rival commentaries on the *Physics* from the 1490s through to the1510s do not dwell on the latitude of forms, and do not begin to use similar visual images in the margins until much later.[71] While there are early editions of fourteenth-century calculators printed at Padua in the 1470s, besides Lefèvre there is no new tradition of calculatory treatises at Paris until 1509, when Alvarus Thomas published an extensive study of the use of ratios in understanding motion, drawing on Bradwardine and Albert of Saxony as well as on their commentary tradition.[72] This tradition continued with the circle around John Mair at the Collège de Montaigu, in the works of Gaspar Lax. Full comparison of these works awaits study, but a swift visual inspection already suggests a very different approach. Thomas's sophisticated mathematics presents regular series of ratios, but fewer geometrical analogies of change along a line. Lax offers almost no diagrams. Their attitude to pedagogy is very different; Thomas presents the work explicity modelled on disputation, not dialogue, while offering advice for how to win disputations, even suggesting that, when ridiculing opponents, students ask for pen and paper to display mathematical prowess—far from the rhetoric of friendship Lefèvre offered. The only place comparable geometrical lines are frequent is in the

[69] Though see Di Liscia, *Zwischen Geometrie und Naturphilosophie*, 31–58.

[70] Indeed, Sylla suggests that Galen's work, a standard textbook in medical faculties, stimulated the Oxford calculators: Edith Dudley Sylla, 'Medieval Concepts of the Latitude of Forms: The Oxford Calculators', *Archives d'histoire doctrinale et littéraire du Moyen Âge* 40 (1973): 226. The earliest printed editions of the *Ars medica* are discussed by Timo Joutsivuo, *Scholastic Tradition and Humanist Innovation: The Concept of Neutrum in Renaissance Medicine* (Helsinki: Academia Scientiarum Fennica, 1999), 33. I have checked commentaries on Galen's *Ars medica* by Pietro Torrigiano, Giacomo da Forlì, and Giovanni Sermoneta—none printed before 1500 include the visual diagrams included in mid-sixteenth-century editions and later periods by Ian Maclean, 'Diagrams in the Defence of Galen: Medical Uses of Tables, Squares, Dichotomies, Wheels, and Latitudes, 1480–1574', in *Transmitting Knowledge: Words, Images, and Instruments in Early Modern Europe*, ed. Sachiko Kusukawa and Ian Maclean (Oxford: Oxford University Press, 2006), 137–64.

[71] One of the most popular was Thomas Bricot and Georgius Bruxellensis, *Textus abbreviatus in cursum totius physices et metaphysicorum Aristotelis* (first edn, Lyons: s.p. 1486). Even Lefèvre's printers did not include lines illustrating motion (Paris: Johann Higman and Wolfgang Hopyl, 1492), until after 1500. Neither is the latitude of forms discussed in the natural philosophy of Pierre Tataret, *Commentarii in libros philosophie naturalis et metaphysice Aristotelis* (first edn, Paris: André Bocard, 1495).

[72] Alvarus Thomas, *Liber de triplici motu* (Paris: Guillaume Anabat, 1509). The fullest study is Edith Dudley Sylla, 'Alvarus Thomas and the Role of Logic and Calculations in Sixteenth Century Natural Philosophy', in *Studies in Medieval Natural Philosophy* (Florence: Leo S. Olschki, 1989), 257–98.

margins of manuscript treatises on the latitude of forms; the Fabrist mathematical physics seems a distinctive printed response to this medieval genre.

This dialogue therefore does not carry us directly into Galilean or Cartesian physics. But the dialogue does reveal Renaissance schoolmen practising the moves that are basic to a mathematical physics, finessing Aristotle's account of mathematical abstraction while reworking the conceptual domain explored by the calculators. Even more remarkable is the fact that these moves are exercised in a widely popular introductory textbook. We have come to expect such a revision of Aristotle within the tomes of exceptional natural philosophers—Scaliger, Telesio, or Campanella—but it is all the more suprising among textbook defenders of a humanist Aristotle such as Lefèvre. This recalibration of the tradition happened precisely through those practices highlighted in previous chapters of this book. First, Lefèvre again used analogy to think about the different modes of reasoning in different disciplines. Second, thinking with the senses allowed one to elide different domains: in particular, this dialogue especially focused on imagining mathematical tools—the mathematical continuum, divided in degrees—in a manner that set tensions within Aristotle's account of quantity. In fact, as we shall see with the category of quantity, Lefèvre focused on mathematical principles, unseating the doctrine of abstraction that justified the Aristotelian suspicion of mathematics.

PRINCIPLES

The whole Fabrist reform of the university was built on a fundamental agreement between Lefèvre and Aristotle: each field of learning works in its own way, founded on its own principles. When Lefèvre urged students to flee the opinions of false Aristotelians 'like a great, deadly plague of the mind', he blamed the ills of contemporary philosophy on the hasty flitting from one discipline to another.[73] Modern obsessions with logical puzzles (*sophismata*) had obscured proper attention to principles. The *moderni* usually focused on two of Aristotle's more advanced works, the *Topics* and the *Sophistic Elenchus*, which informed large sections of Peter of Spain's popular *Summule* and a whole genre of late medieval handbooks on how to solve logical puzzles in disputation: *categoremata*, *sophismata*, and *insolubilia*. Lefèvre's first logical introductions of 1496 were explicitly intended as a sort of crib sheet so students could entirely avoid such 'barbarous sophistry'.[74]

As we saw, such language echoes university critics from Jean Gerson to Leonardo Bruni, who represented what we might see as a late medieval 'common sense'. Theologians such as John Wycliff, Jan Hus, and Henry of Langenstein worried that the people were ill served by specialist theories, while humanists such as Lorenzo Valla poked fun at school subtleties.[75] Late medieval spirituality was shot through

[73] Lefèvre, *Totius philosophiae naturalis paraphrases*, b1v: 'Id preterea animadvertere licet Peripateticos opiniones tanquam grandem et pernoxiam mentis pestem fugere.'

[74] Lefèvre, *Introductiones logicales* (1496).

[75] Lodi Nauta, *In Defense of Common Sense: Lorenzo Valla's Humanist Critique of Scholastic Philosophy* (Cambridge, MA: Harvard University Press, 2009).

with desire to taste, see, and feel God, with the plain physical visuality and materiality of spiritual objects that showed the presence of God's working in the world.[76] The Fabrist heroes Ramon Lull and Nicholas of Cusa in a strange way represented this concern with evidence, with seeing the world's causes more directly, albeit transposed into an intellectualist mode. Lefèvre defended the fourteenth-century Majorican Ramon Lull and his rough Latin, because he 'far excels the wise men of our age'.[77] Lull wrote an art which promised to give universal truth by combinatory reasoning on first principles, where logical structures were assumed to be rooted in true experience of nature.[78] Similarly, another of the Fabrist heroes was Nicholas of Cusa, who wrote dialogues in which an artisan *idiota* offered the topics and principles for understanding—largely mathematical—thus relating the principles for understanding the nature of the mind, of the soul, and of numbers and weights in nature.

For Lefèvre, the key was not to avoid Aristotle, but to unknot the interdisciplinary tangle, pursuing 'physics physically, metaphysics divinely, and logic logically'. For this reason, 'they require one to search out each thing in its proper place and out of its proper principles'.[79] Near the beginning of most of his works Aristotle insisted on the 'proper' modes of reasoning distinctive to each discipline, and Lefèvre echoed the point throughout his introductions. Indeed, as I argued in Chapter 4, the synoptic tables of axiomatic definitions that introduce many Fabrist books reflected this care for elementary principles. Therefore, when he published his second set of paraphrases and introductions to logic in 1503 he did not merely counsel avoiding the discipline altogether, but advised instead to focus on logic's own foundational principles. He presented the whole range of Aristotle's logical works, the *Organon*, editing the standard medieval translation of Boethius with reference to new Greek scholarship. But his paraphrases made short work of the later, technical studies of the *Sophistic Elenchus* and the *Topics*—the texts that late medieval schools had evolved into the genres of paradoxes and modes of disputation known as the *logica moderna*. Lefèvre hastened past those practices, instead lavishing lengthy commentaries and paraphrases on the first part of logic, the *logica antiqua* that dealt primarily with Aristotle's *Categories* and Porphyry's brief *Isagoge*. These dealt, according to Lefèvre's own classification, with the formation of propositions—*proloquia*.[80] At various places, *proloquia* referred to axioms or the

[76] Caroline Walker Bynum, *Christian Materiality: An Essay on Religion in Late Medieval Europe* (New York: Zone, 2011); on a similar 'hunger for reality', see Heiko Augustinus Oberman, *The Dawn of the Reformation: Essays in Late Medieval and Early Reformation Thought* (Edinburgh: T. & T. Clark, 1986), 55.

[77] Jacques Lefèvre d'Étaples (ed.), *Hic continentur libri Remundi pij eremite* (1499), a1v (PE, 77): 'qua [superna infusio] sapientes huius saeculi longe praecelleret'.

[78] Vittorio Hösle and W. Büchel (eds), *Raimundus Lullus, Die neue Logik. Logica Nova*, Philosophische Bibliothek 379 (Hamburg: Felix Meiner Verlag, 1985), introduction.

[79] Lefèvre, *Totius Aristotelis*, b1v (PE, 5): 'Et singula secundum subiectam materiam volunt esse intelligenda, et omnia Physica physice intelliguntur, et Metaphysica divine, et Logica logice. Voluntque singula in propriis locis, et ex propriis esse disquirenda.'

[80] Following a classification shared with Agricola and Ramus, Lefèvre classified the *Organon* into three parts: *pars proloquendi, pars iudicandi, pars inveniendi*. See Cesare Vasoli, *La dialettica e la retorica dell'Umanesimo: 'Invenzione' e 'Metodo' nella cultura del XV e XVI secolo* (Milan: Feltrinelli, 1968), *passim*.

first principles of a discipline.[81] Good logic focused on how logic formed its own first principles.

Since antiquity, the basic elements or principles of logic had been set by Aristotle's *Categories* (usually known as the *Praedicamenta*) and Porphyry's *Isagoge* (or *Quinque Voces*). Porphyry taught how to use five words—definition, genus, differentia, property, and accident—as a basic induction to logic, to the problems of predication, to how definitions and proof work.[82] In sum, Porphyry provided a framework for the Aristotelian philosophy of substance. Aristotle's *Categories* then gave ten categories for thinking about the world, most importantly the first four of substance, quantity, quality, and relation. It is here we find the deepest assumptions of the Aristotelian approach to nature as a set of substances; here Lefèvre and other teachers extending back to Boethius and before wrestled with the problem of turning objects in the world into words and concepts.

In particular, these works set the terms in which philosophers addressed the three speculative disciplines (natural philosophy, mathematics, and first philosophy or metaphysics). This triad traditionally sealed mathematics off from natural philosophy. It was a medieval commonplace, echoing Aristotle's *Metaphysics* E, that physics studies moving things with universals inseparable from matter, while first philosophy studies universals in themselves, absolutely distinct from matter. Aristotle underscored the distinction of mathematics and physics by suggesting that, if mathematics was like either of the others, it was more like theology, since it deals with universals that are separable from matter.[83]

Lefèvre was more ambiguous, however, in his approach to universals. His strategy can be found in his comments on Aristotle's first category, substance, where he sidesteps a fundamental division within late medieval philosophy. Realists held the view that universals really exist outside the mind; there exists a universal form or *species* of cat, existing apart from each individual cat. Nominalists argued that universals are simply a matter of convention and words; we agree on 'cat' to name a certain group of mammals. Lefèvre found both of these positions too extreme:

> Certain people consider things by themselves, completely ignoring notions and concepts (*rationes*); others only care for notions and concepts, and entirely flee from

[81] In his commentary on Aristotle's sustained discussion of first principles in the *Posteriora analytica*, Lefèvre called such principles *dignitates, proloquia, communesque scientie* (axioms, first propositions, common principles), terminology that Aristotle explicitly borrowed from geometry.

[82] See the commentary of Jonathan Barnes, *Porphyry: Introduction* (Oxford: Oxford University Press, 2003).

[83] David Albertson helpfully reconstructs the origins of this commonplace in *Mathematical Theologies: Nicholas of Cusa and the Legacy of Thierry of Chartres* (Oxford: Oxford University Press, 2014), 35–9. This formula is repeated in various contexts: e.g. Clichtove's commentary on Lefèvre's introduction to the *Metaphysica*; Clichtove on *De anima* (both in *Totius philosophiae naturalis paraphrases* (1502), respectively at 403r and 341v–3v); Gérard Roussel, *Divi Severini Boetii arithmetica, duobus discreta libris: adiecto commentario, mysticam numerorum applicationem perstringente, declarara* (Paris: Simon de Colines, 1521), 5rff.

things themselves and their considerations. Neither the former nor the latter can ever genuinely understand Aristotle and other authors.[84]

Neither particular nor universals are enough on their own. (The word 'concepts' here translates *rationes*, a distinctively Boethian term for universals of a certain kind.[85]) In fact, Lefèvre argued, 'those who penetrate analogies' know that they must consider both things and concepts to understand. He criticized logicians and natural philosophers for each taking only one part:

> For a logician is compared to someone who contemplates things through a mirror's images—who deals sometimes with the things themselves on account of their images and the names of things and again the names of those names. A physicist is the reverse, like the one looking on things themselves.[86]

By considering both words and things, Lefèvre sought a *via media* on the matter of universals.[87] In a short dialogue, inserted into his commentary on Aristotle's category of substance, Lefèvre presented his own version of a 'moderate realism'. Besides thing and word, he posited a third conceptual thing:

> For when I give the name 'man' and look at the thing itself, I do not properly name the thing itself, or the species, or the universal, simply and without addition. Instead, I do so with an addition, namely of the universal, conceptually (*ratione*).[88]

Universals exist—but conceptually, inside the mind. The mind constructs a real universal and builds it into the concept. This moderate realist position let Lefèvre avoid the binary of nominalism and realism on the question of universals.[89]

[84] Lefèvre, *Libri logicorum*: 22r–v: 'Quidam enim res solas considerantes, notiones rationesque prorsus abnuunt; alii solas notiones rationesque curantes, res ipsas rerumque considerationes penitus refugiunt. Et neque hi neque illi poterunt unquam sincere Aristotelem ceterosque auctores intelligere.'

[85] Lefèvre explicitly associates Boethius with notion and concept, saying that 'it is to be assumed that Boethius considered the notion and concept, whereas Porphyry considered the thing and subject itself' (*Igitur notionem rationemque Boetius, rem autem ipsam et ipsum subiectum considerasse, ponendus esset Porphyrius*; *Libri logicorum*, 6v). On this terminology of *notio* and *ratio*, see Alain de Libera, *La querelle des universaux. De Platon à la fin du Moyen Age* (Paris: Édition du Seuil, 1996), 48.

[86] Lefèvre, *Libri logicorum*, 22v: 'qui penetrant analogias aperte cognoscant…Assimilatur enim logicus alicui, rerum ex imaginibus speculi contemplatori, et res ipsas interdum, propter imagines illas et rerum nomina et iterum illorum nominum nomina tractanti. Physicus ex opposito: rerum assimilatur contemplatori.'

[87] I suspect Lefèvre here followed Boethius. In his commentary on Porphyry's *Isagoge*, Boethius presented a moderate realist position in which the mind abstracts universals from concrete things. In a process of collecting the likeness (*similitudo*) of each individual object, the mind grasps a universal as a notion or a concept that has a separate, incorporeal reality. Boethius [second commentary on Porphyry]: 'genus and species subsist in one way, they are grasped by the intellect in another way, and they are incorporeal, but they are joined to sensible things and subsist in sensible things. But they are grasped by the intellect as subsisting in themselves and not having their being in anything else' (quotation from John Marenbon, *Boethius* (New York: Oxford University Press, 2003), 30; see also Nauta, *In Defense of Common Sense*, 38).

[88] Lefèvre, *Libri logicorum*, 23v: 'Proinde cum hominem nomino et rem ipsam respicio, non proprie rem illam aut speciem aut universalem appello et simpliciter et sine addito, sed cum addito—scilicet universalem, ratione.'

[89] The position is significant not least because Lefèvre has often been crammed into one or another of the late medieval schools. The story begins with Carl Von Prantl, *Geschichte der Logik im Abendlande*, 4 vols (Leipzig: Hirzel, 1855), 4:278–80, who suggested Lefèvre was nominalist in sympathies,

This moderate realism had profound implications for Lefèvre's philosophy of number, which Lefèvre immediately worked out in the longest of his scholia in the whole work, on Aristotle's second major category, quantity. The category of quantity was the first place premodern students encountered the foundational disjunct between continuous and discrete quantities that we explored in Chapter 5; later they would encounter snippets in the *Physics*, *Posterior Analytics*, and *Metaphysics*. Lefèvre awarded special attention to quantity, over six large pages. (He gave the chief category of substance only four pages.) In outline, the discussion largely seems to confirm Aristotle's distinctions: discrete numbers are integers, while continuous quantities are lines, surfaces, and also describe the flow of time from past to future.[90] As we have already seen, Aristotle had set a chasm between physical qualities and quantity in a series of arguments: since lines are continuous, they can be infinitely divided and do not have minimal parts (i.e. atoms); quantities do not admit of more and less (one triad is not more threeish than another, though qualities like whiteness can be more or less); and quantities have no contraries, since 'small' and 'large' are not absolute but relative.[91]

Subtle shifts in this edifice are significant, given how deeply these assumptions had set Aristotle's natural philosophy on a track veering away from a mathematical physics. Lefèvre's own scholium makes small suggestions about mathematics' role in other sciences, listing astronomy, dialling (i.e. timekeeping), the *Physics*, and the intension and remission of forms. This last science is set in a striking place, near the beginning of the scholium, and returns elsewhere as an example of how to think about the mathematization of qualities. He concludes by differentiating the quantities of extension in space and succession in time from the qualitative changes of intension and remission.[92]

Lefèvre's most distinctive contribution in this scholium was his reorientation of the mental basis of numbers. Where Aristotle's account of quantity remained focused on mathematical objects and the words used to describe them, Lefèvre stepped back to the larger context of the human soul and its relation to the world. His 'moderate realist' account of universals borrowed the Boethian idea that people bring a mental something—a universal—to the objects they encountered and named in the world. Here, Lefèvre integrated this moderate realism with another distinction of Boethius. In an influential passage, Boethius had distinguished between two kinds of numbers: those which number other things (*numerantia*)

based on his introduction to Peter of Spain's *Summule*. Renaudet, 131, argued from one of Beatus' annotations that Lefèvre was nominalist. Both were cited to the same effect by Vasoli, 'Lefèvre e le origini del "Fabrismo"', 233.

[90] Cat. VI, 4b20–6a35.

[91] Cat. VI, 5b10–6a10. Many commentators observe that Aristotle's argument here is far from obvious: he begins with the point that a grain (e.g. of rice) is smaller than a mountain, but larger than other things, so it is not *per se* small, and so does not have some distinct (contrary) quality of smallness.

[92] Lefèvre, *Libri logicorum*, 27r: 'Item albedo suapte natura et per se in partes intensionis diducibilis est, at non iccirco quantitas... In qualitate vero intensio atque remissio.'

and those which are themselves numbered (*numerata*).[93] Lefèvre nuances this distinction by considering human experience. Humans are unique among animals because they count things, he reasons. Only in the human soul, therefore, are there numbers for numbering. Thus there must be not two, but three kinds of numbers: *numerantia*, *numerata*, and then *numeri* themselves. For '*numerantia* belong to souls applying their numbers to things, while *numerata* are those things to which the soul properly and aptly applies numbers. *Numeri* are the discrete concepts of numbers.'[94] To Boethius' scheme, Lefèvre adds numbers *in themselves* (*numeri*, or *numeri formales*)—numbers as they exist purely in contemplation, within the human mind. Here we find the same move made earlier for universals, where the soul has 'concepts' (*rationes*) in the mind that are separate from, but applicable to, the world. Lefèvre retains a wholly pure realm of mathematical objects distinct from numbers which, when mapped onto the world, are a number 'with an addition', an applied principle of counting (*numerantia*).

Here Lefèvre wrestles with the problem of aptness, how numbers can be wholly separate entities while still reliably measuring the physical world. For example, what of a singing tenor voice? How do its pitches neatly fit our intervals, our rhythms? The usual explanation was the Aristotelian theory of abstraction, in which the physicist considered pitch and rhythm *as if* drawn away (literally, to abs-tract is to draw away) from the physical phenomena. But Lefèvre avoids the language of abstraction, instead using language of recognition and approach: 'But reason comes to the object and determines, affirms, and approaches it'.[95] He compares this approach to authors and readers struggling to meet in the same sense:

> Authors especially look to those things that are apt and appropriate, and which exist in themselves—and if they are missing, they look as much they are able, and approach them. Therefore one should pay little attention to those who everywhere ascribe to the authors a false sense, that they wanted to say certain unsuitable things. People of that sort do not explain but instead torture, making up the text's thrust, leading away to every wrong understanding.[96]

The author strives to approximate a meaning; the reader strives to approximate the same meaning. Minds do not meet outside of matter, but both apply themselves to the same sense in the matter of the page.

[93] Boethius, *De trinitate*, III.10–15: 'Numerus enim duplex est, unus quidem quo numeraum, alter vero qui in rebus numerabilibus constat.' This distinction became commonplace in medieval philosophy, especially those areas more closely dependent on Boethius, such as the twelfth-century Chartrians: e.g. Nikolaus M. Häring, *Commentaries on Boethius by Thierry of Chartres and His School* (Toronto: Pontifical Institute of Mediaeval Studies, 1971), *passim*.

[94] Lefèvre, *Libri logicorum*, 27r: 'Unde sunt numerantia, sunt numerata, sunt numeri. Numerantia sunt anime numeros suos rebus applicante; numerata sunt ea quibus anima numeros apte accommodeque applicat. Numeri sunt discrete ille rationes numerandi.'

[95] Lefèvre, *Libri logicorum*, 28v: 'Sed ratio adveniens illum determinat, addicit, atque appropriat'.

[96] Lefèvre, *Libri logicorum*, 28v: 'Auctores autem ea que propria, accommodataque sunt, et que per se sunt, maxime considerant, que si desunt, ea considerant que quammaxime queunt, accedunt ad illa. Unde sit ut audiri parum debeant qui passim auctorum propria sensa, falsa esse astruunt, et impropria quedam auctores voluisse. Qui huiusmodi sunt non exponunt, sed extorquent, vimque littere faciunt, et ad intelligentiam queque pertrahunt repugnantem.'

Measuring therefore does for numbers as reading does for a text. Lefèvre developed the distinction between the two kinds of numbers in the soul, *numeri formales* and *numerantia*, around the practice of measurement. The analogy is that of a standard and a measuring instrument. A man might use his hand to measure out so many palms of material; but his palm is the instrument to convey a certain breadth, which is itself the principal standard. In just this way, the soul possesses a 'principalis' formal number, which in turn is applies to particular cases in the world. But just what are those particular cases? The example of reading requires both reader and author. The unspoken assumption in this philosophy of number is that *there are numbers in the world, waiting to be measured.*

Aristotle had devalued mathematics in physics by arguing that mathematical objects, being surface abstractions, have nothing to say about underlying principles of change in nature. In contrast, Lefèvre and his students argued that universals exist both in the world and in the mind. The ones in the mind are partly given, partly conjured up to match the world. As Bovelles put it in his *Elementa physicalia*, 'man is the one who supplies ideas and species of the sensible world'.[97] Instead of abstraction, we are given a process of measurement, in which a kind of recognition or reading enables these mental principles to approximate the outside world. Because it raised the possibility that numbers do get at deep principles in nature, this modified realism undermined the Aristotelian division of mathematics from natural philosophy.

A MATHEMATICS OF MAKING

Already in his time, Lefèvre's *numeri formales* were recognized as a distinctive alternative to abstraction. The Spanish theologian and cosmographer Pedro Ciruelo flagged the alternative in his *Cursus quatuor mathematicum*, which drew heavily on Clichtove's arithmetic and Lefèvre's music.[98] In a prefatory discussion of the nature of numbers, Ciruelo presented the traditional views of the realists and the nominalists. He then added a third view held by two *neoterici*, the view that 'there exist in the soul of a counting man certain divine and spiritual things, abstracted from time and motion and from all physical qualities and bodily change.'[99] To Lefèvre he assigned the view that 'these mental numbers are certain notions or collective concepts of things... and that such concepts are true species of the predicament of quantity'.[100] He went on to associate Giovanni Pico della Mirandola with the same

[97] Bovelles, *Libri physicorum elementorum* (1512), 57v: 'Quod homo est supplementum idaearum et specierum sensibilis mundi.'

[98] Printed four times, at Alcala (1516, 1523, 1526) and Madrid (1528), the book also included Ciruelo's own augmentation of Clichtove's arithmetic, his paraphrase of Witelo's perspective, and a geometry associated with Bradwardine.

[99] Pedro Ciruelo, *Cursus quatuor mathematicarum artium liberalium* (Madrid: Michaelis de Eguia, 1528), A4r: 'sunt in anima hominis numerantis res quaedam divinae et spirituales abstractae a tempore et motu atque omni phisica qualitate et transmutatione corporali.'

[100] Ciruelo, *Cursus quatuor mathematicarum*, A4r: 'quod numeri illi mentales sunt quaedam notitiae vel conceptus rerum... quod tales conceptus sunt verae species praedicamenti quantitatis'.

application of those mental *numeri formales* to the physical world through the procedure of measurement.[101]

This account of knowing as making responded to fifteenth-century concern about the balance of the active and the contemplative lives. Late medieval reformers argued that education had to fit within the larger *universitas* of knowledge, as well as serve the welfare of society. In Chapter 1 we encountered Jean Gerson, who believed that unfettered natural philosophy distracted clergy from their ecclesiastical duties; contemporaries such as Leonardo Bruni claimed that the kind of physics done by fourteenth-century masters was useless, doing nothing for the *vita activa*. The Fabrists tried to carve a middle way, balancing natural philosophy between its practice as a self-sufficient discipline and its larger framework of wisdom.

They found a source for balancing these requirements in a favourite source, Nicholas of Cusa. Cusanus presented knowing as a process of approximation by iterations of creative, mental measurement.[102] In *De beryllo*, Cusanus tied this directly to the famous dictum of Protagoras:

> You will note the saying of Protagoras that man is the measure of things. For with the senses, he measures sensible objects, with the intellect he measures intelligible ones, and he deals with things beyond intelligible matters by projecting further.[103]

For Cusanus, the measuring mind is fundamentally constructive; it creates ideas within itself in a quasi-divine imitation of God. He repeated the Hermetic observation that 'man is a second god', claiming that

> just as God is the creator of real entities and natural forms, so man is [creator] of rational entities and artificial forms, which are nothing but similitudes of his own intellect, just as God's creatures are similitudes of the divine intellect.[104]

Knowing therefore is a continual process of measurement and making. Just as God created all things by productively measuring them out into the world, man

[101] On Pico and such numbers, see Jean-Marc Mandosio, 'Beyond Pico Della Mirandola: John Dee's "formal Numbers" and "real Cabala"', *Studies in History and Philosophy of Science Part A* 43, no. 3 (2012): 489–97.

[102] The theme of measurement is evident many places. For an account focused on the *Idiota* dialogues, Michael Stadler, 'Zum Begriff der Mensuratio bei Cusanus: Ein Beitrag zur Ortung der Cusanischen Erkenntnislehre', in *Mensura: Mass, Zahl, Zahlensymbolik im Mittelalter*, ed. Albert Zimmermann (Berlin: de Gruyter, 1983), 118–31. On *De coniecturis*, which is constructed around the metaphor of measurement, see Jocelyne Sfez, *L'Art des Conjectures de Nicolas de Cues* (Paris: Beauchesne, 2012).

[103] Cusanus, *De beryllo*, §6: 'notabis dictum Protagorae hominem esse rerum mensuram. Nam cum sensu mensurat sensibilia, cum intellectu intelligibilia, et quae sunt supra intelligibilia in excessu attingit.'

[104] Cusanus, *De beryllo*, §7: 'Quarto adverte Hermetem Trismegistum dicere hominem esse secundum deum. Nam sicut deus est creator entium realium et naturalium formarum, ita homo rationalium entium et formarum artificialium, quae non sunt nisi sui intellectus similitudines sicut creatorae dei divini intellectus similitudines. . . . Unde mensurat suum intellectum per potentiam operum suorum et ex hoc mensurat divinum intellectum, sicut veritas mensuratur per imaginem.' Cusanus also embeds this similitude deep in *De coniecturis*.

creatively conjectures an approximation of that world within his mind. This framework justified the conceit we encountered in Chapter 5, where Cusanus likened the thinking intellect to the mapmaker practising his craft.

The theme of productive measurement informed the account of wisdom that the Fabrists presented whenever they stepped back to discuss the broader goals of university learning.[105] Introducing Aristotle's *Metaphysics*, Lefèvre presented the wise man or *sapiens* as characterized by the *habitus* of knowing the causes of all things. To know all things, the wise man seeks the knowledge of all the arts, from architecture to theology. Such knowledge is ordered, just as nature itself is ordered; and the best forms of knowledge are those which are most certain. As a result, the *sapiens* is less concerned with particulars than with the principles, causes, and being that lie underneath particulars. Around 1500, Clichtove's brief account of all the arts *De divisione artium* followed the standard list of medieval arts, beginning with low mechanical arts such as agriculture and soldiering, moving through the liberal arts of the trivium and quadrivium, and ending with the moral arts (ethics, politics, and economics), natural philosophy, and finally metaphysics and theology.[106] Rules set out the tasks of various arts, and left their goal clear: 'natural philosophy and the rest of the lower arts unfold the way and direct the entry to metaphysical contemplations... Metaphysics is the bound and final end of all the sciences, to which all the others are ordered in a kind of wondrous beauty, as the highest peak of philosophy.'[107] Similarly, Charles de Bovelles organized his masterwork of 1511 around the central treatise *De sapiente*.[108] Here Bovelles similarly conceived wisdom as a collecting all knowledge of the world through the complete cycle of the arts.[109]

To be sure, Lefèvre and his students were not trying to undo the structure of learning they inherited. By setting contemplation or the study of universals at the peak of learning, the Fabrist project reflects the basic framework of medieval learning. Most elements can be traced through Aristotle's *Metaphysics*, where Aristotle had categorized physics, mathematics, and metaphysics together as speculative sciences. The emphasis on *theoria* as the aim of knowledge and the identification

[105] Bovelles, *De sapiente* (in *Liber de intellectu etc.*, 1511), often figures in accounts of the Renaissance vision of man. The first modern edition was published with the German original: Cassirer, *Individuum und Kosmos in der Philosophie der Renaissance* (Leipzig: Teubner, 1927). Landmark studies include Eugene F. Rice, *The Renaissance Idea of Wisdom* (Cambridge, MA: Harvard University Press, 1958); Emmanuel Faye, *Philosophie et perfection de l'homme: De la Renaissance à Descartes* (Paris: Vrin, 1998). More recently, see Michel Ferrari and Tamara Albertini (eds), *Charles de Bovelles'* Liber de Sapiente, *or* Book of the Wise, special issue of *Intellectual History Review* 21, issue 3 (2011).

[106] One might compare works by Hugh of St Victor, Gundissalinus, or Kilwardby. See further studies of this handbook cited in Chapter 2, note 104.

[107] Clichtove, *De divisione artium* ([*c*.1500]; Paris, 1520), c2v: '[Regula] octava: philosophia naturalis necnon caetere discipline inferiores ad metaphisicas contemplationes viam pandunt aditumque ministrant. [Regula] nona: Metaphysica omnium scientiarum meta est et finis ultimus: ad quam ut summm philosophie apicem cetere omnes miro quodam decore ordinantur.'

[108] Bovelles, *Liber de intellectu, etc.* (1511). Other students adopted similar language, e.g. Varennes, *De amore dialogus, etc.* (1512).

[109] On collecting as a key metaphor in this work, see Inigo Bocken, 'The Pictorial Treatises of Charles de Bovelles', *Intellectual History Review* 21, no. 3 (2011): 341–52.

of metaphysics with theology exemplified the broadly Boethian approach to the disciplines that shaped the emerging university.[110] The image of the world as emanating from the *Summum Ens*, in which the natural order presents traces of the goodness and beauty of its origin, picks up themes found in the Victorines as well as Bonaventure's own account of the *Metaphysics*.[111]

Yet, even if the basic framework is standard, the details are distinctive. As we have seen, the Fabrists thought long and hard about new media in teaching, genres of text, modes of speaking—the embodied procedures by which students are cultivated and habituated in knowledge.[112] For the Fabrists, the soul is constructive, its inner powers actively purifying material sensations.[113] Lefèvre and Clichtove offered a suggestive reading of *On the Soul* considering how it actively creates universals out of sensations. Mathematical images and diagrams were particularly important examples of this process, since they most easily evacuated their material content and therefore achieved universal status. Bovelles put it clearest in *De sapiente*, where he described man as a conjecturing double of God. The wise man knows by imitative, creative conjecture within. For this reason, the *sapiens* has two faces (*bifrons*), seeing both inward and outward. Bovelles defines the *sapiens* as a 'little god' who pursues knowledge for the sake of achieving perfect contemplative knowledge, mirroring God's creative action by recreating the order of nature within his own soul. By practising the complete cycle of the arts, the man mirrors the whole universe within himself—speculation, *theoria*, literally makes him the complete human being. Bovelles regards imagination as an active mirror; it plays gatekeeper to the intellect, selecting, purifying, and presenting sensations to the intellect's judgement.[114]

[110] Though there are delicate positions Lefèvre makes too: for example, on the equivocal status of the term *ens* as applied to God (contra Boethius). Contrast various positions in Kent Emery, Russell Friedman, and Andreas Speer (eds), *Philosophy and Theology in the Long Middle Ages: A Tribute to Stephen F. Brown* (Leiden: Brill, 2011).

[111] The original study of the 'book of nature' trope remains invaluable: Ernst Robert Curtius, *European Literature and the Latin Middle Ages*, trans. Willard R. Trask (Princeton, NJ: Princeton University Press, 1953), 319–25. See now Kellie Robertson, *Nature Speaks: Medieval Literature and Aristotelian Philosophy* (University Park, PA: University of Pennsylvania Press, 2017), 60–4; Eric Jorink, *Reading the Book of Nature in the Dutch Golden Age, 1575–1715* (Leiden: Brill, 2010). Clichtove often used the trope in this tradition, e.g. in his introductory notes on Lefèvre's *Paraphrases totius philosophiae naturalis*.

[112] I here support Klinger-Dollé, in Anne-Hélène Klinger-Dollé, *Le* De sensu *de Bovelles: conception philosophique des sens et figuration de la pensée* (Geneva: Droz, 2016), who argues that Bovelles's account of the various senses serves an ideal of university pedagogy as the path to wisdom.

[113] Leen Spruit, *Species Intelligibilis: From Perception to Knowledge, Vol. II: Renaissance Controversies, Later Scholasticism, and the Elimination of the Intelligible Species in Modern Philosophy* (Leiden: Brill, 1995), 39–45; Rebecca Zorach, 'Meditation, Idolatry, Mathematics: The Printed Image in Europe around 1500', in *The Idol in the Age of Art: Objects, Devotions and the Early Modern World*, ed. Michael Wayne Cole and Rebecca Zorach (Farnham: Ashgate, 2009).

[114] Bovelles, *De sensu*, in *Liber de intellectu etc.* (1511), 30r: 'Est igitur ratio in anime centro collocanda, ut verus anime oculus suam spectans circumferentiam, in cuius concavo rationabiliter sive per rationales species omnia continentur. In anime autem circumferentia, alius (convex ex parte) pingendus est oculus, qui corpus attingit, in eoque pariter omnia vidit, quam vocamus imaginationem, cuius actus est ut anime convexa circumferentia. Potentia vero corporis superficies concava, ut cerebrum, cui omnia phantasmata insunt.'

Parallels between mental and bodily making permeate Bovelles's writings. Bovelles uses the image of Prometheus to symbolize this power of invention. The wise man is fertile, and can produce 'the actual from potential, the whole from the principle, the work from the implanted power, understanding from nature, the whole thing from the beginning, the whole from the part, and finally the fruit from the seed'.[115] Just as Prometheus was fabled to have stolen divine fire from above to animate man from clay below, so the *sapiens* uses his godlike power of contemplation to bring the mind to life.[116]

But the physical craft of writing is the means to such wisdom. In *De sensu*, Bovelles developed a comparison between three powers or 'works of the soul, on which hang three forms of life: the contemplative, the active, and the factive'.[117] The first work remains within the soul itself, reflecting on universals. The second addresses the comportment of the body, which Bovelles especially associates with speech and the management of words.[118] The third work of the soul, its factive role, pertains to the mind's direction of activity in making things apart from the body. It is about the mechanical arts; unlike the other two, it is 'illiberal'. The close and constant attention Bovelles awards the mechanical arts is unusual in the history of medieval thought.[119] Although he does not overturn the whole classical division of liberal from mechanical arts, his example of what counts as such a 'factive' art does trouble the polar division of mind and hand. 'Teaching is a practical and active thing', he wrote, upholding moral philosophy as a practical part of the *vita activa*, but 'writing is factive'.[120] Writing, for Bovelles, involved the mind's ability to create in the external world, and so is a mechanical art. Yet as a medium that serves the mind, writing is the first step in learning, since it fixes evanescent sounds for regular review. 'Indeed, writing is a kind of collecting, iteration, memory, and recovery of the word, and it is entirely necessary for forming within us the

[115] *De sapiente*, in *Liber de intellectu etc.* (1511), 123v (wrongly paginated 121v): 'E tenebris [sapiens] emendicat elicitque splendorem, ex potentia, actum; ex principio finem; ex insita vi, opus; ex natura, intellectum; ex inchoatione, perfectum; ex parte, totum; et ex denique semine fructum.' See Faye, *Philosophie et perfection*, 103–9, as well as Jean-Claude Margolin, 'Le mythe de Prométhee dans la philosophie de la Renaissance', in *Histoire de l'exégèse au XVIe siècle*, ed. Olivier Fatio and Pierre Fraenkel (Geneva: Droz, 1978), 241–69.

[116] *De sapiente* 123v (not 121v): 'Hac enim in parte celebrem illum Prometheum imitatur, qui (ut poetarum fabule canunt) aut divum permissione aut mentis et ingenii acumine admissus, nonnunque in ethereos thalamos, postea quam universa celi palatia, attentiore mentis speculatione lustravit, nichil in eis igne sanctius preciosius ac vegetius reperit. Ita et sapiens vi contemplationis sensibilem mundum liquens penetransque in regiam caeli, conceptum ibidem lucidissimum sapientie ignem immortali mentis gremio, in inferiora reporta, eaque sincera ac vegetissima flamma naturalis ipsius tellureusve homo viret, fovetur, animatur.'

[117] Bovelles, *De sensu*, 49v: 'Ab his tribus anime opificiis, trinam hominis pendere vitam: contemplativam, activam, factivam.'

[118] e.g. Bovelles, *De sensu*, 50r: 'Sermo sive oratio et negociose sermonis artes, ad activam [vitam spectat].'

[119] Cf. classifications in Elspeth Whitney, 'Paradise Restored: The Mechanical Arts from Antiquity through the Thirteenth Century', *Transactions of the American Philosophical Society* 80 (1990): 1–169; George Ovitt, Jr., *The Restoration of Perfection: Labor and Technology in Medieval Culture* (New Brunswick, NJ: Rutgers University Press, 1987).

[120] Bovelles, *De sensu*, 50r: 'Docere, practicum et activum [est]; scribere factivum.'

posture of learning.'[121] While here Bovelles focuses on the spoken word (*vox*), his obsessive use of diagrams and images extends this principle to other visual media.

In another work, Bovelles applied this constructionist view of knowledge to mathematical physics. The *Elementorum physicorum libri* (1512) introduces the main topics of Aristotelian physics that Beatus learned under the acronym NaCaMILUT. Bovelles frames these topics, however, with an overt interest in mathematics. His typically playful approach to the metaphysics of numbers is already evident in the treatise's organization: it is composed of ten books, each with ten propositions. Far from being a strict axiomatic assembly, this schema gives Bovelles plenty of space to take excurses. At one point he offers an extended reflection on the 'order of teaching' (*eruditionis ordo*) as a movement from memory to memory. Knowledge moves from a teacher's memory, through their voice, into the student's ear, where it becomes a concept in their mind, which the student then keeps in memory.[122] Throughout the book, Bovelles refers to the creative power of the mind, notably on the topic of the vacuum. He first repeats the Aristotelian view that a vacuum is impossible in nature—like a chimera, it can only be imagined. But this raises the sceptical worry about how we can trust the mind's products. After all, if our minds conceive, invent, or construct everything, how do we tell truth from error? In a chapter *De veritate humanae mentis et concordia eiusdem cum rerum natura*, he returns to the image of the mind as a mirror. Both mirrors and minds can reflect impossible things. In response, Bovelles asks the reader to 'fashion the human mind as a perfect and whole mirror, furnished with two surfaces, the first one outward, the other inward and deeper. I say that images and explanations of impossible entities only go so far as the first exterior surface of the mind, not to the inner and deeper one.'[123] To justify this image, he speaks about a 'practical intellect' (*intellectus practicus*), positioned on the outside of the intellect, which recognizes when inner images clash with information from the senses. This practical intellect acts like the tongue with its gag reflex, expelling unhealthy food from the mouth before it enters the stomach for rumination. We may not quite be convinced. But the image reveals Bovelles at pains to use physical intuitions to defend his constructive account of mind.

This reliance on physical intuition ultimately is the justification for Bovelles's promise that mathematics offered the best way forward in natural philosophy. As for Lefèvre, the discussion of time seemed a good place to consider mathematics, since Aristotle had found time a main example of continuous quantity. 'Many philosophers', Bovelles suggests, 'have said that it is advantageous to see the soul as a self-moving number.'[124] Here Bovelles repeats a Pythagorean view of number as

[121] Bovelles, *De sensu*, 51r: 'Scriptura vero quaedam est vocis collectio, numerus, memoria, resumptio, ad formandum in nobis disciplinae statum pernecessaria.'

[122] Bovelles, *Libri physicorum elementorum*, 31v.

[123] Bovelles, *Libri physicorum*, 37v–8r: 'Finge siquidem humanam mentem speculum esse perfectum atque integrum, geminis praeditum superficiebus, una anteriore et extima, alia vero interiore et profunda. Dico imagines rationesque impossibiliter entium, dumtaxat anteriorem ferire mentis superficiem, non autem interiorem ac profundam.'

[124] Bovelles, *Libri physicorum*, 48v: 'Dixerunt [plerique philosophorum] enim haud ab re illam [animam] esse numerum seipsum moventem.' Aristotle records this view of 'some thinkers' at the

a source, which multiplies itself, as a special example of the Pythagorean view that all things are numbers. But instead of mocking the view as Aristotle had done, Bovelles accepts it, contradicting Aristotle's dismissal of quantities in favour of sensory qualities. The human rational soul is called 'rational' because it gives the 'reasons' or 'ratios' (the context leaves the Latin ambiguous) 'by which all things are counted, divided, distinguished, openly knowing the essences of each, judging their quiddities, determining their degrees and differentiae'.[125] Not only does number account for the 'essences' of things, but 'the senses are held to be slow and unable to count, and do not suffice to penetrate the deepest essences of things, hiding under the variety of accidents'.[126] In a near-perfect reversal of Aristotle's priorities, Bovelles assumes that the mind's numbers and extensions capture even the essences of nature.

* * *

The mathematical turn of Lefèvre's circle was bound up in a fascination with natural change and its relation to human making—a complex of interests all the more evocative because diffuse and often imprecise. Ciruelo noted the company such ideas could keep when he commented on the *numeri formales* shared by Lefèvre and Pico.

> Such men add something rather marvellous, that numbers have certain natural powers and inseparable properties by which they perform external events more perfectly than active qualities of the elements. Those who know such properties of numbers can apply those numbers to make marvellous effects, which they say belongs to natural magic. They confirm their opinion by the authority of the one who gathered together the Psalms in one volume, not in the order in which they were produced [i.e. edited], but in a varied order according to the harmonising properties of numbers and Psalms.[127]

In fact, it was Lefèvre who had provided in his *Quincuplex Psalterium* (1509) the 'harmony' of the various Latin traditions of the Psalms. Ciruelo here presents Lefèvre as an authority among natural magicians who seek causal power in numbers.[128] Regarding the operative power of numbers, Ciruelo perhaps has in mind a

beginning of *De anima*, 404b27–30. See also Ryszard Stachowski, *The Mathematical Soul: An Antique Prototype of the Modern Mathematisation of Psychology* (Amsterdam: Rodopi, 1992), chapter 2.

125 Bovelles, *Libri physicorum*, 48v: 'humana anima rationalis appellatur quia rationes ipsa cunctorum appraehendit, per quas cuncta dinumerat, discernit, distinguit, signulorum noscens essentias eorum in propatulo diiudicans quidditates, gradusque eorum ac differentias d[e]terminans.'

126 Bovelles, *Libri physicorum*, 48v: 'Et nonmodo sensus hactenus hebes et numerandi impotens esse depraehenditur, quatenus ad intimas rerum essentias sub accidentium varietate delitentes penetrare non sufficit.'

127 Ciruelo, *Cursus quatuor mathematicarum*, a4r: 'Addunt isti quod mirabilius est numeros virtutes quasdam naturales habere, et proprietates inseparabiles quibus agant in res exteriores perfectius quam qualitates activae elementorum. Et qui tales numerorum proprietates agnoscunt, possunt applicationes illorum numerorum ad res naturales facere mirabiles effectus, quod dicunt ad magiam naturalem pertinere. Confirmant hanc suam opinionem auctoritate eius qui psalmos Davidicos non eo ordine quo editi fuerant in unum volumen recollegit, sed variato ordine iuxta numerorum et Psalmorum concordantes proprietates'.

128 Note that Ciruelo, in his commentary on Sacrobosco's *Sphere*, named Lefèvre as offering a 'causal' description by modelling the sphere on a lathe. See Chapter 5, note 122.

few points in Lefèvre's commentary on the Psalms, where Lefèvre hinted at the powers of the names of God, as well as the numerology on which such Cabala was based.[129] It is also possible that Ciruelo is obliquely referring to one of the most intriguing texts Lefèvre produced, the *De magia naturali*.

The unpublished *De magia naturali* brings to surface the preoccupations that drive this chapter. Its first book casts the productions of nature in pro-Socratic terms of friendship and hatred, attraction and repulsion, coded in the categories of astrology and sometimes alchemy.[130] The second book supplies a Pythagorean numerology, under the title *de Pythagorica philosophia que ad Magiam introducit*. A full analysis of this work awaits a study of the manuscripts.[131] For now, one of its opening lines may suggest the larger fruits that Lefèvre, Bovelles, and their students expected from both mathematics and natural philosophy: 'The magic of the eastern Chaldeans seems to have been nothing but a certain experienced, practical learning of natural philosophy, applied to works.'[132] With such possibilities on the horizon, it seemed worth applying numbers to nature. This tradition motivated some of the most charismatic propagandists for mathematics of the sixteenth century, such as the Elizabethan intelligencer John Dee, who repeated Pico and Lefèvre's account of 'formal numbers' in his famous *Mathematicall Praeface* to the first English translation of Euclid.[133] A few pages further, Dee hints at what such a mathematics might mean when he cites Nicholas of Cusa's phrase *scientia experimentalis*, coining the English phrase 'Experimentall Science'.[134] Despite such echoes, as an outlet for the idea that mathematical objects have causal power this form of natural magic had limited influence. Lefèvre already had grown wary of how magic could deceive and be misused; the *De magia naturali* had small circulation, now extant in only one full and three partial manuscripts.[135]

Instead, what endured was what the Fabrists made together: their printed books, and the mathematical culture cultivated within them. I have tried to show in this chapter that even one of Aristotle's great Renaissance champions could still systematically consider the causal power of mathematical objects. Lefèvre's dialogues took aim at a favourite target of humanist ridicule: the divisive culture of disputation in which adversarial styles of learning drove hyperspecialization, wedging disciplines apart. In contrast, he presented an alternative style of discourse, in which friends strive together, united in the pursuit of truth—using mathematics to establish a harmony of learning rather than a fragmented cycle of arts. The conceptual tool

[129] Brian P. Copenhaver, 'Lefèvre d'Étaples, Symphorien Champier, and the Secret Names of God', *Journal of the Warburg and Courtauld Institutes* 40 (1977): 189–211.

[130] L. Pierozzi and Jean-Marc Mandosio, 'L'interprétation alchimique de deux travaux d'Hercule dans le "De magia naturali" de Lefèvre d'Étaples', *Chrysopoeia* 5 (1992–6): 190–264.

[131] This is done by Jean-Marc Mandosio, introduction to the forthcoming edition (Paris: Les Belles Lettres). I thank the author for sharing this work in draft.

[132] Olomouc, University Library, MS M.I.119, 174r: 'Chaldeorum orientaliumque magia nihil nisi quedam naturalis philosophie practica operis exercitiva disciplina fuisse videatur.'

[133] John Dee, 'Mathematicall Praeface', in *The Elements of Geometrie of the most auncient Philosopher Euclide of Megara* (London: John Daye, 1571), *1r. See Mandosio, 'Beyond Pico della Mirandola'.

[134] Dee, 'Mathematicall Praeface', A3v. Fritz Nagel, 'Scientia experimentalis: Zur Cusanus-Rezeption in England', *Mitteilungen und Forschungsbeiträge der Cusanus-Gesellschaft* 29 (2005): 95–109.

[135] On the declining interest in magic, see Chapter 2, note 53.

underlying this harmony was analogy, which respects difference while searching out what is shared among objects, ideas, or levels of reality.

Crossing boundaries was also viable because of a constructivist approach to mathematical objects, in which the fictions of the mind's eye can rightly map reality. Thus lines on a page, divided by degrees, could encourage reasoning across conceptual boundaries, testing how mathematical principles might act causally in natural philosophy. Writing for Jesuit colleges a few decades later, Christoph Clavius would justify this constructivist view of mathematical objects by citing antiquity, the theory of postulates found in Proclus' *Commentary on the First book of Euclid's Elements*. A student of Clavius' textbooks, René Descartes, would share this intuition when using mechanical modes of construction to classify mathematical curves.[136] Lefèvre made similar modes of reasoning out of medieval sources.

[136] Peter Dear, *Discipline and Experience: The Mathematical Way in the Scientific Revolution* (Chicago: University of Chicago Press, 1995), 217–20; William R. Shea, *The Magic of Numbers and Motion: The Scientific Career of René Descartes* (Canton, MA: Science History Publications, 1991), chapter 3.

7
Epilogue

The mathematical culture of Lefèvre and his circle took form in a moment of powerful possibility, as the Renaissance university assimilated print. In this moment, mathematics seemed an attractive tool for organizing the university arts around shared goals, virtues, and habits—for forming a certain kind of person. Such a formation relied on old practices: preserving authoritative texts, and making them live again for new generations. To some degree, these practices served to discipline students as docile servants of Europe's governing elite. Testing old pieties about how the humanist liberal arts make free thinkers, Anthony Grafton and Lisa Jardine once argued that Renaissance educational ideals instead served the growing needs of early modern rulers.[1] Mathematics seems an ideal example for their case. Over the next centuries, mathematics certainly was remade in the image of growing empires and their navigators, architects, and accountants.[2] This is the prevailing account of mathematical culture (and the account least critical of modern market-driven science)—it is a story seen from the court, its aims and its habits, rather than from the university and its practices.

Instead, I have told the story of the Fabrist moment from within the university. This perspective is undervalued within the history of science, especially in view of the university's fundamental and growing influence on early modern culture. Courts increasingly depended on the university, not just for bureaucrats; Galileo's credit as courtier is unthinkable without Galileo, professor. I have argued that Lefèvre's books originated in university classrooms, stretched over the frame of the university cursus, in collaboration with students. University practices of disputation, conversation, and reading defined the shape of their mathematical humanism. The university offered a contemplative vision of knowledge, ordering the interlocking disciplines into a whole cycle that aimed towards a singular goal: to form students who could contemplate within themselves the whole of knowledge. While such learning could benefit each part of knowledge, and society, this practice is explained by this university goal. To become such a person was a good in

[1] Anthony T. Grafton and Lisa Jardine, *From Humanism to the Humanities: Education and the Liberal Arts in Fifteenth- and Sixteenth-Century Europe* (Cambridge, MA: Harvard University Press, 1986).

[2] e.g. Britain's burgeoning empire is embedded within the tidal observations Newton used to construct his mathematical physics: Simon Schaffer, 'Newton on the Beach: The Information Order of *Principia Mathematica*', *History of Science* 47 (2009): 243–76. For broader literature see Lesley B. Cormack, Steven A. Walton, and John A. Schuster (eds), *Mathematical Practitioners and the Transformation of Natural Knowledge in Early Modern Europe* (New York: Springer, 2017).

itself. The Fabrists redefined Aristotelian 'true friendship' (*vera amicitia*) as 'studious friendship' (*amicitia studiosa*) among intellectual companions, harmoniously seeking this kind of shared knowledge.[3] Practices always fall short of ideals. But in the short, idealistic moment unfolded in this book, the Fabrists strove to aim university life towards this good.

The moment could not last. The Fabrist circle itself dissolved as members dispersed, their friendships strained by competing allegiances. In the 1510s, Clichtove, as an important representative of the Faculty of Theology, still took Lefèvre's side in debates over the Three Maries. Erasmus attacked him over a footnote.[4] In the 1520s, Lefèvre himself turned to religious reform in the diocese of Meaux, and communication between himself and Clichtove ceased entirely.[5] Such choices also strained relationships with his former students in the Rhineland, such as Beatus Rhenanus, who shared Erasmus' suspicion of early Protestantism.[6]

Mathematical interests changed in these moves from university to monastery, diocese, and finally the court of the king's sister, Marguerite de Navarre. In the monastery, mathematics did not disappear entirely, but now served biblical exegesis. Guillaume Briçonnet was Lefèvre's main patron after 1508, when Lefèvre retired to the monastery of Saint-Germain-des-Prés. Briçonnet had already been a proponent of monastic reform, and Lefèvre's own interests turned increasingly to the texts at the core of monastic life: commentaries on scriptures, first the Psalms (1509), then the Epistles of Paul (1512), and eventually the Gospels (1522). This was not a break, but rather an extension of his philosophical interests. He prepared an edition of a medieval manuscript on scriptural numbers by Odo of Morimond. Josse Clichtove too wrote a modest book, *De mystica numerorum* (1513), exploring the meanings of particularly resonant biblical numbers, such as 7 (days of creation), 12 (number of Israel's tribes and the number of apostles), and 10 (number of the Commandments). As he put it, 'if the philosophical tradition of number has such great capacity (*energia*) for humane disciplines, should sacred letters, whose first author is God, lack such mysteries or contain in themselves nothing so high or hard?'[7] Bovelles wrote in a similar vein, notably a meditation on the *Verbum* of the Gospel of John in relation to the Trinity, and a history of the world divided into

[3] See Chapter 6, the section on 'Friendship and Physics'; Richard J. Oosterhoff, 'Lovers in Paratexts: Oronce Fine's Republic of Mathematics', *Nuncius* 31, no. 3 (2016): 549–83, at 559–66. One might compare what kinds of people later universities intended to make, e.g. Chad Wellmon, *Organizing Enlightenment: Information Overload and the Invention of the Modern Research University* (Baltimore, MD: Johns Hopkins University Press, 2015).

[4] See Chapter 1, note 37.

[5] Henry Heller, 'Reform and Reformers at Meaux, 1518–1525', PhD dissertation, Cornell University, 1969; Henry Heller, 'The Evangelicalism of Lefèvre d'Étaples: 1525', *Studies in the Renaissance* 19 (1972): 42–77. The Meaux Circle predates consolidation of the Protestant movement, clearly shown by Frans Pieter van Stam, 'The Group of Meaux as First Target of Farel and Calvin's Anti-Nicodemism', *Bibliothèque d'Humanisme et Renaissance* 68, no. 2 (2006): 253–75.

[6] Peter G. Bietenholz, *Basle and France in the Sixteenth Century* (Geneva: Droz, 1971).

[7] Josse Clichtove, *De mystica numerorum* (Paris: Henri Estienne, 1513), a1v: 'Quod si in humanis disciplinis philosophica traditione numeri tantam habent energiam, symbolique sunt rerum cognitu dignissimarum, putandum ne est in sacris litteris, quarum primitivus author deus est, illos vacare mysteriis, aut nichil altum arduumve in se continere?'

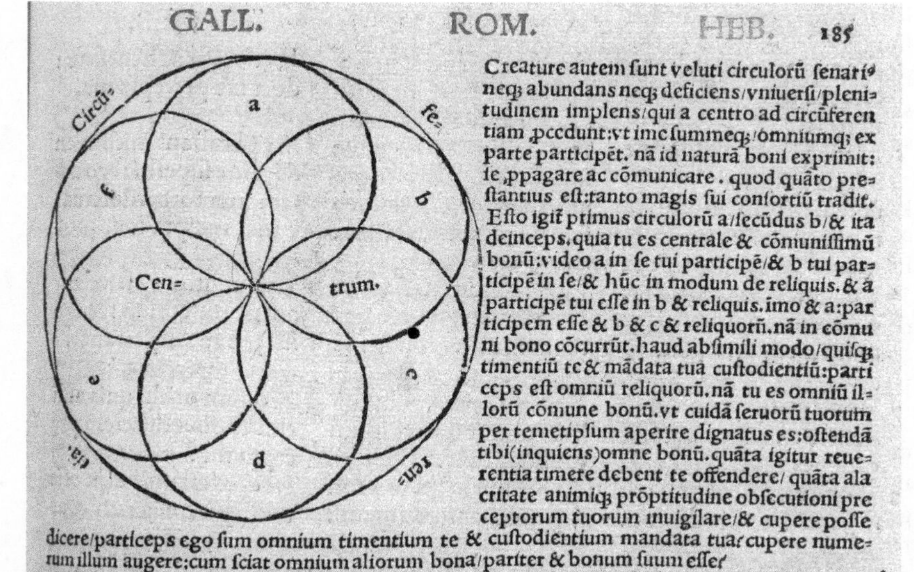
GALL. ROM. HEB. 185

Creature autem ſunt veluti circulorū ſenarius
neq; abundans neq; deficiens/vniuerſi/pleni=
tudinem implens/qui a centro ad circūferen
tiam pcedunt:vt imeſummeq;/omniumq; ex
parte participēt. nā id naturā boni exprimit:
ſe ppagare ac cōmunicare. quod quāto pre=
ſtantius eſt:tanto magis ſui conſortiū tradit.
Eſto igit primus circulorū a/ſecūdus b/& ita
deinceps. quia tu es centrale & cōmuniſſimū
bonū:video a in ſe tui participē/& b tui par=
ticipē in ſe/& hūc in modum de reliquis. & a
participē tui eſſe in b & reliquis. imo & a:par
ticipem eſſe & b & c & reliquorū. nā in cōmu
ni bono cōcurrūt. haud abſimili modo/quiſq;
timentiū te & mādata tua cuſtodientiū:parti
ceps eſt omniū reliquorū. nā tu es omniū il=
lorū cōmune bonū. vt cuidā ſeruorū tuorum
per temetipſum aperire dignatus es:oſtendā
tibi(inquiens)omne bonū. quāta igitur reue=
rentia timere debent te offendere/ quāta ala
critate animiq; prōptitudine obſecutioni pre
ceptorum tuorum inuigilare/& cupere poſſe
dicere/particeps ego ſum omnium timentium te & cuſtodientium mandata tua: cupere nume=
rum illum augere;cum ſciat omnium aliorum bona/pariter & bonum ſuum eſſe.
§ ℭMiſericordia tua dñe plena eſt terra:iuſtificationes tuas doce me. Miſerator imenſe qui omne h
bonū es/qui pars es & merces timentiū te & cuſtodientiū mādata tua/nō debito:ſed merces ſo=

Figure 7.1. God as a circle, enfolding the creation, in which creatures revolve from centre to circumference. Lefèvre, *Quincuplex Psalterium* (1509), 185r. By kind permission of the Syndics of Cambridge University Library, Shelfmark: Adams.4.50.6.

significant chronological segments.[8] Lefèvre himself occasionally returned to mathematical interests as a mode of rumination on scripture. On Psalm 119:63 [Vulgate 118], he marvelled at God as the Cusan circle and point, enfolding all creation within itself (Figure 6.1). The creatures move in circles within, from centre to circumference, imitating the Creator who bounds them. Most strikingly, the image is labelled and linked to the text almost like a diagram, but the text itself addresses God in the second person. The text is a ruminative dialogue rather than a dialectic proof: the diagram is a prayer.

Even such glimpses of number sink further from view after 1521, when Briçonnet called Lefèvre and several colleagues to Meaux, near Paris.[9] As bishop of the diocese, Briçonnet had tried to get his prebend-holders to stay in their parishes and preach; when that failed, he invited Lefèvre and close fellows such as Gérard Roussel and Guillaume Farel to retrain his clergy. Their primary task was to educate preachers in thirty-two preaching circuits, besides preaching themselves. Removed from the leisure of Paris colleges and abbeys, the Meaux circle now spoke to a working diocese. They began to preach and write in French.[10]

[8] Bovelles, *Commentarius in primordiale Evangelium*; Charles de Bovelles, *Aetatum mundi septem supputatio* (Paris: Josse Bade, 1520).

[9] The most up to date account is Jonathan A. Reid, *King's Sister – Queen of Dissent: Marguerite of Navarre (1492–1549) and Her Evangelical Network*, 2 vols (Leiden: Brill, 2009).

[10] On the shift to the vernacular in preaching, see Larissa Taylor, *Soldiers of Christ: Preaching in Late Medieval and Reformation France* (Oxford: Oxford University Press, 1992), 55–6.

Lefèvre published his last Latin commentary on the Bible in 1522, and his next grand project was a French translation of the whole of the Bible. To this new lay audience, extended reflections on abstract mathematics would have seemed futile. In 1521 Gérard Roussel produced a splendid last burst of the old mathematics in an extensive commentary on Boethius' *Arithmetica*, full of reflections on Lefèvre, Cusanus, and the tradition they represented.[11] But the work was the product of an earlier decade, rooted in the university. A preaching handbook from this period betrays not a whisper of those earlier preoccupations.[12]

Lefèvre did survive Meaux, but the experience sublimated his mathematical legacy. Marguerite de Navarre was a close supporter of Briçonnet's experiment—over 120 letters survive from their correspondence during the Meaux years.[13] Briçonnet styled her the 'Captain' of the group, and the known correspondence begins with her request for Briçonnet to send her Michel d'Arande, one of Lefèvre's disciples, to continue his teaching at her court.[14] Despite her support, the Meaux project crumbled early in 1525 under the pressure of the conservative faculty of theology at Paris, having failed to win the king's favour. Retaliation turned ugly: a weaver from Meaux was among the first burned in the French Reformation. Lefèvre fled to Strassburg, where he stayed until he was pardoned the following year. Marguerite arranged for Lefèvre to become Francis I's royal librarian and tutor to his children, a post he held until his death in 1536.

At court, mathematical themes took on new shapes. In one guise, it underwent what we might call a 'literary turn'. Jonathan Reid has carefully reconstructed the extent to which the Navarrian network depended on the Meaux circle, and particularly on Lefèvre. In turn, forced out of the diocesan context, the poets and musicians whom Marguerite drew around herself in the late 1520s became the seminal vernacular voices of the French Renaissance: Clément Marot, François Rabelais, Charles de Sainte-Marthe, the Du Bellay brothers with their clients, and of course Marguerite herself.[15] A full account of this literary turn would recognize how mathematics moved with the *litterae humaniores*, as intellectuals established dominance at court and in the new literary academies such as the Pléiade during Lefèvre and Roussel's last days toward the middle of the century.[16]

[11] Besides Michael Masi, 'The Liberal Arts and Gerardus Ruffus' Commentary on the Boethian De Arithmetica', *Sixteenth Century Journal* 10, no. 2 (1979): 23–41, see especially Henry Heller, 'Nicholas of Cusa and Early French Evangelicalism', *Archiv Für Reformationsgeschichte* 63 (1972): 6–21.

[12] Jacques Lefèvre d'Étaples, *Épistres & Évangiles pour les cinquante & deux sepmaines de l'an: fac-simile de la premiere edition, Simon du Bois*, ed. Michael Andrew Screech (Geneva: Droz, 1964).

[13] Guillaume Briçonnet and Marguerite de Navarre, *Correspondance, 1521–1524*, ed. Christine Martineau, Michel Veissière, and Henry Heller, 2 vols (Geneva: Droz, 1979).

[14] Reid, *King's Sister*, 1:187.

[15] Reid, *King's Sister*, *passim*, but especially 2:447–95.

[16] Such a work does not yet exist. On these academies, the standard work is still Frances A. Yates, *The French Academies of the Sixteenth Century* (London: Warburg Institute, 1947). On mathematical language as influenced by humanist literary standards, see Giovanna Cifoletti, 'Mathematics and Rhetoric: Peletier and Gosselin and the Making of the French Algebraic Tradition', PhD dissertation, Princeton University, 1993; Giovanna Cifoletti , 'From Valla to Viète: The Rhetorical Reform of Logic and its Use in Early Modern Algebra', *Early Science and Medicine* 11, no. 4 (2006): 390–423. For later in the century, see Isabelle Pantin, *La poésie du ciel en France sans la seconde moitié du seizième siècle* (Geneva: Droz, 1995). Cf. S. K. Heninger, Jr., *Touches of Sweet Harmony: Pythagorean Cosmology and Renaissance Poetics* (San Marino, CA: Huntington Library, 1974).

THE LEARNED LEGACY

As the Fabrist community moved away from the university and its encyclopaedic ideals of learning, their mathematical culture changed with it. Yet the project offered at least two legacies. The first has to do with Euclid, who has been a hovering but indistinct presence in this book. Although Euclid's *Elements* were a point of reference in debates over method, e.g. in relation to Aristotle's discussion of learning at the beginning of the *Posterior Analytics*, it seems university students rarely encountered Euclid directly. Lefèvre, Clichtove, and Bovelles never directly mention teaching him. As we have seen, Lefèvre included the Boethian fragments of Euclid as an appendix to the 1500 edition of the *Textus de sphera*. In a similar fashion, Bovelles offered an introduction in the mathematical compendium of 1503, later excerpted and reprinted in various handbooks. It is possible that students were expected to consult manuscript copies, or perhaps a copy following the Latin printings of 1482 and 1491 by Erhard Ratdolt at Venice.

It was not until later that Lefèvre finally published an edition of the *Elements*, and then not for the sake of university learning but rather for purposes of textual scholarship. In 1517, nearly a decade after retiring from the Collège du Cardinal Lemoine to devote himself to biblical studies, Lefèvre tried to make peace between two competing approaches to mathematics. The debate had begun in 1505, when the Venetian humanist Bartholomeo Zamberti published Euclid's *Elements of Geometry*, translated anew from the Greek in seven long years of labour.[17] Pouring scorn on the hubris and inaccuracies of Euclid's 'most barbarous' thirteenth-century translator Campanus of Novara, Zamberti believed he had restored Euclid's propositions to their primitive purity of around 300 BC, and their demonstrations to their state of original creation by Theon of Alexandria, in the fourth century AD.[18]

But philological eloquence meant nothing without mathematical elegance, or so thought the Tuscan Franciscan, Luca Pacioli, whose *Summa mathematica* (Venice, 1494) had squeezed all late medieval mathematics between two book-boards. Fra Pacioli was outraged by Zamberti's cavalier treatment of Campanus. Within four years Pacioli published a newly edited version of the medieval Euclid, retorting that Campanus, not Zamberti, was Euclid's 'most faithful interpreter'—at least once Pacioli himself had expurgated Campanus from the errors he had suffered at the sloppy pens of his medieval copyists. Besides different standards of Latin, Pacioli's Euclid did not include Theon's demonstrations. Campanus had replaced most of these with his *own*, and Pacioli thought them better.

In 1517, Lefèvre resolved this dispute of how to make mathematical progress by simply forcing the two texts to live together: he published Campanus alongside

[17] Zamberti, trans., *Euclidis . . . elementorum libros XIII. cum expositione Theonis* (Venice: Joannes Tacuinus, 1505). This episode is given by Paul Lawrence Rose, *The Italian Renaissance of Mathematics* (Geneva: Droz, 1975), 51–2. See also Hermann Weissenborn, *Die Übersetzungen des Euklid durch Campano und Zamberti eine mathematisch-historische Studie* (Halle a/S.: Schmidt, 1882), 56–64.

[18] For the standard attribution of the demonstrations to Theon of Alexandria in the Renaissance, see Robert Goulding, *Defending Hypatia: Ramus, Savile, and the Renaissance Rediscovery of Mathematical History* (New York: Springer, 2010), xviii, 36, 104, and esp. 150.

Zamberti's translation of Euclid, alternating them. As late as 1558 the Italian mathematician Francesco Maurolico described the benefit of this arrangement, noting that Lefèvre had enabled the two traditions to complement each other, 'since where one errs, the other corrects. For Campanus was faithful to [mathematical] ability and training, and wrongly changed many definitions, and often added something in order to help. Zamberti, despite translating faithfully, did not know how to do [mathematics] and so did not notice where the Greek exemplar was mistaken.'[19]

This is a very different context than the cycle of learning and the pedagogical commitments that animated Lefèvre's earlier mathematical works. For Maurolico, who with Federico Commandino was one of the great Renaissance editors of ancient mathematics, the goal was not for students to internalize certainties, as a way to mathematical literacy, but for experts to resolve uncertainties, textual and mathematical. Precisely because Lefèvre's Euclid includes none of the summaries, tables, and simplifying devices of his previous mathematical works, it was well suited to advanced study. Lefèvre's other collated translations were only of advanced works, beyond the BA curriculum.[20] The layout of each page makes clear the hierarchy and responsibilities of authors and commentators. Each book opens by proclaiming Euclid 'by far the most illustrious philosopher and prince of mathematicians', and propositions are organized 'first from the commentator Campanus, then from the Greek commentator Theon, as translated by the Venetian Bartholomaeo Zamberti'.[21] Euclid's own words—assumed only to be the definitions and propositions—are in larger type, and labelled with its translator. In smaller type, every demonstration or comment is labelled with its assumed author, named in upper case letters: *CAMPANUS*, or *THEON, ex Zamberto*. This care extends even to the diagrams, where often as not the edition offered wholly new diagrams for the different scholia of Campanus and Theon. The page becomes a scaffold for dismantling, scouring—and restoring—the authoritative façade of Euclid's Latin text.

[19] Maurolico to Juan de Vega, Messina, 8 August 1556: 'Celebris erat in euclideis libris apud nos Campani traditio; transtulit inde Zambertus Theonis editionem. Jacobus Faber hos in unum iunxit; utique melius facturus si e duobus unum opus coaptasset, ne idem bis repeteret. Nam, cum uterque peccasset, uterque corrigendus erat. Campanus enim, ingenio ac professioni confisus, multa in diffinitionibus perperam mutavit, nonnunquam aliquid ad usum adiicit. Zambertus, dum omnia fideliter transfert, ignarus negocii ne quidem mendas graeci exemplaris animadvertit, totusque in Campanum et ultra modestiae terminos excandescit, atque ibi ut plurimum eum carpit, ubi reprehendendus non est.' From BnF ms Lat. 7473, published in R. Moscheo (ed.), *I Gesuiti e le matematiche nel secolo XVI. Maurolico, Clavio e l'esperienza siciliana* (Messina: Società Messinese di Storia Patria, 1998), 42/sig. a4r–v, http://www.dm.unipi.it/pages/maurolic/edizioni/epistola/propriae/prop-12.htm (accessed 13 October 2017). In 1536, Francesco Maurolico had decried Zamberti's translation, claiming it only showed the need for new ones: see Maurolico, *Cosmographia* (Venice, 1543), sig. A2, cit. Rose, *The Italian Renaissance of Mathematics*, 74.

[20] Important examples are the three translations of the *Nicomachean Ethics* in the *Decem librorum moralium, tres conversiones* (1497), the *Quincuplex Psalterium* (1509), or the edition of Aristotle's *Metaphysica* in the translations of Bessarion and Argyropolus (1515).

[21] Lefèvre (ed.), *Euclidis elementa*, e.g. at 3r: 'Euclidis Megarensis clarissimi philosophi mathematicorumque facile principis: primum ex Campano, deinde ex Theone Graeco commentatore, interprete Bartholomaeo Zamberto Veneto, Geometricorum elementorum liber primus.'

This careful structure betrays sustained attention, even if Lefèvre only supplied a prefatory letter and no additional commentaries of his own. In that letter, Lefèvre thanked his old student François Briçonnet, cousin to his patron Guillaume, for asking him to publish the edition. It seems he had already edited the first ten books of Campanus' version some years earlier. When he was interrupted in 1514, he turned the task over to Michael Pontanus—that year Pontanus had proven his abilities by seeing Lefèvre's edition of Nicholas of Cusa's works through the press of Josse Bade.[22]

The first Euclid printed in France fits well within Henri Estienne's profile.[23] The printer had built his business around Lefèvre's works ever since he took over the press from Johann Higman in 1502—nearly all his earliest editions are by Lefèvre's circle, beginning with the compendium of Lefèvre, Clichtove, and Bovelles in 1503. Estienne seems to have specialized somewhat in a few significant, large, difficult volumes, in comparison to Bade who published more than twice as many works per year, often in collaboration with other Paris printers. In 1516 and 1517 Bade printed at least fifty-seven books, many in smaller formats; in those same years, Estienne published seventeen, of which eleven were in folio and six in quarto formats.[24] In retrospect, Estienne's press achieved particular triumphs with Lefèvre's *Quincuplex Psalterium* (1509), and his commentaries on the epistles of Saint Paul (1512, 2nd edition 1515). But despite this appetite for first editions of scholarly prestige, Estienne continued to publish university books for Lefèvre. Just in 1516 and 1517, Lefèvre's textbooks accounted for nearly a third of Estienne's output: Lefèvre's commentary on the *Nicomachean Ethics* (10 April 1516), the *Textus de sphera* (10 May 1516), the *Astronomicon* (with Clichtove's commentary, 1516), the *Artificialis introductio* to Aristotle's *Ethics* (February 1517), and the *Introductiones logicales* (with Clichtove's commentary, 30 April 1517).

Estienne specialized in large, challenging books, but the Euclid was an outstanding feat: 522 folio pages of irregular, layered typography, filled with enough woodcuts to accompany *two* versions of the *Elements*.[25] Briçonnet's prominent place on the title page and in the prefatory letter may reflect his financial support, though it is impossible to be sure. One clue to the resources bound up in printing the book comes from the price lists of Henri's son Robert Estienne, at least a decade later.

[22] Lefèvre (ed.), *Euclidis elementa*, 2r–v (PE, 379–80).

[23] On this edition, see Charles Thomas-Stanford, *Early Editions of Euclid's Elements* (London: Bibliographical Society, 1926), no. 6; PE, 378–83; Fred Schreiber, *The Estiennes* (New York: E.K. Schreiber, 1982), 35–7.

[24] Philippe Renouard, *Bibliographie des impressions et des œuvres de Josse Badius Ascensius, imprimeur et humaniste, 1462–1535*, 3 vols (New York: Burt Franklin, 1980), 1:83–5; Antoine August Renouard, *Annales de l'imprimerie des Estienne; ou, Histoire de la famille des Estienne et de ses editions* (Paris: J. Renouard et cie, 1843), 18–19.

[25] The first edition was already a feat. It is possible these diagrams were not all woodcuts, but that some were steel strips set in a matrix—though I have not noticed the telltale distortion of circles that such steel strips might leave. See Renzo Baldasso, 'La stampa dell' Editio princeps degli Elementi di Euclide (Venezia, Erhard Ratdolt, 1482)', in *The Books of Venice* (Venice: La Musa Talia/Oak Knoll Press, 2007), 61–100.

Robert listed the Euclid at 25 *sols*, by far the most expensive book.[26] At the time, the sum represented half a month's wages for a labourer in Paris.[27] The price may indicate the book was rare, but a comparison with other books also suggests that it was because of its expensive woodcuts. Even large folio volumes like the threefold translations of Aristotle's *Ethics*, without images, are priced cheaply at 2 *sols* 6 *déniers*. Meanwhile, books with more images cost several times more: Bovelles's works (1510) and an edition of Lefèvre's commentaries on the logical organon (1510) both cost 15 *sols*, while a copy of the much slimmer *Textus de sphaera* from 1516 costs 12 *sols*.

The apparatus and expense of the book takes Euclid far from the university classroom—it is a very different mathematical enterprise than we have found in the textbooks. Indeed, Estienne's edition was always present in sixteenth-century Euclidean scholarship. In 1533 Simon Grynaeus, professor of Greek in Basel, published the *editio princeps* of the Greek Euclid. This text would become important for a few elite mathematicians reading Greek, such as Commandino and Maurolico. But most scholars still depended on the Latin. In 1537 Herwagen also republished Lefèvre's edition, and despite the expense and difficulty of the enterprise he published it again in 1546 and 1558. These three editions also recycled a prefatory letter on geometry by Grynaeus' old school friend, Philip Melanchthon. Before Commandino's landmark translation of 1573, even adepts with excellent Greek such as Maurolico relied on the Euclid collated by Lefèvre and Pontanus.

The influence of this edition can be sensed perhaps in the more pedagogical versions of Euclid in the sixteenth century. One telling example is Oronce Fine's edition of the first six books of the *Elements*. To his patron, Francis I, Fine bewailed how students were left, 'through the carelessness of parents or teachers, without any taste of Euclid'.[28] But on the title page and in the letter to the reader, Oronce advertised the work's scholarly integrity, having inserted Euclid's Greek text (i.e. the propositions) and reworked Zamberti's translation of both propositions and demonstrations himself. By 1536, it seems, one could imagine that students might be interested in Euclid's Greek and up-to-date translations. The text itself, however, clearly owes a debt to Campanus. In the definitions that preface book five, where Zamberti simply translated the definitions without comment, Campanus had added pages of scholia, detailing the nature of *rationes*, *proportiones*, *analogiae*, and *habitudines*, building on Aristotle's discussion of quantity in the *Categories*.[29] Fine adopted Zamberti's translation of Euclid's definitions—explained with Campanus' commentary, redressed in more fashionable Latin.

[26] These prices are given by Renouard, *Annales*. At xiv–xv Renouard mentions the catalogue sources, but I have not been able to track them down.

[27] Around 1510 a building labourer earned 2s 6d per day. Micheline Baulant, 'Le salaire des ouvriers du bâtiment à Paris de 1400 à 1726', *Annales* 26, no. 2 (1971): 479.

[28] Oronce Fine, *In sex priores libros geometricorum elementorum Euclidis Megarensis demonstrationes: quibus ipsius Euclidis textus Graecus, suis locis insertus est; una cum interpretatione Latina Bartholomaei Zamberti Veneti* (Paris: Simon de Colines, 1536).

[29] Lefèvre (ed.), *Euclidis elementa*, 57r–62r; Fine (ed.), *In sex priores libros elementorum Euclidis* (1536), 107–15.

THE UTILITY OF MATHEMATICAL CULTURE

A second legacy of the Fabrist project has to do with its defence of mathematics. What is mathematics good for? Lefèvre's answer, I have argued, largely related mathematics to the contemplative aims of the university, as a propaedeutic to forming the universal scholar. This adjusts the narrative that mathematical culture spontaneously generated as European courts and merchants came to recognize the self-evident utility of mathematics as a tool of state and commerce. Mathematical culture did not leap fully formed from Galileo's head: it was made. Mathematics had to defend itself against cultural suspicion. It was perceived as difficult, obscure, and uncomfortably close to dubious magic and astrological prognostication.[30] Those we might properly call mathematical practitioners, who forged a career around their mathematical skills, were few and often unappreciated. When Francis I made Oronce Fine the first professor of mathematics in the Collège Royal in 1531, Fine became the sixteenth century's most visible mathematician, standing on the pedestal of both Paris's university and France's court. He was perhaps the most influential mathematical author of sixteenth-century Europe.[31] Yet his career was defined by financial hardship, which drove him to extraordinary productivity in crafting instruments, making maps, writing, editing, and illustrating books.[32] Jean Fernel, one of Fine's friends, also attempted to build a life around mathematics, but eventually was nudged by his father into the more lucrative career of medicine.[33] We will be hard pressed to understand the eventual profile of mathematical culture if we only focus on the material successes of its central practitioners. To understand how their contemporaries evaluated the uses of mathematics, we must consider the mathematics they encountered in the university.

It is true, however, that over the course of the sixteenth century there developed a strong rhetoric of the utility of mathematics. An extreme case has often been found in the figure of Gabriel Harvey.[34] We are indebted to Harvey for the dense annotations he left in his many books, which reveal his ambitions to transform himself into a courtly intelligencer with a special penchant for practical, technical knowledge that might serve the realm.[35] Harvey's roving eye fell on mathematics

[30] See Chapter 1.

[31] The dynamics of Fine's eminence are explored by Henrique Leitão, 'Pedro Nunes against Oronce Fine: Content and Context of a Refutation', in Alexander Marr (ed.), *The Worlds of Oronce Fine: Mathematics, Instruments and Print in Renaissance France* (Donington: Shaun Tyas, 2009), 156–71. For an example of Fine's prominence in mathematical libraries, see Alexander Marr, 'A Renaissance Library Rediscovered: The "Repertorium Librorum Mathematica" of Jean I Du Temps', *The Library* 9, no. 4 (2008): 428–70. The *Repertorium* includes eight works by Oronce Fine, the most of any modern author: nos 7a–d; 17a; 45a–b; 82a.

[32] Richard Ross, 'The Mathematical Works of Oronce Finé', PhD dissertation, Columbia University, 1971, 26–9.

[33] John Henry, '"Mathematics Made No Contribution to the Public Weal": Why Jean Fernel (1497–1558) Became a Physician', *Centaurus* 53, no. 3 (2011): 193–220.

[34] Jessica Wolfe, *Humanism, Machinery, and Renaissance Literature* (Cambridge University Press, 2004), chapter 5.

[35] Biographical details available in Jason Scott-Warren, 'Harvey, Gabriel (1552/3–1631)', *Oxford Dictionary of National Biography*, Oxford University Press, 2004; online edn, Jan 2016.

in the late 1570s, when he began to look beyond his post as university praelector at Cambridge to the royal court. Mathematics promised the kind of power of use at court, as Jardine and Grafton showed of Harvey's contemporaneous reading of Livy.[36] Harvey continually preached the value of the common-sense practicality of 'our Robert Recorde' and John Dee, keeping records of the artisans best exemplifying such knowledge.[37] But he recorded such thoughts in the archetypic mathematical text of the medieval university—in a 1527 edition of Lefèvre's *Textus de sphaera*. He first turned to the book in 1580 on the recommendation of Sir Philip Sidney, apparently during his short tenure as secretary to Robert Dudley, Earl of Leicester.[38] From the first annotation of the book, Harvey framed mathematics as best rooted in practical craft, contrasting the ignorance of scholars with the knowledge they might gain from 'architects, workers in wood and metal, gem-cutters, painters, measurers, sailors, and whoever deals with mathematical craft'.[39] But Lefèvre and his circle were exceptional among university scholars, for Harvey classed them with the *praegnantes polytechni* who contributed to the advancement of knowledge.

Harvey had picked up—and narrowed—a traditional defence of mathematics. Since Regiomontanus' oration in Padua in 1464, pedagogues had recycled tropes about the antiquity, dignity, and utility of mathematics. But between Lefèvre and Peter Ramus, another of Harvey's heroes, promises of utility became ever more prominent.

To be sure, Harvey's critique of university scholars somewhat missed the point. The contemplative vision of university knowledge for its own sake had always expected practical fruit. Clichtove explained that the *res publica* in fact mirrored the underlying order or *unitas* of creation itself.[40] A self-conscious society therefore should reflect on this unity, which would inculcate virtues that benefit the public order. Mathematics mattered partly for its own sake, as a segment of general learning. But this encyclopaedic remit always included the broader practical possibilities of the disciplines. A discipline such as mathematics would have intrinsic intellectual goods, but advantages would spill over into the commonweal too.

Still, Harvey was not entirely wrong to identify a stronger emphasis on utility among the Fabrists. Charles de Bovelles especially retained an interest in practical mathematics late into life. In 1511 Henri Estienne published his *Geometrie practique*, the first vernacular geometry printed in French.[41] Bovelles defended his use

[36] Lisa Jardine and Anthony Grafton, '"Studied for Action": How Gabriel Harvey Read His Livy', *Past and Present* no. 129 (1990): 30–78.

[37] Nicholas Popper, 'The English Polydaedali: How Gabriel Harvey Read Late Tudor London', *Journal of the History of Ideas* 66, no. 3 (2005): 351–81.

[38] Lefèvre, *Textus de sphaera* (Paris: Simon de Colines, 1527). Shelfmark: BL 533.k.1 [hereafter cited as Harvey's Sacrobosco]. A note on the title page reads 'Arte, et virtute, 1580'.

[39] Harvey's Sacrobosco, a1r, *in margine*: 'architectos fabros lignarios, et metallicos, lapicidae, pictores, mensores, nautas, et quicumque mathematicum aliquod exercent artificium'.

[40] PE, 367–8.

[41] Jean-Claude Margolin, 'Une Géométrie fort singulière: la Géométrie pratique de Charles de Bovelles (Paris, S. de Colines, 1542)', in *Verum et Factum: Beiträge zur Geistesgeschichte und Philosophie der Renaissance zum 60. Geburtstag von Stephan Otto* (Frankfurt am Main, 1993), 437–51. Practical geometry had existed throughout the middle ages, though rarely in *studia*. See B. L. Ullman,

of the vernacular in a Latin dedicatory letter to Étienne Petit, the king's treasurer. He pointed out that all human arts have both theoretical and practical aspects—even theology, 'the apex of all disciplines and highest theory' (*omnium disciplinarum apicem, summeque speculationis*), has a practical application in teaching men how to act, the deeds leading to God. By analogy, then, all disciplines lead from speculation to operation, from theory to practice.

> But all contemplation and internal reason is prior to the work and to the external or material remainder, just as whatever you meditate on is prior to what you will say or do. Thus it flows from the mind within and rises then into the mouth or the hands. Therefore, since there is no human art so liberal that it cannot also be mechanical . . . I printed this book in the vernacular tongue.[42]

Bovelles gave architecture as the example of the unity of liberal and practical arts, and he closed the letter by noting that Petit, as royal treasurer, was expected to build great buildings.

Bovelles came to practical geometry from theory at a time when such studies were getting good press in Paris. The Italian architect Fra Giocondo had come to France with Charles VIII, and he designed garden fountains and the Pont Notre-Dame (after it collapsed in 1499), and lectured on Vitruvius before he returned to Italy in 1505. These lectures were popular in Bovelles's circles. In 1503 Lefèvre recorded that he attended Giocondo's lectures, possibly at the house of Germain de Ganay. Guillaume Budé drew on those lectures when explaining Roman buildings in his *Annotationes in pandectas* (1508).[43]

Harvey could have found intimations of more practical geometry even within Lefèvre, Clichtove, and Bovelles's mathematical compendium of 1503. There Bovelles offered a defence of practical geometry that moved from the hierarchy of disciplines to the world of work. Bovelles retained a lifelong fascination with material objects, and his *oeuvre* uniquely combined tactile examples with abstract reflection. The lines, surfaces, and bodies of geometry epitomized this mysterious tangle of touchable and thinkable. He compared geometrical objects to numbers:

> Indeed number is more hidden than magnitude; even if you wanted to, you could not tell where it should be placed. Now, when a sound is made, it disappears without a

'Geometry in the Medieval Quadrivium', in *Studi di bibliografia e di storia in onore di Tammaro de Marinis*, vol. 4 (Verona, 1964), 263–85; Stephen K. Victor, *Practical Geometry in the High Middle Ages: Artis Cuiuslibet Consummatio and the Pratike de Geometrie* (Philadelphia, PA: American Philosophical Society, 1979).

[42] Charles de Bovelles, *Géométrie en françoys. Cy commence le Livre de l'art et Science de Geometrie: avecques les figures sur chascune rigle au long declarees par lesquelles on peut entendre et facillement comprendre ledit art et science de Geometrie* (Paris: Henri Estienne, 1511), [1]v: 'Ceterum contemplatio omnis et interna ratio, prior est opere et extero materialive complemento, quemadmodum prius est quicquid meditaris, eo quod loqueris, aut operaris. Quandoquidem ab interna mente, fluit oriturque, quicque deinceps in ore aut in manibus est. Cum igitur nulla sit humana ars tam liberalis quam mechanica esse nequeat … hanc vernacula lingua Geometriam cudimus.'

[43] Jacques Lefèvre d'Étaples, *Libri logicorum ad archteypos recogniti cum novis ad litteram commentariis ad felices primum Parhisiorum et communiter aliorum studiorum successus in lucem prodeant ferantque litteris opem* (Paris: Wolfgang Hopyl and Henri Estienne, 1503), 78v. See also P. de Nolhac, 'Recherches sur Fra Giocondo de Vérone', *Courrier de l'art* 8 (1888): 77–9; Vladimir Juřen, 'Fra Giovanni Giocondo et le début des études vitruviennes en France', *Rinascimento*, 2nd ser., 14 (1974): 101–15; Adolfo Tura, *Fra Giocondo et les textes français de géométrie pratique* (Geneva: Droz, 2008), 41–50.

> trace, and by that fact is less certain to us than magnitude. For the senses of magnitude are very certain, the former being touch, and the latter vision. And the heavens are separated from us by a nearly inaccessible space (*medium*), so thus it is that number, sounds, and the things of heaven are uncertain, while magnitudes are more certain to us, and their knowledge is prior with respect to us and less difficult in comparison with the other parts of mathematics.[44]

Geometrical objects are available to more senses, more permanently, than other features of the world—and therefore geometry is more certain and easier than even other mathematical disciplines.

Bovelles also promised technological advancement with historical anecdotes of geometrical invention: Egyptian priests had developed geometry to measure the flooding Nile.[45] He repeated this in the prologue to the *Livre singulier et utile*, a new practical geometry he published in 1542, expanding on how the Nile's annual flooding caused popular unrest. Egyptian pharaohs mandated their priests to control these 'controversies of the people' (*controversies populaires*), and so they devised the art of measurement to make adequate barriers to control the flooding Nile. While Bovelles did note that Greeks such as Pythagoras, Archimedes, and Euclid greatly refined geometry in a form that was now printed and so *divulgué*, he relativized their contribution by adding that 'there is no science so perfect that it cannot be increased everyday by many inventions, and so be given greater perfection'.[46] He cited Daedalus, said to have created the Labyrinth using mathematical tools; Perdix, Daedalus' nephew, who invented the geometer's emblematic tool, the metal compass; and Chorebus (according to Pliny) and Scytam Anacharsim (according to Strabo), who discovered the circular power of the wheel to make pottery.

Bovelles's geometrical dreams were about future development, and instead of serving an idyllic past of perfect knowledge, his history bespoke future promise. Bovelles capped the list with a story incessantly repeated by Renaissance pedagogues, of Archimedes, whose mathematical weapons held an invading force of Romans at a distance for many days.[47] This last story was an especially useful anecdote, no doubt, because it combined the promise of military technology—utility needing no defence—with an element of mystery, as it was not clear exactly what Archimedes had devised with such lethal power. Mathematical invention was *potentially* powerful, if only you applied yourself.

[44] Jacques Lefèvre d'Étaples, Josse Clichtove, and Charles de Bovelles, *Epitome compendiosaque in libros arithmeticos divi Severini Boetii, adiecto familiari [Clichtovei] commentario dilucidata. Praxis numerandi certis quibusdam regulis (auctore Clichtoveo). Introductio in geometriam Caroli Bovilli. Astronomicon Stapulensis* (Paris: Wolfgang Hopyl and Henri Estienne, 1503), 49r–v: 'Numerus quippe magnitudine secretior, quem vel si velimus, minime liceat ubi sit assignari. Sonus vero factus, evestigio disperit, ipsa ideo nobis magnitudine incertior. Magnitudinis namque certissimi sunt sensus, hic quidem tactus ille vero visus. Et a nobis celi, inaccessibili ferme medio distant, unde sit ut numeri, soni, et celi incertiores, magnitudines vero certiores nobis sint, et earum scientia nobis ceteris mathematice partibus prior, arduaque minus.'

[45] Goulding, *Defending Hypatia*, 1.

[46] Charles de Bovelles, *Livre singulier et utile, touchant l'art et practique de Geometrie, composé nouvellement en Francoys* (Paris: Simon Colines, 1542), 3r–v.

[47] Lefèvre, Clichtove, and Bovelles, *Epitome, etc.*, 49v.

Bovelles repeated the story in a lengthy addition to the 1547 edition of the *Livre singulier et utile*, with a new explanation of why he thought modern mathematics could support such promises. Here the story of Archimedes follows a list of practical 'secrets of nature' that demonstrate what geometry might be good for. Bovelles clearly believed that the example of Archimedes could be repeated, for he added that Archimedes was certainly a 'grand and subtle Geometer'—but he never managed to 'find and discover' the quadrature of the circle. This was a widely known flaw in Archimedes' otherwise heroic record. In his *Categories*, Aristotle had made the quadrature of the circle an example of objects of knowledge that existed, but had not yet been found. Bovelles triumphantly recorded that the quadrature of the circle had been 'in our time, discovered and affirmed without much effort'.[48] There were a series of recent claimants to this honour, from Nicholas of Cusa to Bovelles himself, as well as his friend Oronce Fine.[49] Bovelles made a popularized version of his quadrature available in the *Livre singulier et utile*, presenting it as practical, since it allowed conversions between square and round columns. Rhetorically, the solution of quadrature backed up technological promises: by squaring the circle, modern mathematicians had proved they could progress beyond their ancient exemplars in theory—*a fortiori*, they could also progress in technological (if not military) affairs.

We can hear all of these arguments echoed by the next generations of mathematicians at Paris. The closest inheritor was Oronce Fine. He enters the historical record in an edition of Lefèvre's *Astronomicon*, where he names himself in a distich on the final page.[50] Fine supplied engravings and perhaps the new annotations found in later editions of Lefèvre's *Textus de Sphaera* from 1516 onward; he also reworked Lefèvre's commentary in his own *De mundi sphaera, sive cosmographia*, as we saw in Chapter 5.[51] In 1535 he adapted large sections of Lefèvre, Clichtove,

[48] Charles de Bovelles, *Geometrie practique, nouvellement reveue, augmentee et grandement enrichie* (Paris: Reginald Chauderon, 1547), 69v.

[49] Aristotle, *Categories*, 7b30–4. Boethius, in his authoritative commentary on the *Categories*, claimed that the quadrature of the circle was known in his own day, though not in the earlier age of Archimedes. Lefèvre referred to Boethius' observation as well as Cusanus' apparent success, in his commentary: Lefèvre, *Libri logicorum*, 33r. All such proofs, and Fine especially, were disproved with great pleasure by the increasingly rigorous community of mathematicians in the second half of the sixteenth century. For literature on the debates over the quadrature of the circle, see Robert Goulding, 'Polemic in the Margin: Henry Savile against Joseph Scaliger's Quadrature of the Circle', in *Scientia in margine: études sur les marginalia dans les manuscrits scientifiques du Moyen Age à la Renaissance*, ed. Danielle Jacquart and Charles Burnett (Geneva: Droz, 2005), 241–59.

[50] Oronce Fine (ed.), *Theoricarum nouarum Textus Georgii Purbachii, cum expositione Capuani, Item de Prierio; Insuper Jacobi Fabri Stapulensis astronomicon: Omnia nuper summa diligentia emendata cum figuris ac commodatisssimis longe castigatius insculptis quam prius suis in locis adiectis* (Paris: Jean Petit, 1515), final page. On Fine's career as illustrator, see Robert Brun, 'Un illustrateur méconnu: Oronce Finé', *Arts et métiers graphiques* 41 (1934): 51–7; Isabelle Pantin, 'Oronce Finé mathématicien et homme du livre: la pratique éditoriale comme moteur d'évolution', in *Mise en forme des savoirs à la Renaissance: A la croisée des idées, des techniques et des publics*, ed. Isabelle Pantin and Gérald Péoux (Paris: Armand Colin, 2013), 19–40; Isabelle Pantin, 'Altior incubuit animus sub imagine mundi: L'inspiration du cosmographe d'après un gravure d'Oronce Finé', in *Les méditations cosmographiques à la Renaissance* (Paris: Presses de l'université Paris-Sorbonne, 2009), 69–90.

[51] First published within the *Protomathesis* (Paris: G. Morrhy, 1532), but also separately in 1542 (abbrev.), 1551, 1555, and (in French) 1551 and 1552.

and Bovelles's *Epitome compendiosa* (1503) on arithmetic, geometry, and perspective, adding them to his edition of the *Margarita philosophica*.[52] Fabrist university texts were the raw material for Fine's mathematical culture.

The kind of mathematics that actually occupied his time and made his reputation was practical. Beyond his exquisite maps, most of Fine's published works deal with measuring, calculating, and even instructing others on how to make their own sundials and other instruments; his one brief work on music deals with the use of various scales, not harmonic theory; he published several canons for interpreting ephemerides, the basis of astrological work; his masterwork the *Protomathesis* included a 'practical' arithmetic, instructions for applying cosmography to the mapping of land and sea, as well as a final treatise on dialling, the design and use of sundials and other measuring instruments.

Perhaps because Fine's exceptional craft skills left him in danger of being identified solely with the workshop, his prefaces especially focus on the propaedeutic value of mathematics.[53] Fine justified all of this practical mathematics by drawing attention to the way mathematics shaped one's capacity for invention. Geometry, he wrote to the King in 1530,

> Makes our wits lively
> Learned, subtle, taught, and fit
> To invent all manner of new thing.
> Do we not have the towns and cities
> Churches, houses, and castles
> Thus embellished by [geometry]?[54]

The use of *decorez* (embellish) in the last line disregards the uses of geometry in masonry and carpentry, instead focusing on design. He thus links geometry with the more decorous Vitruvian tradition and the liberal arts.[55] Even in his 'practical arithmetic'—the genre most closely associated with abacus and commerce—Fine takes pains to ensure readers do not expect him to consider 'the host of common questions', because he thought that was 'not only useless, but also unworthy of a mathematical man'. This book offered only 'the purer and universal practice of

[52] Reisch, *Margarita philosophica*, ed. Fine (Basel: Henricus Petrus for Conrad Resch, 1535). Also published 1583, and 1599 (in Italian).

[53] For detailed analysis of this rhetoric, see Isabelle Pantin, 'Oronce Fine's Role as Royal Lecturer', in *The Worlds of Oronce Fine. Mathematics, Instruments and Print in Renaissance France*, ed. Alexander Marr (Donington: Shaun Tyas, 2009), 13–30; Angela Axworthy, 'The Epistemological Foundations of the Propaedeutic Status of Mathematics according to the Epistolary and Prefatory Writings of Oronce Fine', in *The Worlds of Oronce Fine*, 31–51; Angela Axworthy, *Le Mathématicien renaissant et son savoir: le statut des mathématiques selon Oronce Fine* (Paris: Classiques Garnier, 2016), esp. 127–78.

[54] *Epistre exhortative* §7. Edited in Axworthy, *Le Mathématicien renaissant*, 365–80, at 374: 'Elle nous rend les espritz excitez | Doctes, subtilz, instruycts, habillitez | pour inventer toute chose nouvelle, | Ne fait on pas les villes & citez | Temples, maisons, & chasteaux habitez | Si decorez par le moyen d'icelle?'

[55] The scholarship on the reception of Vitruvius is very large. For an overview of responses by artisans and humanists, see Pamela O. Long, *Artisan/Practitioners and the Rise of the New Sciences, 1400–1600* (Corvallis, OR: Oregon State University Press, 2011), 62–93.

arithmetic'.[56] The next edition of the book made the point in the title: *Practical Arithmetic: Greatly Useful and Necessary for Those Who Aspire Not Merely to Mathematics, but to Philosophy*.[57] A bit defensively, Fine pursued the propaedeutic defence of mathematics in even the least likely places.

The possibility of progress beyond the ancients, as Bovelles suggested had happened in geometry, tantalized Paris intellectuals at the middle of the sixteenth century. Moving in the orbit of Marguerite de Navarre, poets such as Pierre Ronsard and Joachim Du Bellay explored the possibilities of French as a language equal in abundance and expression to ancient Latin, in part spurred on by the accomplishments of Petrarch in Italy. Their friend Jacques Peletier du Mans was as talented a mathematician as he was a poet, and applied his literary tastes to algebra and arithmetic.[58] Peletier circled the theme of progress in a series of letters interspersed through his *L'Aritmetique* (1549), dedicated to his old friend Theodore Beza. The age was one of progress, Peletier exclaimed: 'Has another time ever been found in which Philosophy, Poetry, Painting, Architecture, and new inventions for all sorts of commodities necessary for human life have flourished more?'[59] Print was not making people stupid, as some claimed, but nurturing curiosity. Even practical arithmetic was a sign of flourishing times. After all, the ancients had only written down the basics of theoretical arithmetic (i.e. number theory); they had left practical problems of commercial abacus to the vicissitudes of time. His new arithmetic therefore was yet another example of how moderns had surpassed antiquity.

As Natalie Zemon Davis once argued, Peletier's arithmetic belongs in the trajectory of the changing evaluation of practical mathematics. The Pléiade of Peletier in the 1540s and the *Académie* of Jean-Antoine de Baïf in the 1570s certainly took up the contemplative mathematics of heavenly harmonies and disciplining the mind, as Frances Yates argued at length.[60] At the same time, they also supported practical geometry for military purposes, and supported the authors of commercial arithmetics.[61] Peletier and Baïf represented court and academy, outside the university. But the most vociferous defence of mathematics' utility came from a voice for university reform, Peter Ramus.

[56] Oronce Fine, *Arithmetica practica*, 3rd edn (Paris: Simon de Colines, 1542), 67r: 'Ne miretur quispiam, aut nobis leviter imponat, si hanc nostram Arithmeticae praxin, innumera regularum seu vulgarium quaestionum multitudine, honorare distulerimus; utpote, quoniam id non inutile tantum, sed viro etiam mathematico censuimus indignum... Hac igitur de causa, puriorem ac universalem Arithmeticae praxin his quatuor libris perstringendam fore duximus.'

[57] Oronce Fine, *Arithmetica practica... Iis qui ad liberam quamvis, nedum Mathematicam, adspirant philosophiam perutilis, admodumque necessaria* (Paris: Simon de Colines, 1544).

[58] Cifoletti, 'Mathematics and Rhetoric'; Isabelle Pantin, 'La représentation des mathématiques chez Jacques Peletier du Mans: Cosmos hiéroglyphique ou ordre rhétorique?' *Rhetorica* 20, no. 4 (2002): 375–89.

[59] Jacques Peletier, *L'aritmetique, departie en quatre livres à Theodore Debesze* (Paris: Marnef, 1549), 3r. See Natalie Zemon Davis, 'Peletier and Beza Part Company', *Studies in the Renaissance* 11 (1964): 188–222.

[60] Yates, *The French Academies*.

[61] Natalie Zemon Davis, 'Mathematicians in the Sixteenth-Century French Academies: Some Further Evidence', *Renaissance News* 11, no. 1 (1958): 3–10; Natalie Zemon Davis, 'Sixteenth-Century French Arithmetics on the Business Life', *Journal of the History of Ideas* 21, no. 1 (1960): 18–48.

Like the Fabrists, Ramus hoped to reform university dialectic from the inside. Already in his early critique of Aristotelian dialectic, he argued that the deep, methodical structure that God gave both world and mind are, ultimately, mathematical. When he was banned from teaching philosophy, then, he turned his full attention to mathematics, convinced it held the key to method. Eventually he gained a chair at the Collège Royal, as professor of 'Philosophy and Eloquence', and when Oronce Fine's successor died, informally took it on himself to teach mathematics. Ramus therefore cultivated a circle of talented young mathematicians—almost by necessity, for his talent in mathematics was slender, and his biographer tells us he needed colleagues to coach him in the mornings on the mathematics he was to teach in the afternoons. This community of protégés would have great influence on mathematics in France, for they dominated the next generation of mathematical teaching in Paris: Jean Pena, Pierre Forcadel, Friedrich Risner, and Henri de Monantheuil.[62]

Reliance on others to teach his favoured discipline seems not to have bothered Ramus greatly. After all, he shared the attitude towards mathematical authorship that I have described earlier in this book, seeing propositions as doing the real work of mathematics; the demonstrations are not 'proofs' but simply explanations with which a teacher can always be flexible.[63] Besides, why should mathematics be quick or solitary? Unlike Peletier, Ramus held that progress in mathematics was slow and communal. His mature views were built around a historiography in which Euclid had destroyed the pristine geometry of antiquity by over-theorizing it.[64]

In fact, Ramus came to think mathematics was all about useful knowledge. Its origins were in practice. Introducing his *Arithmetica* (1555), he argued that in antiquity mathematics was the property of schoolboys as well as craftsmen such as painters and architects.[65] Indeed, the best mathematics was natural, and craftsmen represented the primitive unity of mathematics and nature. Therefore the way to advance the discipline was through the untutored insight of sailors, architects, and farmers. In his *Scholae mathematicae* (1569) he praised mathematics abroad, in England, Scotland, Italy, Spain, Portugal, and especially Germany, where mathematics was prominent in trade corporations as well as in universities.[66] France was conspicuously absent from the list. Ramus smarted from being banned from teaching philosophy in Paris, and then being rejected from the chair in mathematics—he explained that his natural mathematics was the true 'Socratic philosophy' that bypassed useless trivialities of the schools, and identified his sufferings with the trials of Socrates.[67] The French universities needed the dose of Socratic common sense underlying all useful mathematics.

[62] Isabelle Pantin, 'Teaching Mathematics and Astronomy in France: The Collège Royal (1550–1650)', *Science & Education* 15 (2006): 189–207.

[63] Robert Goulding, 'Method and Mathematics: Peter Ramus' Histories of the Sciences', *Journal of the History of Ideas* 67, no. 1 (2006): 63–85, at 71.

[64] Goulding, *Defending Hypatia*, 19–74.

[65] Peter Ramus, *Arithmeticae libri tres* (Paris: Andreas Wechel, 1555), a2v.

[66] Peter Ramus, *Scholarum mathematicarum, libri unus et triginta* (Basel: Eusebius Episcopius and Nicolas Pratre, 1569). Ramus digressed on the state of mathematics in England and Scotland in book I, Germany in book II, and Italy, Spain, and Portugal in book III.

[67] Goulding, *Defending Hypatia*, 42–3.

Ramus' own followers were not quite so dismissive of Paris's mathematical tradition. Gabriel Harvey had turned to what has been called 'pragmatic humanism' by way of Ramus. Early on, he identified Ramus as his guide in an oration he gave in 1575 as the praelector of rhetoric at Cambridge—the same year he published an *Ode natalitia*, an elegy to Ramus and his first publication. In following years, Ramus became ever more central in his reading notes.[68] Ramus decried his predecessors at Paris as useless. So it is especially ironic that Harvey found Ramus' notion of utility when he opened the mathematical books of Lefèvre's circle.

* * *

Harvey, Ramus, Peletier, and later Bacon, Descartes, and Galileo adopted a rhetoric of utility, progress, and novelty that deliberately obscured the textbooks they themselves had used, and they excoriated the very universities that cultivated the readers they expected to persuade. Historians have often recognized this duplicity, yet rarely have they paid serious attention to that broader mathematical culture, unconsciously replicating the prejudices of their chosen sources. Even those who have considered the universities—the German and English universities, the Italian studia, or the Jesuit colleges—have ignored the roots of mathematical culture in the late medieval university.[69] There is no doubt that the seventeenth century presented a new coherence of mathematical visions of the world—as does our own age. But we will not understand these developments if we ignore how learning and media, university and printing press came together to make such a mathematical culture.

In this book, I have gone to the origins of the printed textbook in Europe, in the circle of Lefèvre d'Étaples. For them, mathematics harmonized the cycle of the arts; in time, Lefèvre's works would populate bibliographies by Possevino and Vossius and others for the 'complete', 'perfect', or 'general scholar'.[70] We may be tempted to see this harmony of knowledge as an outmoded piety, an educational commonplace for mealy-mouthed educationalists. So it was. But it was also a sociological fact. Until the late seventeenth century, the 'general scholar' defined the mainstream ideal of learning for early modern Europeans. This fact gives the argument presented in this book weight beyond Lefèvre's circle. We have seen mathematical culture first germinating within the classroom, and then being disseminated in the experimental print projects of students and masters together. This was not a culture limited to mathematical practitioners, who constructed their careers around the various mathematical arts from architecture to astrology. Rather, these books cultivated a form of mathematical learning that was far more widespread and much more representative of mathematics' place in early modern culture. The books made in Lefèvre's circle set the background of assumptions that readers activated when they encountered more specialized mathematical texts. They set *doxa*, 'what everyone knows'. They established what is now common sense.

[68] Grafton and Jardine, *From Humanism to the Humanities*, 183–96.

[69] See the discussion in Chapter 1.

[70] See Chapter 1, note 36. On the general scholar, see Richard Serjeantson, introduction to *Generall Learning: A Seventeenth-Century Treatise on the Formation of the General Scholar by Meric Casaubon* (Cambridge: RTM Publications, 1999).

APPENDIX

Handlist of Books Annotated by Beatus Rhenanus, 1502–7

This appendix lists the schoolbooks Beatus Rhenanus annotated while studying at Paris. These are only books plausibly connected to university education, not all the books Beatus acquired during this time, which number more than 254 and were meticulously listed by a schoolmaster of nineteenth-century Sélestat, Gustav Knod.[1] His study remains the point of entry to this period of Beatus' life. This handlist supplements Knod's bibliography by taking account of annotations in Beatus' books; Knod only recorded annotations that index the awareness Beatus revealed of classical antiquity. Knod did mark Beatus' progress through the Paris arts curriculum, but omitted any account of which fascicles were bound together, leading to certain omissions. I have organized this list by current shelfmark and binding, which are in most cases original with Beatus Rhenanus (e.g. pp. 70–9, 162–4 above), recording some material conditions of his reading.

Joseph Walter compiled the main catalogue of printed books in the Bibliothèque humaniste de Sélestat in the early twentieth century.[2] Some of Walter's entries are highlighted in the third volume of Paul Oskar Kristeller's *Iter Italicum*, but the researcher must beware that all Sélestat shelfmarks cited by Kristeller are out of date.[3] The Bibliothèque humaniste de Sélestat is currently undergoing major renovations and revising its online catalogue, preventing me from rechecking certain entries.

To identify Beatus' level of interest, in this table I have distinguished between 'heavily' and 'lightly' annotated pages. Generally, heavily annotated pages have five or more different annotations or marks on the page (including underlining), though I have also considered a single long paragraph to be heavily annotated. Lightly annotated pages have four or fewer of Beatus' marks. Unless otherwise indicated, all notes are in the hand of Beatus (BR). This very rough quantification does not indicate *how* Beatus read, but does identify *which* books he read carefully while in school. While textbooks written by Lefèvre and his colleagues are heavily annotated with the signs of class lectures and intensive study, the others bear many fewer marks and perhaps were used chiefly for reference.

[1] Knod, *Aus der Bibliothek des Beatus Rhenanus*. This must be supplemented with the product of James Hirstein's extraordinary diligence, *Epistulae Beati Rhenani*. Anne-Hélène Klinger-Dollé is beginning a large collaboration on these annotations.

[2] Joseph Walter, *Catalogue général de la Bibliothèque municipale, primière série, les livre imprimés, troisième partie: Incunables & XVIme siècle* (Colmar: Imprimerie Alstasia, 1929).

[3] Paul Oskar Kristeller, *Iter Italicum: A Finding List of Uncatalogued or Incompletely Catalogued Humanistic Manuscripts of the Renaissance in Italian and Other Libraries* (London: Warburg Institute/ Brill, 1963).

Year location	**Shelfmark** Book	**Comments**
1501 Schlettstadt	**K 867** Peter of Spain, *Dicta circa summulas* (Heidelberg: Misch, 1490). Knod 20. Not in USTC/ISTC.	Title page has BR's ex libris, as well as some notes, and marks from previous owners. Few markings by BR.
	K 981 a. Ps-Albert the Great, *De secretis mulierum* (Speyer: Johann and Hist, 1483). Not in Knod. USTC 742402 b. Agostino Dati, *Elegantiae minores* (Speyer: Hist, 1491). Knod 10. Not in USTC/ISTC. c. Pseudo-Seneca, *Proverbia Senece secundum ordinem Alphabeti* (Strassburg? 1496?). Knod 17. d. Matheolus Perusinus, *Artis memoratiue Matheoli Perusini medicine Doctoris praestantissimi, Tractatus vtilissimus: com nonullis Plinij et Gordani documentis* (Strassburg: Schott, 10 Oct 1498). Knod 21. USTC 746934 e. [missing?] f. Dionysius Afer [Pomponius Mela], *Cosmographie sue de situ orbis Dionysii [Afer] Per Priscianum e greco in latinum metrica traductio* (Cologne, 1499). Knod 15. Not in USTC.	Annotations are relatively rare and light in these books; the most are found in the *Ars memorativa* (d).
	K 1051 Alexander de Villa Dei and Johannes Synthen, *Dicta super prima et secunda parte Alexandri* (Strassburg: Prüss, 1499). Knod 9. USTC 742634	This volume is noteworthy as one of the few volumes in the BHS that was in part written by Hegius, headmaster in Deventer, at the school of the Brethren of the Common Life.
1502 Schlettstadt	**K 829** Boethius, *De consolatione Philosophie cum commentis Thome [Aquino]* (Strassburg: Prüss, 1491). Knod 54. USTC 743512	A few scattered annotations in Beatus' hand, but not obviously systematic.
	K 948 Thomas Bricot, George of Brussels, and Peter of Spain, *Expositio Magistri Georgii super summulas* (Paris: Baligault, 1495). Knod 55. USTC 761183	This book contains quite a few annotations, some heavy.
	K1159 Peter of Spain and Marsilio of Padua, *Commentum emendatum et correctum in primum et quartum tractatus* (Hagenau: Gran, 1495). Knod 56. USTC 746469	Aside from the note on the inside of the front board, 'Est Beati Rhinow Slettstatensis', this exemplar is devoid of notes.

	K 810 a. Marsilio Ficino, *De triplici vita* (Bologna: Faelli, 1501). Knod 29. USTC 829418 b. Aquitanus Prosper, *Epigrammata sancti Prosperi episcopi regiensis de viciis et virtutibus ex dictis Augustini* (Mainz: Friedberg, 1494). Knod 48? USTC 748337 c. Jacobus Wimpfeling, *Germania ad rempublicam Argen.*, d. Jacobus Wimpfeling, *Isidoneus Germanicus* (Strasbourg: Grüninger, 1497). Knod 33–4. USTC 749858 *N.B.: (c) and (d) appear to be two works, published together as one volume; they do not appear to be the edition indicated by Knod.* e. Baptista Guarino, *De modo et ordine docendi ac discendi* (Heidelberg: Knoblochtzer, 1489). Knod 25. Not in USTC f. = **MS 325.** Incipit *Publii Vergilii carmen. Contra luxuriam et ebrietatem.* g. Filipo Beroaldo, *De felicitate opusculum* (Bologna: Plato de Benedictis, 1495). Knod 28. USTC 999724 i. Theoderic Gresemundus, *Lucubraciuncule bonarum septem artium liberalium Apologiam eiusdemque cum philosophia dialogum et orationem ad rerum publicarum rectores in se complectentes* (Mainz: Petrus Fribergensis, 1494). Knod 65; Walter 231. USTC 745387 k. Ps-Bede, *Repertorium sive tabula generalis autoritatem Aristotelis et philosophorum, cum commento per modum alphabeti* (Cologne: Quentell, 1495). USTC 743332	a. A handful of notes. e. Carefully annotated, followed by a paragraph comment on Cicero and education. i. Only a few notes, not in BR's hand.
1503 Schlettstadt	**K 1135** Boethius, *De philosophico consolatum sive de consolatione philosophie... cum figuris* (Strasbourg: Grüninger, 1501). Knod 65. USTC 616871	
1503 Paris	**K 875** a. Ermolao Barbaro, *Libri paraphraseos Themistii peripatetici acutissimi* (Venice, 1499) USTC 990523 b. Giles of Rome, *Expositio domini Egidii Romani super libros de Anima* (Venice, 1500). Not in USTC c. Walter Burley. *Super artem veterem Porphyrii et Aristotelis* (Venice, 1493). USTC 996485	Original binding. Diagrams on the flyleaves. All three works in the volume have been rubricated in the same hand. a. Of 224 pages, at least 20 are heavily annotated, and 15 lightly. b. Of 218 pages, 4 heavily annotated, 7 lightly. c. Of 164 leaves, 8 are heavily annotated, 2 lightly.

(*Continued*)

Year location	**Shelfmark** Book	**Comments**
	BHS K 908 a. Lucretius, *De rerum natura* (Venice, 1495). Knod 62; Walter 303. USTC 993475	Very few notes. a. 'Est Beati Rhenani Schletstattini Anno Domini millesimo quigentesimo tertio.'
	K 1079 a. Martinus Magistri, *Expositio perutilis et necessaria super libro predicabilium Porphyrii* (Paris, 1499). USTC 201556 b. Thomas Bricot, *Tractatus insolubilium* (Paris, s.d. [1498?]) c. *Practica sophismatum Buridani* (s.n.s.d.) d. Jean Buridan, *Consequentie magistri Joannis Buridani emendate per Albertum* (Paris: Marchant, 1499). USTC 202053 e. Andreas Limos, *Dubia in insolubilibus noviter emendata* (Paris: Gourmont, 1499). USTC 202053 f. Symphorien Champier, *Janua logice et phisice* (Lyon, 1498). USTC 201418	No annotations.
	K 1046 a. Jacques Lefèvre d'Étaples, Josse Clichtove, and Charles de Bovelles, *Epitome compendiosaque in libros arithmeticos Boetii* (Paris: Estienne, 1503). Knod 76. USTC 142874 b. Jordanus Nemorarii, *Elementa arithmetica, etc.* (Paris: Hopyl and Higman, 1496). Knod 77. USTC 201338 c. Sacrobosco, *Textus de sphera* (Paris: Estienne, 1500). Knod 78. USTC 201872	a. Of 224 pages, 64 are heavily annotated; pages with less than four notes are 36. b. No annotations. c. Out of a total of 32 folios (64 pages), pages heavily annotated are 17; 14 pages with less than four notes.
	K 1047 Jacques Lefèvre d'Étaples, *Libri logicorum* (Paris: Hopyl and Estienne, 1503). Knod 79. USTC 142880	Of 594 pages (297 folios), 121 are heavily annotated (5 or more notes per page), and 135 are lightly annotated. The guard- and paste-down pages are covered with annotations, and the final folios of the volume include a treatise titled 'Elementorum logicalium liber primus diffinitiones'.
	BHS K 1111 Politianus, *Omnia opera* (Venice, 1498). Knod 66. USTC 991842	Some lightly annotated pages, including on Poliziano's *Lamia*.

	BHS K 1123 a. Quintilianus, *Institutiones* (Venice, 1494). Knod 67. USTC 991632 b. Martianus Capella, *Opus de Nuptiis Philologie et Mercurii librio duo* (Venice: Henr. de Sancto Urso, 1499). USTC 996343	a. Some pages lightly annotated, especially in the first book. Title page includes various quotations from classical authors. b. Some light notes, especially in the section on dialectic. One of the final flyleaves includes some quotations from Plutarch.
	BHS K 1199 Jacques Lefèvre d'Étaples, *In hoc opere continentur totius philosophie naturalis paraphrasis... hoc ordine digeste* (Paris, Hopyl, 1501/2. USTC 142754	Of 864 pages (432 folia), about 300 are heavily annotated, and over 228 lightly.
	BHS K 1276 Aristotle, *Opera omnia* (Venice: Gregorio di Gregorios, 1496). USTC 997543	Some notes, in BR's school hand.
1504 Paris	**BHS MS 58** (cahier d'étudiant)	
	BHS K 822 a. Desiderius Erasmus, *Lucubratiunculae aliquot Erasmi* (Antwerp: Martens, 1503). Knod 85. USTC 400246 b. Desiderius Erasmus, *Adagiorum collectanea* (Paris: Philippe, 1500). Missing from Knod. USTC 201899 c. Bigus Pictorius Ferrariensis, *Opusculorum Christianorum libri tres* (Modena: Domenico Rocociala, 1498). Knod 83. USTC 996860 d. Giovanni Pico della Mirandola, *Auree epistole* (Paris: Michael le Noir, 1499). Knod 89. USTC 202041 e. [Rudolphus Agricola Frisius], *Axiochus Platonis* (Milan: Haliatte, s.d.). Knod 115. e1. Charles de Bovelles, *Metaphysicum introductorium* (Paris: Mercator, 1503). Knod 103. USTC 180087 e2. Charles de Bovelles, *In artem oppositorum introductio* (Paris: Hopyl, 1501). Knod 100. USTC 186511 f. Dio [Chrysostomous?], *Prusensis philosophus* (Paris: Denidel, s.d.). Knod 84. USTC 201705? g. = **MS 327**. *Eodipi Interpretationes.*	a. Title page includes 'Est Beati Rhenani Schletstattini. Anno Humanae reparationis 1.5.0.4. ¶ parrhisiis. ma. pro.' b. 'Est Beati Rhenani Schletstattini Parrhissiis m[anu]. pr[opria]. 1504.' e. Rhenanus scratched out the colophon, noting that the real author was not Plato but Rudolph Agricola. e1. A few notes indicate that Beatus read this. e2. Beatus wrote on the title page 'De intellectuali philosophandi modo, Isagoge.' Annotated, including a note on analogy. f. Several notes in Beatus' youthful and older hands. g. Several handwritten quotations, in Beatus' older hand, from various Latin poetry.
	K 853 Jean Buridan, *Commentum magistri Johannis Dorp super textu summularum* ([Lyons]: Carcain, 1499). Knod 111?	No notes found.

(*Continued*)

Year location	**Shelfmark** Book	**Comments**
	K 936 Boethius, *Opera* (Venice: Forlivio, 1499). Knod 96.	
	K 950 MS [no shelfmark] *Invenies sequentem Questionem a Georgio determinatam solio xxxiiii libri physicorum tertii* a. Thomas Bricot and George of Bruxelles, *Quaestiones super philosophiam Aristotelis* (Paris: Hopyl, 1491). Knod 99. USTC 201110 b. Thomas Bricot, *Metaphysica* (continuous with previous) c. Pedro Ciruelo, Pierre D'Ailly, *Uberrimum de sphaera* (Paris, 1498). Knod 106. USTC 202166 d. Pedro Ciruelo (ed.), Thomas Bradwardine, *Geometrica speculativa* (Paris: Marchant, 1495). Knod 105. USTC 201284 e. MS incipit *Circa principium octavi physicorum.*	MS. In Beatus' school hand, this appears to be a response to George's commentary on Aristotle. a. Heavy annotation on the title page, as well as on a few pages within the commentary itself. b. Very minor underlining, though the final page of the volume includes two paragraphs in Beatus' hand. c. Lightly annotated d. No notes. e. Beatus' hand, for a dozen folia.
	MS 15	
	K 984 a. = **MS 339.** b. Jacques Lefèvre d'Étaples (ed.), *Mercurij Trismegisti Liber de potestate et sapientia dei per Marsilium Ficinum traductus ad Cosmum Medicem* (Paris: Higman and Hopyl, 1494). Knod 114. USTC 202354 c. Lucianus, Brutus, Diogenes, *Palinurus, scipio romanus, etc.* (Avignon: Tepe, 1497). Knod 94. USTC 761274 d. Robert Gaguin, *Epistole, orationes, etc.* (Paris: Bocard, 1498). Knod 86. USTC 201431 e. Robert Gaguin, *De variis vite humane incommodis* (Paris). Knod 87. USTC 201430? f. Robert Gaguin, *Ars versificatoria* (Paris). Knod 88. Unclear which of the many Paris editions. g. Athanagoras, Xenocrates, and Cebes, *Athanagoras de resurrectione. Xenocrates platonis auditore de morte. Cebetis thebani Aristotelis auditoris tabula, miro artificio vite instituta continens* (Paris: Marchant, 1498). Knod 93. USTC 201404	Some notes on flyleaves, front and back. a. Mansucript of notes on Apuleius, 'in Libro de dogmate platonis'. b. Lightly annotated. g. Notes in Beatus' older hand.

K 1005

a. Giles of Rome, *Expositio Domini Egidii Romani super Libro priorum analeticorum Aristotelis cum textu eiusdem* (Venice: Toresano de Asula, 27 Sept 1499). Knod 97. USTC 998198
b. Laurentianus Florentinus, *In librum Aristotelis de elocutione* (Venice: Octavian Scot, 1495). Knod 101. Walter 291.
[c]. Giles of Rome, *Expositio Egidii romani super libros posteriorum* (Venice: Boneto Locatello for Ottaviano Scoto, 1495). Knod 98. USTC 998200

Just a few notes of loci, renvois, in the first few folios of b.
a. Title page: 'Est Beati Rhenani Sletstattini | Anno Sal. 1.5.0.4 | Parrhisiis. Ma.Pro.' No notes observed.
b. 'Est Beati Rhenani Sletstattini | Anno salutis 1504 | Ma. Pro. parrhisiis.'

K 1214

a. *Repertorium dictorum, Aristotelis, Ave | roys, aliorumque philosophorum* (Bologna: Bazalerius de Bazaleriis, 1491). Not in Knod. USTC 999310
b. Albert the Great, *Philosophia naturalis* (Basel: Furter, 22 June, 1506). Not in Knod. USTC 610612
c. Pierre d'Ailly, *Tractatus de anima* (Paris: Roce, 1503). Knod 107. USTC 142911
d. Fantinus, *Liber terminorum* (Paris: Mercator, 4 Oct 1499). Knod 112. USTC 202075
e. Pierre d'Ailly, *Tractatus exponibilium magistri Petri de Allyaco* (Paris: Caillaut, s.d.). Knod 108. USTC 760136?
f. Pierre d'Ailly, *Conceptus et insolubilia* (Paris, 1501). Knod 109. USTC 182440
g. Albertus [Saxonia], *Obligationum tractatus* (Paris 1500). Knod 91.
h. *Destructio naturarum communium contra reales* (Strasbourg, *c.*1495) Knod 113[b].
i. Michaelis, *Argumenta communia ad inferendum sophistice* (s.d.s.l.). Knod 113[a].
k. d'Ailly, *Destructiones modorum significandi* (s.d.s.l.).
l. [ps. Albert the Great], *Pulcerrimus tractatus de modo opponendi et r[espondi] necessarius valde omnibus volentibus acutissime arguere*... (Cologne: Quentell, 1498). Not in Knod.
m. Johann Widmann, *Tractatus de pustulis | que vulgato nomine dicuntur mal de | franzos Doctoris Ioannis widmam* (Strassburg: Grüniger, 1497). Not in Knod. USTC 749819

Knod dated this volume to 1504; Beatus appears to have written 1.4.0.4. on the title page of 1214d; likely the first 4 was meant to be a 5, but not the second, judging from (d).
This volume was rebound later, and appears to have gathered a number of school ephemera, which do not bear printer marks or dates.
d. Lots of notes, with the ex libris of BR, 'Est Beati Rhynow Schletstattini | Anno Salutis 1.4.0.4, | in alma Parrhisio | rum Academia | Ma. Pro.'

(*Continued*)

Year location	**Shelfmark** Book	**Comments**
1505 Paris	**K 852** a. Jacques Lefèvre d'Étaples, *Artificialis introductio per modum Epitomatis in decem libros Ethicorum Aristotelis adiectis elucidata commentariis* (Paris: Estienne, 1502). Knod 137. USTC 180032 b. Aristotle, *Decem librorum Moralium Aristotelis, tres conversiones* (Paris: Estienne, August 1505). Knod 166? USTC 143015	Pastedowns and flyleaves full of notes. a. 'Est Beati Rhenani Schletstattini ? Anno Domini Virtutum. Supra sesquimilesimum quinto, Parhisiis, Ma. P.' Of 58 folia (116 pages), 11 lightly, 15 heavily annotated. b. Of nearly 400 pages, 33 are lightly annotated and 106 are heavily annotated. (Knod dates this to 1506.)
	K 861 Giovanni Pico della Mirandola, *Commentationes Joannis Pici... quibis* (Bologna: Benedictus Hectoris, 1496). Knod 127. USTC 992054	
	K 880 Sacrobosco, *Opus sphaericum cum commentis, Esculani, Capuani, Stapulensis* (Venice, 1499). Knod 139. USTC 993974	BR's ownership notes on the title page, but virtually no marks within.
	K 983b Vergil, Polydore, *De inventoribus rerum* (Paris: Augrain and Bignet, 1501). Knod 132.	Not in the Sélestat online catalogue.
	K 1014 Peter Lombard, *Sententiae* (Paris: Bocard, 23 Nov 1497). Knod 142.	Rebound in the early 20th century. On the verso of the title page, a couple of lists, including 'In analogia correspondet' 'Est Beati Rhenani Seleustattini. Beniignissimi | redemptoris Nostri Anno . 1.5.0.5. | Ma. Pro. in celeberima Parrhisiorum Lutetia.'
	K 1076 a. Anselm Meianus, *Enchiridion naturale* (Paris, 1500). Knod 144. USTC 201755 b. Petrus Tartaretus, *Questiones morales in octo capita distincte* (Paris: Demarnet, 1504). Knod 146. USTC 182606	A few notes on the paste-down. a. 'Est Beati Rhenano | Schletstattini. Ma- | nu propria. Par- | rhisiis . 1.5.0.5.' b. 'Est Beati Rhenani Selestatini' [written later in his older hand]
	K 1083 a. Jacques Lefèvre d'Étaples, *Ars moralis in magna moralia introductoris* (Paris: Mercator, 1500). Knod 136. b. Aristotle, ed. Lefèvre, *Aethica seu moralia Aristotelis* (Paris: Roce, 1503). Knod 135? USTC 182519	a. Only the title page and one quotation from Giorgio Valla, on *prudentia* (fol. B3r), is annotated. b. Of 115 folia (230 pages), at least 61 pages are heavily annotated, and at least 2 lightly annotated.

	K 1134 a. Raymond Lull, *Contemplationes Remundi; Libellus blaquerne* (Paris: Petit Mercator, 10 Dec 1505). Knod 138. USTC 142995 b. Various Church Fathers, *Illustrium virorum opuscula* (Paris: André Bocard for Jean Petit, 1500). Not in Knod. USTC 201945 c. Tertullian, *Apologeticus adversus gentes* (Venice: Benali, 1494?). Knod 131. USTC 990532 d. Raymond Sebond, *Theologia naturalis s. liber creaturarum* (Nuremberg: Koberger, 1502). Knod 145. USTC 696769	BR's ownership on each volume. a. Few notes. Lovely rubrication in blue and red. b. No notes. c. Blank page in place of title page; at the top: 'Est Beati Rhenani Schletstattini ? Anno post natum Salvatorem | 1.5.0.5: Parisiis | Man. Prop.' BR wrote the title in his own hand on this page. Beautiful rubrication, as in (a). d. No notes.
	K 1180 a. Cleomedes, Aristides, Dion, Plutarch, *De contemplatione orbium excelsorum. Aelius Aristides: Ad Rhodienses de concordia oratio; Dio Chrysostomus: Ad Nicomedenses oratio; De concordia oratio; Plutarchus: De virtute morali; Coniugalia praecepta* (Brixen: Misinta, 1497). Knod 140. USTC 995663 b. Censorinus, *De die natali. Cebes: Tabula; Plutarchus: De invidia et odio* (Venice: Vitali, *c.*1498–9). Knod 147. USTC 996108	Title page: 'Est Beatis Rhenani | Parisiis. 1.5.0.5. | M.P.' a. Occasional light marginalia, in BR's older hand. b. Light marginalia in a neater hand, but probably after BR's school period. In (b), most marginalia near the end of the volume, on hymns, the 'mens collecta', in letter between Basil the Great and Gregory Nazianzen.
1506 Paris	**K 951** Nicholas of Cusa, *Opuscula varia* (Strassburg: Flach, 1488). Knod 173.	No title page, but on the first guard page, in BR's hand: 'OPERA NICOLAI DE CUSA | Est Beati Rhenani Selestatini, Anno pietatis | M.D.VI. | In preclara ac nobili | Parisiorum Lute- | tia. | M.P.' Light annotations throughout, but heavy annotations in certain works, especially the *Compendium* and *De beryllo*.
	K 1017 a. ?? b. Aristotle, *Clarissima singularisque totius philosophie Tatareti* (Lyon: Wolff, 1506). Knod 176. USTC 143126 c. Aristotle, *Expositio magistri Petri Tatareti super textu logices Aristotelis* (Lyon: Wolff, 1506). Knod 164. USTC 154973	No marks noticed.

(*Continued*)

Year location	Shelfmark Book	Comments
	K1071 a. Hermes, Lazarelus, *Pimander, Asclepius, Crater Hermetis*, ed. Jacques Lefèvre d'Étaples (Paris: Estienne, 1505). Knod 176. USTC 143065 b. Ephrem Syrus, *Sermones* (Paris: Marchand, 1505). USTC 143089	Both (a) and (b) are rubricated similarly in blue and red, with leaded lines around the textblock, indicating that these were prepared together.
	K 1077 a. Josse Clichtove and Jacques Lefèvre d'Étaples, *Introductiones in terminos* (Paris: Estienne, 1505). Knod 161. USTC 182627 b. Charles de Bovelles and Jacques Lefèvre d'Étaples, *Ars suppositionum Jacobi Fabri Stapulensis adiectis passim Caroli Bovilli viromandui annotationibus* (Paris: Petit, 1500). Knod 163. USTC 201898 c. Charles de Bovelles, *De artium constitutione et utilitate* (Paris: Petit/Philippe, 1506). Knod 158. d. Thomas Bradwardine, d. Pedro Ciruelo, *Arithmetica* (Paris: Lambert, 1505). Knod 159. USTC 180127 e. Pedro Ciruelo, *Tractatus Arithmetice practice* (Paris: Lambert, 1505). Knod 160. USTC 180127	a. Lightly annotated throughout. b. No notes. c. This is a very rare edition, later than the copy in the BL (which is from 1500, printed by Estienne). No notes. d. Only 'B&R' initialled on title page, and the initials repeated within the printer's mark. e. As for (d), 'B&R' is initialled on title page, and the initials repeated within the printer's mark; but also with 'Petri Cirueli' added below title.
	K 1139 Jacques Lefèvre d'Étaples, *In hoc opere continentur totius phylosophie naturalis paraphrases, hoc ordine digeste* (Paris: Estienne, 2 Dec. 1504). Walter 651; but not in Knod. USTC 142937	Title page: 'Est Beati Rhenani | Selestatensis | M.D.VI | Parisijs.' No notes.
1507 Paris	**K 352** a. Horace, *Carmina* (with comm. of Josse Bade) (Paris: Petit, Sept 1503). USTC 142886 b. Horatio, *Sermones et epistolae, etc.* (with comm. of Josse Bade) (Paris: Petit, 1503). c. Valla, *Elegantiae* (Paris: Josse Bade, 1505). Knod 233. USTC 181583 d. Aristotle, *Artificialis introductio per modum Epitomatis in decem libros Ethicorum Aristotelis* (Paris: Hopyl and Estienne, 1502). Knod 137? USTC 180032	Folio, early binding, but with signatures/ex libris from another owner—quite likely Beatus Rhenanus bought this volume secondhand. a. Considerable annotation, in Beatus' older hand. b. While there are occasional markings throughout, there are many notes in the epistolae. Probably BR's older cursive hand, though some might be by another. c. Free of notes. d. Considerable number of notes, in the hand of Francisci Murarii, signed on the end guard page.

K 856	
a. Macrobius, *De somno scipionis* (Brescia: Angelus Britannicus, 1501). Knod 186. USTC 839408 b. Valla, Ptolemy, *Commentationes. In quadripartitum* (Venice: Simon de Papiensis Gabis, 1502). Not in Knod. USTC 861867 c. Giovanni Antonio Panteo, *Annotationes Ioannis Antonii Panthei Vernonensis ex trium dierum confabulationibus* (Venice: Moretto, 1505). USTC 846520	Only a couple of minor annotations within the whole volume.
K 877	
Aristotle, *Habentur hoc volumina haec Theodoro Gaza interprete. Aristotelis de natura animalium* (Venice: Manuzio,1504). Knod 236. USTC 810862	Large folio. The volume looks in pristine condition, with the habitual 'Est Beati Rhenani Slecstattini. Parisiis, M.D.VII'. But mostly clean of marginalia. There is enough in *De part. anim.* to indicate that he read it. In the prefatory letter to *De historia plantarum*, he notes the reference to 'tu diue Nicolae'—Nicholas of Cusa?
K 934	
a. Symphorien Champier, *Liber de quadruplici vita, Theologia Asclepij hermetis trismegisti* (Lyon: Gueynardi et Buguetanni, 1507). Knod 237. USTC 143270 b. Aulus Persius Flaccus, *Satyrae. Comm: Joannes Britannicus and Jodocus Badius Ascensius* (Lyon: Wolff, 27 Jan. 1499/1500). Not in Knod. USTC 202042	No marginalia.
K 985	
e. Pomponius Mela, *Cosmographia sive de situ orbis* (Venice: de Pensi, 1498). Knod 223. USTC 993029	It is not clear how Knod dated BR's acquisition of this volume. I have found no markings in it. Knod notes, as if he found within: 'Est Beati Rhenani Parisiis MDVII'.
K 1000	
Lorenzo Valla, *Reconcinnatio totius dialectice & fundamentorum universalis* (Milan: Guillaume le Signerre, 1496–1500). Knod 232. USTC 990123	Title page: 'Est Beati Rhenani. | Parisiis. M.D. VII.' Several notes in the first pages.

(*Continued*)

Year location	Shelfmark Book	Comments
	K 1041 a. Marsilio Ficino (trans.), *Iamblichus de mysteriis Aegyptiorum, Proclus in Platonicum, etc.* (Venice: Manuzio, 1497). Knod 239. USTC 994070 b. Ammonius, *Ammonius in quinque voces Porphyrii*, trans. Pomponius Gauricus (Venice: Battista and Sessa, 1504). Not in Knod. USTC 809136 c. Johannes Reuchlin, *De verbo mirifico* (Basel: Amerbach, 1494). Knod 250. USTC 748528	There are scattered but suggestive notes on all three, especially in (b), Ammonius's commentary on Porphyry.
	K 1044 Heraclides, *Pro piorum recreatione et in hoc opere contenta: Epistola Jacobi Stapulensis editoris ante indicem. Index contentorum. Ad lectores. Paradysus Heraclidis. Epistola Clementis. Recognitiones Petri apostoli. Complementum epistole Clementis. Epistola Anacleti*, ed. Jacques Lefèvre d'Étaples (Paris: Guy Marchand for Jean Petit, 1504). Knod 238. USTC 142952	Few annotations.
	K 1078 a. Johannes Damascenus, *Theologia damasceni s. de orthodoxa fide*, trans. Jacques Lefèvre d'Étaples (Paris: Estienne, 1507). Knod 246. USTC 180285 b. Ammonius, *Ammonius in quinque voces Porphyrii, trans. Pomponius Gauricus* (Venice: Battista and Sessa, 1504). Not in Knod. USTC 809136	Published with verses by Beatus Rhenanus to Robert Fortune (fol. 114v). No annotations.
	K 1161 a. Guillaume d'Auvergne, *De claustro anime*; Hugh of St. Victor, *De claustro anime libri quatuor* (Paris: Estienne, 1507). Knod 249. USTC 143227 b. Ricoldus de Monte Crucis, *Confutatio Alcorani seu legis Saracenorum* (Basel: Kessler, 1507). Not in Knod. USTC 624688 c. *Tractatus de martyrio sanctorum* (Basel: Wolff, 1492). Not in Knod. USTC 746918 d. Alanus ab Insulis, *De maximis theologiae* (Basel: Wolff, 1492). Knod 244. USTC 742322 e. Savonarola, *Revelatio de tribulationibus nostrorum temporum* (Paris: Mercator, 1496). Knod 251. USTC 201352	Very occasional notes.

Bibliography

MANUSCRIPTS

Lefèvre d'Étaples, Jacques, *De magia naturali*
Olomouc University Library M I 119
BnF lat. 7454
Rhenanus, Beatus
Cahier écolier, BHS MS 50
Cahier étudiant, BHS, MS 58
BHS K 810f [=MS 325], Incipit *Publii Vergilii carmen. Contra Luxuriam et ebrietatem*

PRIMARY LITERATURE

Allen, Michael J. B. (ed.), *Nuptial Arithmetic: Marsilio Ficino's Commentary on the Fatal Number in Book VIII of Plato's 'Republic'* (Berkeley, CA: University of California Press, 1994).

Bade, Josse, *Rosetum exercitiorum spiritualium* of Jean Mombaer (Paris: Josse Bade for Jean Petit and Jean Scabelarius, 1510).

Baldi, Bernardino, *Cronica de Matematici* (Urbino: A.A. Monticelli, 1707).

Barbaro, Ermolao (trans.), *Paraphraseos Themistii* (Venice, 1481).

Barbaro, Ermolao, *Castigationes Plinianae* (Rome: Eucharius Silber, 1492).

Barker, Andrew (ed.), *Greek Musical Writings*, vol. II: *Harmonic and Acoustic Theory* (Cambridge: Cambridge University Press, 1990).

Beza, Theodore, *Icones* (Geneva: Ioannes Laonius, 1580).

Biancani, Giuseppe, *De mathematicarum natura dissertatio* (Bologna: Bartholomaeus Cochius, 1615).

Biancani, Giuseppe, *Aristotelis loca mathematica* (Bologna: Bartholomaeus Cochius, 1615).

Boethius, *Fundamentals of Music*, trans. Calvin M. Bower, ed. Claude V. Palisca (New Haven, CT: Yale University Press, 1989).

Bovelles, Charles de, *In artem oppositorum introductio* (Paris: Wolfgang Hopyl, 1501).

Bovelles, Charles de, *Liber de intellectu; Liber de sensu; Liber de nichilo; Ars oppositorum; Liber de generatione; Liber de sapiente; Liber de duodecim numeris; Epistole complures. Insuper mathematicum opus quadripartitum: De numeris perfectis; De mathematicis rosis; De geometricis corporibus; De geometricis supplementis* (Paris: Henri Estienne, 1511).

Bovelles, Charles de, *Géométrie en françoys. Cy commence le Livre de l'art et Science de Geometrie : avecques les figures sur chascune rigle au long declarees par lesquelles on peut entendre et facillement comprendre ledit art et science de Geometrie* (Paris: Henri Estienne, 1511).

Bovelles, Charles de, *Libri physicorum elementorum* (Paris: Josse Bade, 1512).

Bovelles, Charles de, *Quaestionum theologicarum libri septem* (Paris: Josse Bade, 1513).

Bovelles, Charles de, *Commentarius in primordiale evangelium divi Joannis; Vita Remundi eremitae; Philosophicae et historicae aliquot epistolae* (Paris: Josse Bade, 1514).

Bovelles, Charles de, *Livre singulier et utile, touchant l'art et practique de Geometrie, composé nouvellement en Francoys* (Paris: Simon Colines, 1542).

Bovelles, Charles de, *Geometrie practique, nouvellement reveue, augmentee et grandement enrichie* (Paris: Reginald Chauderon, 1547).

Briçonnet, Guillaume, and Marguerite de Navarre, *Correspondance, 1521–1524*, ed. Christine Martineau, Michel Veissière, and Henry Heller, 2 vols (Geneva: Droz, 1979).
Bricot, Thomas, and Georgius Bruxellensis, *Textus abbreviatus in cursum totius physices et metaphysicorum Aristotelis* (1st edn, Lyons: s.p. 1486).
Bruni, Leonardo, *De disputationum exercitationisque studiorum usu* (Basel: Heinrich Petri, 1536).
Caesarius, Johannes (ed.), *Introductio Jacobi Fabri Stapulensis in Arithmeticam; Ars supputandi Clichtovei; Epitome rerum geometricarum Bovilli* (Deventer: R. Pafraet, 1507).
Campanus of Novara and Euclid's Elements, ed. H. L. L. Busard (Stuttgart: Franz Steiner Verlag, 2005).
Capella, *Martianus Capella and the Seven Liberal Arts*, vol. II: *The Marriage of Philology and Mercury*, trans. William Harris Stahl and E. L. Burge (New York: Columbia University Press, 1977).
Champier, Symphorien, *Ianua logice et physice* (Lyons: Guillaume Balsarin, 1498).
Champier, Symphorien, *Duellum epistolare: Galie et Etalie antiquitates summatim complectens* (Lyons: Froben et al., 1519).
Ciruelo, Pedro, *Cursus quatuor mathematicarum artium liberalium* (Alcalá: A.G. de Brocar, 1516).
Ciruelo, Pedro, *Cursus quatuor mathematicarum artium liberalium* (Madrid: Michaelis de Eguia, 1528 [the title page reads 1526]).
Clichtove, Josse, *De divisione artium* ([*c.*1500], Paris, 1520).
Clichtove, Josse (ed.), *Opera [Hugonis de Sancto Victore]: De institutione novitiorum. De operibus trium dierum. De arra anime. De laude charitatis. De modo orandi. Duplex exposito orationis dominice. De quinque septenis. De septem donis Spiritus Sancti* (Paris: Estienne, 1506).
Clichtove, Josse (ed.), *De claustro animae* [by Hughes de Fouilloy] (Paris: Estienne, 1507).
Clichtove, Josse (ed.), *Hugonis de Sancto Victore Allegoriarum in utrunque testamentum libri decem* (Paris: Henri Estienne, 1517).
Clichtove, Josse, and Johannes Caesarius, *Introductio in terminorum cognitionem, in libros Logicorum Aristotelis, authore Iodoco Clichtoveo Neoportuensi, una cum Ioannis Caesarii Commentariis* (Paris: Gabriel Buon, 1560).
Commandino, Federico, *Euclidis Elementorum libri XV. Unà cum scholiis antiquis* (Pisa: Iacobo Chriegher Germano, 1572).
Cuno, Johannes, and Beatus Rhenanus (eds), *Divini Gregorii Nyssae Episcopi qui fuit frater Basilii Magni libri* (Basel: Mathias Schurer, 1512).
Cusanus, Nicolaus, *Opuscula theologica et mathematica* (Strassburg: Martin Flasch, 1488).
Descartes, René, *The Philosophical Writings of Descartes*, vol. I, trans. John Cottingham et al. (Cambridge: Cambridge University Press, 1984).
Ficino, Marsilio, *Opera omnia Platonis* (Venice: Andrea Toresani de Asula, 1491).
Ficino, Marsilio, *Commentaria in Platonem* (Florence, 1496).
Ficino, Marsilio, *Commentary on Plotinus*, vol. 4: *Ennead III, Part 1*, ed. and trans. Stephen Gersh (Cambridge, MA: Harvard University Press, 2017).
Fine, Oronce (ed.), *Theoricarum nouarum Textus Georgii Purbachii, cum expositione Capuani, Item de Prierio; Insuper Jacobi Fabri Stapulensis astronomicon: Omnia nuper summa diligentia emendata cum figuris ac commodatisssimis longe castigatius insculptis quam prius suis in locis adiectis* (Paris: Jean Petit, 1515).
Fine, Oronce, *Protomathesis* (Paris: G. Morrhy, 1532).
Fine, Oronce, *In sex priores libros geometricorum elementorum Euclidis Megarensis demonstrationes: quibus ipsius Euclidis textus Graecus, suis locis insertus est; una cum interpretatione Latina Bartholomaei Zamberti Veneti* (Paris: Simon de Colines, 1536).

Fine, Oronce, *Arithmetica practica* (Paris: Simon de Colines, 1542, 1544).
Fine, Oronce, *De Mundi sphaera, sive Cosmographia, libri V. ab ipso authore denuo castigati, et marginalibus (ut vocant) annotationibus recens illustrati: quibus tum prima Astronomiae pars, tum Geographiae ac Hydrographiae rudimenta pertractantur* (Paris: Michel de Vascovan, 1555).
Gerson, Jean, *Œuvres Complètes*, ed. Palémon Glorieux (Paris: Desclée, 1960–73).
Giovio, Paolo, *Elogia doctorum virorum ab avorum memoria publicatis ingenii monumentis illustrium* (Antwerp: Ioannes Bellerus, 1557).
Gogava, Antonio (ed.), *Aristoxeni Harmonicorum Elementorum libri III. Cl. Ptolemæi Harmonicorum, seu de Musica lib. III. Aristotelis de Obiecto Auditus Fragmentum ex Porphyrij commentariis* (Venice, 1562).
Goulet, Robert, *Compendium on the Magnificence, Dignity, and Excellence of the University of Paris in the Year of Grace 1517*, trans. Robert Belle Bourke (Philadelphia, PA: University of Pennsylvania Press, 1928).
Hartmann, Alfred (ed.), *Die Amerbachkorrespondenz*, 10 vols (Basel: Verlag der Universitätsbibliothek, 1942–95).
Heath, Thomas L. (ed.), *The Thirteen Books of Euclid's Elements*, vol. 1: *Books 1–2* (1908, 2nd edn 1925; New York: Dover Publications, 1956).
Hegius, Alexander, *Dialogi duo de sacro sancte incarnationis mysterio adiuncta pache inveniendi ratione, in quibus continetur ratio totius computi ecclesiastici et ferme totius sphere mundi, Ars supputatoria calcularis, Tractatulus de numero ad alium relato sive numerorum proportionibus*, ed. Jacobus Faber (Cologne: Heinrich Quentell, 1508).
Hirstein, James (ed.), *Epistulae Beati Rhenani. La Correspondance latine et grecque de Beatus Rhenanus de Sélestat. Edition critique raisonnée avec traduction et commentaire*, vol. 1 (1506–17) (Turnhout: Brepols, 2013).
John of Salisbury, *Metalogicon*, ed. Clement C. J. Webb (Oxford: Clarendon Press, 1929).
Lefèvre d'Étaples, Jacques, [*Totius Aristotelis philosophiae naturalis paraphrases*] (Paris: [Johann Higman], 1492).
Lefèvre d'Étaples, Jacques, *Introductio in metaphysicorum libros Aristotelis*, ed. Josse Clichtove (Paris: Johann Higman, 1494).
Lefèvre d'Étaples, Jacques (ed.), *Mercurij Trismegisti Liber de potestate et sapientia dei per Marsilium Ficinum traductus ad Cosmum Medicem*, trans. Marsilio Ficino (Paris: Hopyl, 1494).
Lefèvre d'Étaples, Jacques, *Textus de sphera Johannis de Sacrobosco, cum additione (quantum necessarium est) adiecta: novo commentario nuper edito ad utilitatem studentium philosophice parisiensis academie: illustratus* (Paris: Wolfgang Hopyl, 1495).
Lefèvre d'Étaples, Jacques, *Elementa arithmetica; Elementa musicalia; Epitome in libros arithmeticos divi Severini Boetii; Rithmimachie ludus que et pugna numerorum appellatur* (Paris: Higman and Hopyl, 1496).
Lefèvre d'Étaples, Jacques, *Introductiones logicales in suppositiones, in predicabilia, in divisiones, in predicamenta, in librum de enunciatione, in primum priorum, in secundum priorum, in libros posteriorum, in locos dialecticos, in fallacias, in obligationes, in insolubilia* (Paris: Guy Marchant, 1496).
Lefèvre d'Étaples, Jacques, *Decem librorum Moralium Aristotelis tres conuersiones: prima Argyropili Byzantii; secunda Leonardi Aretini; tertia vero Antiqua per Capita et numeros conciliate: communi familiarique commentario ad Argyropilum ad lectio* (Paris: Higman and Hopyl, 1497).
Lefèvre d'Étaples, Jacques (ed.), *Hic continentur libri Remundi [Lulli] pij eremite* (Paris: Petit, 1499).

Lefèvre d'Étaples, Jacques (ed.), *Theologia vivificans, cibus solidus. Dionysii Celestis hierarchia. Ecclesiastica hierarchia. Diuina nomina. Mystica theologia. Undecim epistole. Ignatii Undecim epistole. Polycarpi epistola una* (Paris: Higman and Hopyl, 1499).

Lefèvre d'Étaples, Jacques, and Josse Clichtove, *Totius philosophiae naturalis paraphrases, adiecto commentario* (Paris: Hopyl, 1502).

Lefèvre d'Étaples, Jacques, and Josse Clichtove, *Artificialis introductio per modum Epitomatis in decem libros Ethicorum Aristotelis adiectis elucidata [Clichtovei] commentariis* (Paris: Wolfgang Hopyl and Henri Estienne, 1502).

Lefèvre d'Étaples, Jacques, Josse Clichtove, and Charles de Bovelles, *Epitome compendiosaque introductio in libros arithmeticos divi Severini Boetii, adiecto familiari [Clichtovei] commentario dilucidata. Praxis numerandi certis quibusdam regulis (auctore Clichtoveo). Introductio in geometriam Caroli Bovilli. Astronomicon Stapulensis* (Paris: Wolfgang Hopyl and Henri Estienne, 1503).

Lefèvre d'Étaples, Jacques, *Libri logicorum ad archteypos recogniti cum novis ad litteram commentariis ad felices primum Parhisiorum et communiter aliorum studiorum successus in lucem prodeant ferantque litteris opem* (Paris: Wolfgang Hopyl and Henri Estienne, 1503).

Lefèvre d'Étaples, Jacques (ed.), *Pro Piorum recreatione et in hoc opere contenta etc.* (Paris: Jean Petit, 1504).

Lefèvre d'Étaples, Jacques (ed.), *Primum volumen Contemplationum Remundi [Lulli] duos libros continens. Libellus Blaquerne de amico et amato* (Paris: Guy Marchant for Jean Petit, 1505).

Lefèvre d'Étaples, Jacques (ed.), *Pimander. Mercurii Trismegisti liber de sapientia et potestate dei. Asclepius. Eiusdem Mercurii liber de voluntate divina. Item Crater Hermetis A Lazarelo Semptempedano*, trans. Marsilio Ficino (Paris: Henri Estienne, 1505).

Lefèvre d'Étaples, Jacques, *Politicorum libro octo. Commentarii. Economicorum duo. Commentarii. Hecatonomiarum septem. Economiarum publ. unus. Explanationis Leonardi [Bruni] in Oeconomica duo* (Paris: Estienne, 1506).

Lefèvre d'Étaples, Jacques (ed.), *Georgii Trapezontii dialectica* (Paris: Estienne, 1508).

Lefèvre d'Étaples, Jacques, *Quincuplex Psalterium* (Paris: Henri Estienne, 1509).

Lefèvre d'Étaples, Jacques (ed.), Richard of St. Victor, *De superdivina Trinitate theologicum opus hexade librorum distinctum. Commentarius artificio analytico* (Paris: Estienne, 1510).

Lefèvre d'Étaples, Jacques (ed.), *Devoti et venerabilis patris Ioannis Rusberi presbyteri, canonici observantiae beati Augustini, de ornatu spiritualium nuptiarum libri tres* (Paris: Estienne, 1512).

Lefèvre d'Étaples, Jacques, *Epistolae ad Rhomanos, Corinthios, Galatas, Ephesios, Philippenses, Colossenses, Thessalonie, Timotheum, Titum, Philemonem, Hebraeos. Epistolae ad Laodicenses, Senecam. Linus de passione Petri et Pauli* (Paris: Estienne, 1512).

Lefèvre d'Étaples, Jacques (ed.), *Haec accurata recognitio trium voluminum, Operum clariss. P. Nicolai Cusae Cardinalis* (Paris: Josse Bade, 1514).

Lefèvre d'Étaples, Jacques (ed.), *Proverbia Raemundi. Philosophia amoris eiusdem* (Paris: Josse Bade, 1516).

Lefèvre d'Étaples, Jacques (ed.), *Euclidis Megarensis mathematici clarissimi Elementorum geometricorum libri xv, Campani Galli transalpini in eosdem commentariorum libri xv, Theonis Alexandrini Bartholamaeo Zamberto Veneto interprete, in tredecim priores, commentariorum libri xiii, Hypsiclis Alexandrini in duos posteriores, eodem Bartholomaeo Zamberto Veneto interprete, commentariorum libri ii* (Paris: Henri Estienne, 1517).

Le Livre des XXIV Philosophes, ed. Françoise Hudry (Paris: Vrin, 2009).

Lucács, L. (ed.), *Monumenta Paedagogica Societatis Jesu* (Rome, 1965–92).

Maurolico, Francesco, *Cosmographia* (Venice, 1543).

Mersenne, Marin, *Traité de l'harmonie universelle* (Paris: Fayard, 2003).
Meyer, Hubert (ed.), Jean Sturm, 'Vie de Beatus Rhenanus', trans. Charles Munier, *Annuaire. Les Amis de La Bibliothèque Humanist de Sélestat* 35 (1985): 7–18.
Milliet de Chales, Claude-François, *Cursus seu mundus mathematicus* (Lyons: Officina Anissoniana, 1674).
Mombaer, Jean, *Rosetum exercitiorum spiritualium* (Paris: Josse Bade for Jean Petit and Jean Scabelarius, 1510).
Moscheo, R. (ed.), *I Gesuiti e le matematiche nel secolo XVI. Maurolico, Clavio e l'esperienza siciliana* (Messina: Società Messinese di Storia Patria, 1998).
Nifo, Agostino, *Aristotelis Stagiritae Topicorum libri octo* (Venice: Girolamo Scoto, 1569).
Peletier, Jacques, *L'aritmetique, departie en quatre livres à Theodore Debesze* (Paris: Marnef, 1549).
Peter of Spain, *Tractatus, Called Afterwards Summule Logicales*, ed. L. M. de Rijk (Assen: Van Gorcum, 1972).
Possevino, Antonio, *Bibliotheca selecta: qua agitur de ratione studiorum* (Rome: Typographia Apostolica Vaticana, 1593).
Ramus, Peter, *Dialecticae institutiones* (Paris: Jacques Bogard, 1543).
Ramus, Peter, *Arithmeticae libri tres* (Paris: Andreas Wechel, 1555).
Ramus, Peter, *Proemium mathematicum* (Paris: Andreas Wechelus, 1567).
Ramus, Peter, *Scholarum mathematicarum, libri unus et triginta* (Basel: Eusebius Episcopius and Nicolas Pratre, 1569).
Reisch, Gregor, *Margarita philosophica*, ed. Oronce Fine (Basel: Henricus Petrus for Conrad Resch, 1535).
Rice, Eugene F. Jr. (ed.), *The Prefatory Epistles of Jacques Lefèvre d'Étaples and Related Texts* (New York: Columbia University Press, 1972).
Ringmann, Matthias, *Grammatica figurata* [St Die, 1509].
Roussel, Gérard, *Divi Severini Boetii Arithmetica duobus discreta libris, adiecto commentario, mysticam numerorum applicationem perstringente, declarata* (Paris: Simon Colines, 1521).
Serjeantson, Richard, *Generall Learning: A Seventeenth-Century Treatise on the Formation of the General Scholar by Meric Casaubon* (Cambridge: RTM Publications, 1999).
Sherburne, Edward, *The Sphere of Marcus Manilius Made an English Poem: With Annotations and an Astronomical Appendix* (London: Nathanael Brooke, 1675).
Silíceo, Juan Martínez, *Arithmetica in theoricen, et praxim scissa*, ed. Oronce Fine (Paris: Henri Estienne, 1519).
Tataret, Pierre, *Commentarii in libros philosophie naturalis et metaphysice Aristotelis* (1st edn, Paris: André Bocard, 1495).
Themistius, *Commentaire sur le traité de l'ame d'Aristote: Traduction de Guillaume de Moerbeke*, ed. G. Verbeke (Leiden: Brill, 1973).
Thomas, Alvarus, *Liber de triplici motu* (Paris: Guillaume Anabat, 1509).
Thuasne, Louis (ed.), *Roberti Gaguini Epistole et orationes*, 2 vols (Paris: É. Bouillon, 1903).
Trithemius, Johannes, *De scriptoribus ecclesiasticis*, ed. anon. (Paris: Bertold Rembolt and Jean Petit, 1512).
Valla, Giorgio, *Cleonidae Harmonicum introductorium [incl. Euclidis sectio canonis] interprete Georgio Valla Placentino. L. Vitruvii Pollionis de Architectura libri decem. Sexti Frontini de Aquaeductibus liber unus. Angelic Policiani opusculum, quod Panepistemon inscribitur. Angeli Policiani in priora analytica praelectio, cui titulus est Lamia* (Venice: Simone Bevilacqua, 1497).
Valla, Giorgio, *De expetendis et fugiendis rebus* (Venice: Aldus Manutius, 1501).
Van Engen, John (ed.), *Devotio Moderna: Basic Writings* (New York: Paulist Press, 1988).

Varennes, Alain de, *De amore dialogus, de luce dialogi, etc.* (Paris: Henri Estienne, 1512).
Vogt, Otto (ed.), *Dr. Johannes Bugenhagens Briefwechsel* (Stettin: Leon Saunier, 1888).
Vossius, Gerardus Joannes, and Franciscus Junius, *De quatuor artibus popularibus, de philologia, et scientiis mathematicis, cui operi subjungitur, chronologia mathematicorum, libri tres* (Amsterdam: Ioannis Blaeu, 1660).
Zamberti, Bartholomaeo (trans.), *Euclidis megarensis philosophi platonici Mathematicarum disciplinarum Janitoris: Habent in hoc volumine quiqunque ad mathematicam substantiam aspirant: elementorum libros XIII. cum expositione Theonis... voluta in Campani interpretatione:... adiuncta : Deputatum... Euclidi volumen XIIII. cum exposione Hypsi. Alex. Itidemque & Phaeno. Specu. & Perspe. cum expositione Theonis. ac mirandus ille liber Datorum cum expositione Pappi Mechanici una cum Marini dialectici protheoria* (Venice: Joannes Tacuinus, 1505).
Zarlino, Gioseffo, *Le institutione harmoniche* (2nd edn, Venice: De Franceschi, 1573).
Zarlino, Gioseffo, *Sopplimenti musicali* (Venice, 1588).
Zwingli, Huldrich, *Zwinglis Sämtliche Werke*, vol. VII [= Corpus Reformatorum 94] (Leipzig: Heinsius, 1911).

MODERN LITERATURE

Abdounur, Oscar J., 'Ratios and Music in the Late Middle Ages: A Preliminary Survey', in *Music and Mathematics in Late Medieval and Early Modern Europe*, ed. Philippe Vendrix (Turnhout: Brepols, 2008), 23–69.
Albertson, David, 'Mystical Philosophy in the Fifteenth Century: New Directions in Research on Nicholas of Cusa', *Religion Compass* 4, no. 8 (2010): 471–85.
Albertson, David, *Mathematical Theologies: Nicholas of Cusa and the Legacy of Thierry of Chartres* (Oxford: Oxford University Press, 2014).
Allen, Michael J. B., 'Ficino, Daemonic Mathematics, and the Spirit', in *Natural Particulars: Nature and the Disciplines in Renaissance Europe*, ed. Anthony Grafton and Nancy G. Siraisi (Cambridge, MA: MIT Press, 1999), 121–37.
Allen, Percy Stafford, 'Erasmus' Relations with His Printers', *The Library* 13, no. 1 (1913): 297–322.
Armstrong [Tyler], Elizabeth, *Robert Estienne, Royal Printer: An Historical Study of the Elder Stephanus* (Cambridge: Cambridge University Press, 1954).
Armstrong Tyler, Elizabeth, 'Jacques Lefèvre d'Etaples and Henri Estienne the Elder, 1502–1520', in *The French Mind: Studies in Honour of Gustave Rudler*, ed. W. Grayburn Moore (Oxford: Sutherland and Starkis, 1952), 17–33.
Ash, Eric H., *Power, Knowledge, and Expertise in Elizabethan England* (Baltimore, MD: Johns Hopkins University Press, 2004).
Avril, François, and Jean Lafaurie (eds), *La Librairie de Charles V* (Paris: Bibliothèque nationale de France, 1968).
Axworthy, Angela, 'The Epistemological Foundations of the Propaedeutic Status of Mathematics according to the Epistolary and Prefatory Writings of Oronce Fine', in *The Worlds of Oronce Fine: Mathematics, Instruments and Print in Renaissance France*, ed. Alexander Marr (Donington: Shaun Tyas, 2009), 31–5.
Axworthy, Angela, *Le Mathématicien renaissant et son savoir: le statut des mathématiques selon Oronce Fine* (Paris: Classiques Garnier, 2016).
Azzolini, Monica, *The Duke and the Stars: Astrology and Politics in Renaissance Milan* (Cambridge, MA: Harvard University Press, 2013).

Bailey, Michael D., *Battling Demons: Witchcraft, Heresy, and Reform in the Late Middle Ages* (University Park, PA: Pennsylvania State University Press, 2003).

Bakker, Paul J. J. M., 'The Statutes of the Collège du Montaigu: Prelude to a Future Edition', *History of Universities* 22, no. 2 (2007): 76–111.

Baldasso, Renzo, 'La stampa dell' *editio princeps* degli *Elementi* di Euclide (Venezia, Erhard Ratdolt, 1482)', in *The Books of Venice* (Venice: La Musa Talia, 2007), 61–100.

Barbierato, Federico, 'Writing, Reading, Writing: Scribal Culture and Magical Texts in Early Modern Venice', *Italian Studies* 66, no. 2 (2011): 263–76.

Barbour, Murray, *Tuning and Temperament: A Historical Survey* (1951; Mineola, NY: Dover, 2004).

Barker, Peter, 'The Reality of Peurbach's Orbs: Cosmological Continuity in Fifteenth and Sixteenth Century Astronomy', in *Change and Continuity in Early Modern Cosmology*, ed. Patrick J. Boner (Berlin: Springer, 2011), 7–32.

Barker, Andrew, *The Science of Harmonics in Classical Greece* (Cambridge: Cambridge University Press, 2007).

Barnes, Jonathan, *Porphyry: Introduction* (Oxford: Oxford University Press, 2003).

Baron, Sabrina Alcorn, Eric N. Lindquist, and Eleanor F. Shevlin (eds), *Agent of Change: Print Culture Studies After Elizabeth L. Eisenstein* (Boston, MA: University of Massachusetts Press, 2007).

Baulant, Micheline, 'Le salaire des ouvriers du bâtiment à Paris de 1400 à 1726', *Annales* 26, no. 2 (1971): 463–83.

Baumann, Christoph C., 'Dictata in quinque predicantes voces. Ein Kommentar zur Isagoge des Porphyrius in der Aufzeichnung des Beatus Rhenanus', PhD thesis, Universität Zürich, 2008.

Baur, Ludwig, *Nicolaus Cusanus und Pseudo-Dionysius im Lichte der Zitate und Randbemerkungen des Cusanus. Cusanus-Texte. III. Marginalien. 1* (Heidelberg, 1940).

Bedouelle, Guy, *Lefèvre d'Étaples et l'intelligence des Écritures* (Geneva: Droz, 1976).

Bedouelle, Guy, *Le Quincuplex Psalterium de Lefèvre d'Étaples: Un guide de lecture* (Geneva: Droz, 1979).

Bertoloni Meli, Domenico, 'Guidobaldo Dal Monte and the Archimedean Revival', *Nuncius* 7, no. 1 (1992): 3–34.

Bertoloni Meli, Domenico, *Thinking with Objects: The Transformation of Mechanics in the Seventeenth Century* (Baltimore, MD: Johns Hopkins University Press, 2006).

Besse, Jean-Marc, *Les Grandeurs de la terre: aspects du savoir géographique à la renaissance* (Paris: ENS Editions, 2003).

Biagioli, Mario, 'The Social Status of Italian Mathematicians', *History of Science* 27, no. 1 (1989): 41–95.

Biagioli, Mario, *Galileo, Courtier: The Practice of Science in the Culture of Absolutism* (Chicago: University of Chicago Press, 1993).

Biagioli, Mario, *Galileo's Instruments of Credit: Telescopes, Images, Secrecy* (Chicago: University of Chicago Press, 2006).

Bianchi, Luca, 'Continuity and Change in the Aristotelian Tradition', in *The Cambridge Companion to Renaissance Philosophy*, ed. James Hankins (Cambridge: Cambridge University Press, 2007), 49–71.

Bianchi, Luca, 'From Lefèvre d'Etaples to Giulio Landi: Uses of the Dialogue in Renaissance Aristotelianism', in *Humanism and Early Modern Philosophy*, ed. Jill Kraye and M. W. F. Stone (London: Routledge, 2000), 41–58.

Bietenholz, Peter G., *Basle and France in the Sixteenth Century* (Geneva: Droz, 1971).

Black, Crofton, *Pico's Heptaplus and Biblical Hermeneutics* (Leiden: Brill, 2006).
Blair, Ann, 'Lectures on Ovid's Metamorphoses: The Class Notes of a 16th-Century Paris Schoolboy', *Princeton University Library Chronicle* 50 (1989): 117–44.
Blair, Ann, 'Humanist Methods in Natural Philosophy: The Commonplace Book', *Journal for the History of Ideas* 53, no. 4 (1992): 541–51.
Blair, Ann, 'Student Manuscripts and the Textbook', in *Scholarly Knowledge: Textbooks in Early Modern Europe*, ed. Emidio Campi, Simone De Angelis, and Anja-Silvia Goeing (Geneva: Droz, 2008), 39–73.
Blair, Ann, 'The Rise of Note-Taking in Early Modern Europe', *Intellectual History Review* 20, no. 3 (2010): 303–16.
Blair, Ann, *Too Much to Know: Managing Scholarly Information before the Modern Age* (New Haven, CT: Yale University Press, 2010).
Bocken, Inigo, 'The Pictorial Treatises of Charles de Bovelles', *Intellectual History Review* 21, no. 3 (2011): 341–52.
Bolzoni, Lina, *The Gallery of Memory: Literary and Iconographic Models in the Age of the Printing Press*, trans. Jeremy Parzen (Toronto: University of Toronto Press, 2001).
Bonitz, H., *Index Aristotelicus* (Berlin: Reimer, 1831).
Bos, Henk J. M., *Redefining Geometrical Exactness: Descartes' Transformation of the Early Modern Concept of Construction* (Springer, 2001).
Boudet, Jean-Patrice (ed.), *Le Recueil des plus celebres astrologues de Simon de Phares* (Paris: Honoré Champion, 1999).
Boudet, Jean-Patrice, 'A "College of Astrology and Medicine"? Charles V, Gervais Chrétien, and the Scientific Manuscripts of Maître Gervais's College', *Studies in History and Philosophy of Science* 41, no. 2 (2010): 102.
Boyer, Carl B., 'Proportion, Equation, Function: Three Steps in the Development of a Concept', *Scripta Mathematica* 12 (1946): 5–13.
Bresc, H., and I. Heullant-Donat, 'Pour une réévaluation de la "révolution du papier" dans l'Occident médiéval', *Scriptorium* 61, no. 2 (2007): 354–83.
Broadie, Alexander, 'John Mair's Dialogus de Materia Theologo Tractanda: Introduction, Text and Translation', in *Christian Humanism: Essays in Honour of Arjo Vanderjagt* (Leiden: Brill, 2009), 419–30.
Brockliss, L. W. B., 'Patterns of Attendance at the University of Paris, 1400–1800', *Historical Journal* 21, no. 3 (1978): 503–44.
Brockliss, L. W. B., *French Higher Education in the Seventeenth and Eighteenth Centuries: A Cultural History* (Oxford: Clarendon Press, 1987).
Brown, Gary I., 'The Evolution of the Term "Mixed Mathematics"', *Journal of the History of Ideas* 52, no. 1 (1991): 81–102.
Brun, Robert, 'Un illustrateur méconnu: Oronce Finé', *Arts et métiers graphiques* 41 (1934): 51–7.
Brush, J. W., 'Lefèvre d'Etaples: Three Phases of His Life and Work', in *Reformation Studies: Essays in Honor of Roland H. Bainton*, ed. Franklin Hamlin Littell (Richmond, VA: John Knox Press, 1962), 117–28.
Buchwald, Jed Z., and Mordechai Feingold, *Newton and the Origin of Civilization* (Princeton, NJ: Princeton University Press, 2012).
Burtt, Edwin Arthur, *The Metaphysical Foundations of Modern Science* (2nd edn, 1924; London: Routledge & Kegan Paul, 1932).
Busard, H. L. L., 'Über die Entwicklung der Mathematik in Westeuropa zwischen 1100 und 1500', *NTM Zeitschrift für Geschichte der Wissenschaften, Technik und Medizin* 5, no. 1 (1997): 211–35.

Bycroft, Michael, 'How to Save the Symmetry Principle', in *The Philosophy of Historical Case Studies*, ed. Tilman Sauer and Raphael Scholl (Dordrecht: Springer, 2016), 11–29.

Bynum, Caroline Walker, *Christian Materiality: An Essay on Religion in Late Medieval Europe* (New York: Zone, 2011).

Cadden, Joan, 'Charles V, Nicole Oresme, and Christine de Pizan: Unities and Uses of Knowledge in Fourteenth-Century France', in *Texts and Contexts in Ancient and Medieval Science: Studies on the Occasion of John E. Murdoch's Seventieth Birthday*, ed. Edith Dudley Sylla and Michael M. McVaugh (Leiden: Brill, 1997), 108–244.

Camille, Michael, 'Visual Art in Two Manuscripts of the Ars Notaria', in *Conjuring Spirits: Texts and Traditions of Medieval Ritual Magic*, ed. Claire Fanger (Stroud: Sutton, 1998), 110–39.

Cantor, Moritz, *Vorlesungen Über Geschichte der Mathematik*, vol. II (Leipzig: Teubner, 1892).

Caroti, Stefano, 'La critica contra l'astrologia di Nicole Oresme e la sua influenza nel Medioevo e nel Rinascimento', *Atti della Accademia Nazionale dei Lincei*, Memorie. Classe di scienze morali, storiche e filologiche, 23 (1979): 543–648.

Carpenter, Nan Cooke, *Music in the Medieval and Renaissance Universities* (1958; New York: Da Capo Press, 1972).

Carrière, V., 'Lefèvre d'Étaples à l'Université de Paris (1475–1520)', in *Etudes historiques dediées à la mémoire de M. Roger Rodière* (Arras, 1947), 109–20.

Carruthers, Mary, *The Craft of Thought: Meditation, Rhetoric, and the Making of Images, 400–1200* (Cambridge: Cambridge University Press, 1998).

Carruthers, Mary, *The Book of Memory: A Study of Memory in Medieval Culture* (1990; 2nd edn, Cambridge: Cambridge University Press, 2008).

Cassirer, Ernst, *Individuum und Kosmos in der Philosophie der Renaissance* (Leipzig: Teubner, 1927).

Cassirer, Ernst, 'The Relationship between Metaphysics and Scientific Method', in *Galileo, Man of Science*, ed. Ernan McMullin, trans. Edward W. Strong (New York: Basic Books, 1967), 338–51.

Cave, Terence, *The Cornucopian Text: Problems of Writing in the French Renaissance* (Oxford: Oxford University Press, 1979).

Cave, Terence (ed.), *Thomas More's Utopia in Early Modern Europe: Paratexts and Contexts* (Manchester: Manchester University Press, 2008).

Chartier, Roger, Dominique Julia, and Marie-Madeleine Compère, *L'education en France du XVIe au XVIIIe siècle* (Paris: Société d'édition d'enseignement supérieur, 1976).

Cifoletti, Giovanna, 'Mathematics and Rhetoric: Peletier and Gosselin and the Making of the French Algebraic Tradition', PhD dissertation, Princeton University, 1993.

Cifoletti, Giovanna, 'From Valla to Viète: The Rhetorical Reform of Logic and its Use in Early Modern Algebra', *Early Science & Medicine* 11, no. 4 (2006): 390–423.

Claessens, Guy, 'Het denken verbeeld: De vroegmoderne receptie (1533–1650) van Proclus' Commentaar op het eerste boek van Euclides' Elementen', PhD dissertation, University of Leuven, 2011.

Clark, Stuart, *Vanities of the Eye: Vision in Early Modern European Culture* (Oxford: Oxford University Press, 2007), 9–20.

Claudin, Anatole, *Histoire de l'imprimerie en France au XVe et au XVIe siècle*, 4 vols (Paris, 1900–14).

Cleary, John J., *Aristotle and Mathematics: Aporetic Method in Cosmology and Metaphysics* (Leiden: Brill, 1995).

Clerval, Alexandre, *De Judoci Clichtovei, Neoportuensis, doctoris theologi parisiensis et carnotensis canonici: vita et operibus (1472–1543)* (Paris: A. Picard, 1894).

Colomer, Eusebius, 'Zu Dem Aufsatz von Rudolf Haubst "Der Junge Cusanus war im Jahre 1428 zu Handschriften-Studien in Paris"', in *Mitteilungen und Forschungsbeiträge der Cusanus-Gesellschaft*, ed. Rudolf Haubst (Mainz: Matthias-Grünewald-Verlag, 1982), 57–70.

Compère, Marie-Madeleine, 'Les collèges de l'université de Paris au XVIe siècle: structures institutionnelles et fonctions éducatives', in *I collegi universitari in Europa tra il XIV e il XVIII secolo: Atti del convegno di studi della commissione internazionale per la storia della Università, Siena-Bologna, 16–19 maggio 1988*, ed. Hilde de Ridder-Symoens and Domenico Maffei (Milan: Guiffre editore, 1991), 101–18.

Compère, Marie-Madeleine, *Les Collèges Français: 16e–18e siècles* (Paris: INRP, 2002).

Constable, Giles, 'The Popularity of Twelfth-Century Spiritual Writers in the Late Middle Ages', in *Renaissance: Studies in Honor of Hans Baron*, ed. Anthony Molho and John A. Tedeschi (Florence, 1970), 3–28.

Constable, Giles, 'Twelfth-Century Spirituality and the Late Middle Ages', in *Medieval and Renaissance Studies* 5, ed. O. B. Hardison Jr. (Chapel Hill, NC: University of North Carolina Press, 1971), 27–60.

Coopland, G. W., *Nicole Oresme and the Astrologers: A Study of His Livre de Divinacions* (Liverpool: Liverpool University Press, 1952).

Copeland, Rita, *Pedagogy, Intellectuals, and Dissent in the Later Middle Ages: Lollardy and Ideas of Learning* (Cambridge: Cambridge University Press, 2001).

Copenhaver, Brian P., 'Lefèvre d'Étaples, Symphorien Champier, and the Secret Names of God', *Journal of the Warburg and Courtauld Institutes* 40 (1977): 189–211.

Copenhaver, Brian P., *Symphorien Champier and the Reception of the Occultist Tradition in Renaissance France* (The Hague: Mouton Publishers, 1978), 56–7.

Copenhaver, Brian P. (ed.), *Hermetica: The Greek Corpus Hermeticum and the Latin Asclepius in a New English Translation, with Notes and Introduction* (Cambridge: Cambridge University Press, 1992).

Copenhaver, Brian P., and Thomas M. Ward, 'Notes from a Nominalist in a New Incunabulum by Symphorien Champier', in *Essays in Renaissance Thought in Honour of Monfasani*, ed. Alison Frazier and Patrick Nold (Leiden: Brill, 2015), 546–604.

Cormack, Lesley B., Steven A. Walton, and John A. Schuster (eds), *Mathematical Practitioners and the Transformation of Natural Knowledge in Early Modern Europe* (New York: Springer, 2017).

Corneanu, Sorana, *Regimens of the Mind: Boyle, Locke, and the Early Modern Cultura Animi Tradition* (Chicago: University of Chicago Press, 2012).

Corsten, Severin, 'Universities and Early Printing', in *Bibliography and the Study of 15th-Century Civilization* (London: British Library, 1987), 83–123.

Courtenay, William J., 'The Bible in the Fourteenth Century: Some Observations', *Church History: Studies in Christianity and Culture* 54, no. 2 (1985): 176–87.

Cranz, F. Edward, and Charles B. Schmitt (eds), *A Bibliography of Aristotle Editions 1501–1600* (2nd edn, Baden-Baden: Koerner, 1984).

Crapulli, Giovanni, *Mathesis universalis: Genesi di un'idea nel XVI secolo* (Rome: Ataneo, 1969).

Crombie, Alistair C., *Styles of Scientific Thinking in the European Tradition: The History of Argument and Explanation Especially in the Mathematical and Biomedical Sciences and Arts* (London: Duckworth, 1994).

Crowther, Kathleen M., and Peter Barker, 'Training the Intelligent Eye: Understanding Illustrations in Early Modern Astronomy Texts', *Isis* 104, no. 3 (2013): 429–70.

Cunningham, Andrew, and Sachiko Kusukawa (ed. and trans.), *Natural Philosophy Epitomised: Books 8–11 of Gregor Reisch's Philosophical Pearl (1503)* (Farnham: Ashgate, 2010).

Curtius, Ernst Robert, *European Literature and the Latin Middle Ages*, trans. Willard R. Trask (Princeton, NJ: Princeton University Press, 1953).

Cypess, Rebecca, *Curious and Modern Inventions: Instrumental Music as Discovery in Galileo's Italy* (Chicago: University of Chicago Press, 2016).

Dalché, Patrick Gautier, *La Géographie de Ptolémée en Occident (IVe–XVIe siècle)* (Turnhout: Brepols, 2009).

D'Amico, John F., *Theory and Practice in Renaissance Textual Criticism: Beatus Rhenanus Between Conjecture and History* (Berkeley, CA: University of California Press, 1988).

Davis, Natalie Zemon, 'Mathematicians in the Sixteenth-Century French Academies: Some Further Evidence', *Renaissance News* 11, no. 1 (1958): 3–10.

Davis, Natalie Zemon, 'Sixteenth-Century French Arithmetics on the Business Life', *Journal of the History of Ideas* 21, no. 1 (1960): 18–48.

Dear, Peter, 'Mersenne's Suggestion: Cartesian Meditation and the Mathematical Model of Knowledge in the Seventeenth Century', in *Descartes and His Contemporaries: Meditations, Objections, and Replies*, ed. Roger Ariew and Marjorie Grene (Chicago: University of Chicago Press, 1995), 44–62.

Destrez, Jean, *La Pecia dans les manuscrits universitaires du XIII et XIV siècles* (Paris: Editions J. Vautrain, 1935).

Di Liscia, Daniel A., *Zwischen Geometrie und Naturphilosophie: die Entwicklung der Formlatitudenlehre im deutschen Sprachraum* (Munich: Universitätsbibliothek, 2002).

Dionisotti, Carlo, 'Ermolao Barbaro e la Fortuna di Suiseth', in *Medioevo e Rinascimento: Studi in onore di Bruno Nardi*, 2 vols (Florence: Sansoni, 1955), 1: 219–53.

Eisenstein, Elizabeth L., *The Printing Press as an Agent of Change: Communications and Cultural Transformations in Early-Modern Europe*, 2 vols (Cambridge: Cambridge University Press, 1979).

Eisenstein, Elizabeth L., *Divine Art, Infernal Machine: The Reception of Printing in the West from First Impressions to the Sense of an Ending* (Philadelphia, PA: University of Pennsylvania Press, 2012).

Ekholm, Karin J., 'Tartaglia's Ragioni: A Maestro d'abaco's Mixed Approach to the Bombardier's Problem', *British Journal for the History of Science* 43, no. 2 (2010): 181–207.

Emery, Kent, 'Mysticism and the Coincidence of Opposites in Sixteenth- and Seventeenth-Century France', *Journal for the History of Ideas* 45, no. 1 (1984): 3–23.

Emery, Kent, Russell Friedman, and Andreas Speer (eds), *Philosophy and Theology in the Long Middle Ages: A Tribute to Stephen F. Brown* (Leiden: Brill, 2011).

Evans, Gillian Rosemary, 'The "Sub-Euclidean" Geometry of the Earlier Middle Ages, up to the Mid-Twelfth Century', *Archive for History of Exact Sciences* 16, no. 2 (1976): 105–18.

Evans, Gillian Rosemary, 'Introductions to Boethius's "Arithmetica" of the Tenth to the Fourteenth Century', *History of Science*, 16, no. 1 (1978), 22–41.

Fabris, Cécile, *Etudier et vivre à Paris au moyen âge: le Collège de Laon, XIVe–XVe siècles* (Paris: École nationale des chartes, 2005).

Farge, James K., *Biographical Register of Paris Doctors of Theology, 1500–1536* (Toronto: Pontifical Institute of Mediaeval Studies, 1980).

Farge, James K., 'Erasmus, the University of Paris, and the Profession of Theology', *Erasmus of Rotterdam Society Yearbook* 19 (1999): 18–46.

Farge, James K. (ed.), *Students and Teachers at the University of Paris: The Generation of 1500. A Critical Edition of Bibliothèque de l'Université de Paris (Sorbonne), Archives, Registres 89 and 90* (Leiden: Brill, 2006).

Faye, Emmanuel, 'Beatus Rhenanus lecteur et étudiant de Charles de Bovelles', *Annuaire des Amis de la Bibliothèque Humanist de Sélestat* 45 (1995): 119–38.

Faye, Emmanuel, *Philosophie et perfection de l'homme: De la Renaissance à Descartes* (Paris: Vrin, 1998).

Faye, Emmanuel, 'Nicolas de Cues et Charles de Bovelles dans le manuscrit "Exigua pluvia" de Beatus Rhenanus', *Archives d'histoire doctrinale et littéraire du moyen âge* 65 (1998), 415–50.

Febvre, Lucien, 'The Origins of the French Reformation: A Badly-Put Question?', repr. in *A New Kind of History: From the Writings of Lucien Febvre* (London: Routledge & Kegan Paul, 1973), 44–107 (orig. 1929).

Feingold, Mordechai, *The Mathematician's Apprenticeship: Science, Universities and Society in England, 1560–1640* (Cambridge: Cambridge University Press, 1984).

Fend, Michael, 'The Changing Functions of Senso and Ragione in Italian Music Theory of the Late Sixteenth Century', in *The Second Sense: Studies in Hearing and Musical Judgement from Antiquity to the Seventeenth Century*, ed. Charles Burnett, Michael Fend, and Penelope Gouk (London: Warburg Institute, 1991), 199–221.

Ferrari, Michel, and Tamara Albertini (eds), *Charles de Bovelles'* Liber de Sapiente, *or* Book of the Wise, special issue of *Intellectual History Review* 21, issue 3 (2011).

Folkerts, Menso, 'Regiomontanus' Role in the Transmission and Transformation of Greek Mathematics', in *Tradition, Transmission, Transformation: Proceedings of Two Conferences on Pre-Modern Science Held at the University of Oklahoma*, ed. F. J. Ragep, Sally Ragep, and Steven John Livesey (Leiden: Brill, 1996), 83–113.

Folkerts, Menso, '*Boethius' Geometrie II: ein mathematisches Lehrbuch des Mittelalters* (Weisbaden: F. Steiner, 1970).

Fournier, Florence, 'L'Enseignement d'Aristote à l'Université dans les années 1490: le rôle et la place de Jacques Lefèvre d'Étaples à travers l'étude de son cours et de son manuel de 1492 sur le De anima (livre I)', MA dissertation, Centre d'Études Supérieurs de la Renaissance, 2006.

Fowler, David H., *The Mathematics of Plato's Academy: A New Reconstruction* (2nd edn, Oxford: Clarendon Press, 1999).

Frangenberg, Thomas, '"Auditus Visu Prestantior": Comparisons of Hearing and Vision in Charles de Bovelles's "Liber de Sensibus"', in *The Second Sense. Studies in Hearing and Musical Judgement from Antiquity to the Seventeenth Century*, ed. Charles Burnett, Michael Fend, and Penelope Gouk (London: Warburg Institute, 1991), 71–94.

Gabriel, Astrik L., *Student Life in Ave Maria College, Mediaeval Paris* (Notre Dame, IN: University of Notre Dame Press, 1955).

Ganay, Ernest de, *Un Chancelier de France sous Louis XII: Jehan de Ganay* (Paris, 1932).

Garin, Eugenio, *L'éducation de L'homme Moderne*, trans. Jacqueline Humbert (1957; Paris: Fayard, 2003).

Garin, Eugenio, 'Echi italiani de Erasmo e Lefèvre d'Étaples', *Rivista critica di storia della filosofia* 24 (1971): 88–90.

Genette, Gérard, *Paratexts: Thresholds of Interpretation*, trans. Jane E. Lewin (Cambridge: Cambridge University Press, 1997).

Gentile, Sebastiano, 'Giano Lascaris, Germain de Ganay e la "prisca theologia" in Francia', *Rinascimento* 2nd ser., 26 (1986): 51–76.

Gerritsen, Johan, 'Printing at Froben's: An Eye-Witness Account', *Studies in Bibliography* 44 (1991): 144–63.

Gingerich, Owen, 'From Copernicus to Kepler: Heliocentrism as Model and as Reality', *Proceedings of the American Philosophical Society* 117, no. 6 (1973): 513–22.

Goff, Jacques Le, *Les intellectuels au Moyen Âge* (1957; 2nd edn, Paris: Éditions du Seuil, 1985).

Goldman, David Paul, 'Nicholas Cusanus' Contribution to Music Theory', *Rivista internazionale di musica sacra* 10 (1989): 308–38.

Gorochov, Nathalie, 'Le collège du Cardinal Lemoine au XVIe siècle', *Mémoires de Paris et l'Ile-de-France* 42 (1991): 219–59.

Goulding, Robert, 'Polemic in the Margin: Henry Savile against Joseph Scaliger's Quadrature of the Circle', in *Scientia in margine: études sur les marginalia dans les manuscrits scientifiques du Moyen Age à la Renaissance*, ed. Danielle Jacquart and Charles Burnett (Geneva: Droz, 2005), 241–59.

Goulding, Robert, 'Method and Mathematics: Peter Ramus' Histories of the Sciences', *Journal of the History of Ideas* 67, no. 1 (2006): 63–85.

Goulding, Robert, *Defending Hypatia: Ramus, Savile, and the Renaissance Rediscovery of Mathematical History* (New York: Springer, 2010).

Grabmann, Martin, *Methoden und Hilfsmittel des Aristotelisstudiums im Mittelalter* (Munich: Verlag der Bayerischen Akademie der Wissenschaften, 1939).

Graf, Charles-Henri, *Essai sur la vie et les écrits de Jacques Lefèvre d'Étaples* (Strasbourg: G.L. Schüler, 1842).

Graf-Stuhlhofer, Franz, *Humanismus zwischen Hof und Universität: Georg Tanstetter (Collimitius) und sein wissenschaftliches Umfeld im Wien des frühen 16. Jahrhunderts* (Vienna: WUV-Universitüts Verlag, 1996).

Grafton, Anthony T., 'Teacher, Text and Pupil in the Renaissance Class-Room: A Case Study from a Parisian College', *History of Universities* 1 (1981): 37–70.

Grafton, Anthony T., 'Textbooks and the Disciplines', in *Scholarly Knowledge: Textbooks in Early Modern Europe*, ed. Emidio Campi, Simone De Angelis, and Anja-Silvia Goeing (Geneva: Droz, 2008), 11–36.

Grafton, Anthony T., *The Culture of Correction in Renaissance Europe* (London: British Library, 2011).

Grafton, Anthony T., and Lisa Jardine, *From Humanism to the Humanities: Education and the Liberal Arts in Fifteenth- and Sixteenth-Century Europe* (Cambridge, MA: Harvard University Press, 1986).

Grafton, Anthony T., and Urs Leu, *Henricus Glareanus's (1488–1563) Chronologia of the Ancient World: A Facsimile Edition of a Heavily Annotated Copy Held in Princeton University Library* (Leiden: Brill, 2013).

Grafton, Anthony, and Glenn W. Most (eds), *Canonical Texts and Scholarly Practices: A Global Comparative Approach* (Cambridge: Cambridge University Press, 2016).

Grendler, Paul F., *The Universities of the Italian Renaissance* (Baltimore, MD: Johns Hopkins University Press, 2002).

Groote, Inga Mai, and Bernhard Köble, 'Glarean the Professor and His Students' Books: Copied Lecture Notes', *Bibliotheque d' Humanisme et Renaissance: Travaux et Documents* 73, no. 1 (2011): 61–91.

Groote, Inga Mai, 'Studying Music and Arithmetic with Glarean: Contextualizing the Epitomes and Annotationes among the Sources for Glarean's Teaching', in *Heinrich Glarean's Books: The Intellectual World of a Sixteenth-Century Musical Humanist*, ed.

Iain Fenlon and Inga Mai Groote (Cambridge: Cambridge University Press, 2013), 195–222.

Grössing, Helmuth, *Humanistische Naturwissenschaft: Zur Geschichte der Wiener mathematischen Schulen des 15. und 16. Jahrhunderts* (Baden-Baden: Valentin Koerner, 1983).

Guerlac, Rita, *Juan Luis Vives against the Pseudodialecticians: A Humanist Attack on Medieval Logic: The Attack on the Pseudialecticians and On Dialectic* (Dordrecht: Springer, 1979).

Hamburger, Jeffrey F., 'The Visual and the Visionary: The Image in Late Medieval Monastic Devotions', *Viator* 20 (1989): 161–82.

Hamburger, Jeffrey F., and Anne-Marie Bouché (eds), *The Mind's Eye: Art and Theological Argument in the Medieval West* (Princeton, NJ: Princeton University Press, 2005).

Hamel, Jürgen, 'Johannes Sacrobosco: Handbuch der Astronomie, Kommenierte Bibliographie der Drucke der "Sphaera" 1472 bis 1656', in *Wege der Erkenntnis: Festschrift für Dieter B. Herrmann zum 65. Geburtstag*, ed. Dietmar Fürst, Dieter B. Herrmann, and Eckehard Rothenberg (Frankfurt am Main: Harri Deutsch Verlag, 2004), 115–70.

Hamel, Jürgen, 'Johannes de Sacroboscos Sphaera', in *Gutenberg-Jahrbuch*, ed. Stephan Füssel (Weisbaden: Harrassowitz Verlag, 2006), 113–36.

Hamel, Jürgen, *Studien zur 'Sphaera' des Johannes de Sacrobosco* (Leipzig: Akademische Verlagsanstalt, 2014).

Hankins, James, 'Ptolemy's Geography in the Renaissance', in *Humanism and Platonism in the Italian Renaissance*, 2 vols (Rome: Edizioni di storia e letteratura, 2003), 1:457–68.

Häring, Nikolaus M., *Commentaries on Boethius by Thierry of Chartres and His School* (Toronto: Pontifical Institute of Mediaeval Studies, 1971).

Haubst, Rudolf, 'Der junge Cusanus war im Jahre 1428 zu Handschriften-Studien in Paris', in *Mitteilungen und Forschungsbeiträge der Cusanus-Gesellschaft*, ed. Rudolf Haubst (Mainz: Matthias-Grünewald-Verlag, 1980), 198–205.

Haugen, Kristine Louise, 'Academic Charisma and the Old Regime', *History of Universities* 22, no. 1 (2007): 203–9.

Hayton, Darin, 'Instruments and Demonstrations in the Astrological Curriculum: Evidence from the University of Vienna, 1500–1530', *Studies in History and Philosophy of Science Part C* 41, no. 2 (2010): 125–34.

Hayton, Darin, *The Crown and the Cosmos: Astrology and the Politics of Maximilian I* (Pittsburgh, PA: University of Pittsburgh Press, 2015).

Heiberg, J. L., *Beiträge zur Geschichte Georg Vallas und seiner Bibliothek* (Leipzig: Harrossowitz, 1896).

Heller, Henry, 'Reform and Reformers at Meaux, 1518–1525', PhD dissertation, Cornell University, 1969.

Heller, Henry, 'The Evangelicalism of Lefèvre d'Étaples: 1525', *Studies in the Renaissance* 19 (1972): 42–77.

Heninger, S. K. Jr., *Touches of Sweet Harmony: Pythagorean Cosmology and Renaissance Poetics* (San Marino, CA: Huntington Library, 1974).

Heninger, S. K. Jr., 'Oronce Finé and English Textbooks for the Mathematical Sciences', in *Studies in the Continental Background of Renaissance English Literature. Essays Presented to John L. Lievsay*, ed. D. B. J. Randall and G. W. Williams (Durham, NC: Duke University Press, 1977), 171–85.

Henry, John, '"Mathematics Made No Contribution to the Public Weal": Why Jean Fernel (1497–1558) Became a Physician', *Centaurus* 53, no. 3 (2011): 193–220.

Hesse, Mary, 'Aristotle's Logic of Analogy', *Philosophical Quarterly* 15, no. 61 (1965): 328–40.

Hirai, Hiro, *Le Concept de semence dans les théories de la matière à la Renaissance: De Marsile Ficin à Pierre Gassendi* (Turnhout: Brepols, 2005).

Hobart, Michael E., and Zachary S. Schiffman, *Information Ages: Literacy, Numeracy, and the Computer Revolution* (Baltimore, MD: Johns Hopkins University Press, 2000).

Hobbins, Daniel, 'The Schoolman as Public Intellectual: Jean Gerson and the Late Medieval Tract', *American Historical Review* 108, no. 5 (2003).

Hobbins, Daniel, *Authorship and Publicity Before Print: Jean Gerson and the Transformation of Late Medieval Learning* (Philadelphia, PA: University of Pennsylvania Press, 2009).

Hoenen, Maarten J. F. M., 'Academics and Intellectual Life in the Low Countries; The University Career of Heymeric de Campo († 1460)', *Recherches de Théologie Ancienne et Médiévale* 61 (1994): 173–209.

Hoenen, Maarten J. F. M., 'Via Antiqua and Via Moderna in the Fifteenth Century: Doctrinal, Institutional, and Church Political Factors in the *Wegestreit*', in *The Medieval Heritage in Early Modern Metaphysics and Modal Theory, 1400–1700*, ed. Russell L. Friedman and Lauge O. Nielsen (Dordrecht: Kluwer, 2003), 9–36.

Hoppe, Brigitte, 'Die Vernetzung der Mathematisch ausgerichteten Anwendungsgebiete mit den Fächern des Quadriviums in der Frühen Neuzeit', in *Der "mathematicus" zur Entwicklung und Bedeutung einer neuen Berufsgruppe in der Zeit Gerhard Mercators*, ed. Irmgarde Hantsche (Bochum: Brockmeyer, 1996), 1–33.

Hösle, Vittorio, and W. Büchel (eds), *Raimundus Lullus, Die neue Logik. Logica Nova* (Hamburg: Felix Meiner Verlag, 1985).

Høyrup, Jens, 'Jordanus de Nemore, 13th Century Mathematical Innovator: An Essay on Intellectual Context, Achievement, and Failure', *Archive for History of Exact Sciences* 38, no. 4 (1988): 307–63.

Høyrup, Jens, *In Measure, Number, and Weight: Studies in Mathematics and Culture* (Buffalo, NY: State University of New York Press, 1994).

Hughes, B. B., 'Toward an Explication of Ambrosiana MS D 186 Inf.', *Scriptorium* 26 (1972): 125–7.

Hughes, Philip Edgcumbe, *Lefèvre: Pioneer of Ecclesiastical Renewal in France* (Grand Rapids, MI: Eerdmans, 1984).

Imbach, Ruedi, 'Das Centheologicon des Heymericus de Campo und die darin enthalten Cusanus-Reminiszenzen', *Traditio* 39 (1983): 466–77.

Jacquart, Danielle, 'Theory, Everyday Practice, and Three Fifteenth-Century Physicians', *Osiris*, 2nd ser., 6 (1990): 148–50.

Janssen, Frans A., 'The Rise of the Typographical Paragraph', in *Cognition and the Book: Typologies of Formal Organization of Knowledge in the Printed Book of the Early Modern Period*, ed. Karl A. E. Enenkel and Wolfgang Neuber (Leiden: Brill, 2005), 9–32.

Janssen, Frans A., *Technique and Design in the History of Printing: 26 Essays* (Houten: Hes & De Graaf, 2004).

Jardine, Lisa, and Anthony Grafton, '"Studied for Action": How Gabriel Harvey Read His Livy', *Past and Present* no. 129 (1990): 30–78.

Jardine, Lisa, *Erasmus, Man of Letters: The Construction of Charisma in Print* (Princeton, NJ: Princeton University Press, 1993).

Johnston, Stephen, 'The Identity of the Mathematical Practitioner in 16th-Century England', in *Der 'Mathematicus': Zur Entwicklung und Bedeutung einer neuen Berufsgruppe in der Zeit Gerhard Mercators*, ed. Irmgarde Hantsche (Bochum: Brockmeyer, 1996), 93–120.

Jorink, Eric, *Reading the Book of Nature in the Dutch Golden Age, 1575–1715* (Leiden: Brill, 2010).

Jourdain, Charles, 'Le collège du Cardinal Lemoine', *Mémoires de la Société de l'histoire de Paris et de l'Ile-de-France* 3 (1876): 42–81.

Joutsivuo, Timo, *Scholastic Tradition and Humanist Innovation: The Concept of Neutrum in Renaissance Medicine* (Helsinki: Academia Scientiarum Fennica, 1999).

Jung, Elzbieta, 'Intension and Remission of Forms', in *Encyclopedia of Medieval Philosophy: Philosophy Between 500 and 1500*, ed. Henrik Lagerlund (Dortrecht: Springer, 2011), 551–5.

Juřen, Vladimir, 'Fra Giovanni Giocondo et le début des études vitruviennes en France', *Rinascimento*, 2nd ser., 14 (1974): 101–15.

Kalatzi, Maria, *Hermonymos: A Study in Scribal, Literary and Teaching Activities in the Fifteenth and Early Sixteenth Centuries* (Athens, 2009), 66–85.

Kaluza, Zenon, *Les querelles doctrinales à Paris: Nominalistes et realistes aux confins du XIVe et du XVe siècles* (Bergamo: Pierluigi Lubrina, 1988).

Keller, Vera, 'Painted Friends: Political Interest and the Transformation of International Learned Sociability', in *Friendship in the Middle Ages and Early Modern Culture*, ed. Marilyn Sandidge and Albrecht Classen (Berlin: de Gruyter, 2011), 661–92.

Kerby-Fulton, Kathryn, and Stephen Justice, 'Langlandian Reading Circles and the Civil Service in London and Dublin, 1380–1427', *New Medieval Literatures* 1 (1997): 59–83.

Kessler, Eckhard, 'Clavius entre Proclus et Descartes', in *Les jésuites à la Renaissance: Système éducatif et production du savoir*, ed. Luce Giard (Paris: Presses universitaires, 1995), 285–308.

Kessler, Eckhard, 'Introducing Aristotle to the Sixteenth Century: The Lefèvre Enterprise', in *Philosophy in the Sixteenth and Seventeenth Centuries: Conversations with Aristotle*, ed. Constance Blackwell and Sachiko Kusukawa (Aldershot: Ashgate, 1999), 1–21.

Kirschner, Stefan, 'Oresme on Intension and Remission of Qualities in His Commentary on Aristotle's Physics', *Vivarium* 38, no. 2 (2000): 255–74.

Klein, Jacob, *Greek Mathematical Thought and the Origin of Algebra*, trans. Eva Brann (1934; New York: Dover, 1992).

Klinger-Dollé, Anne-Hélène, *Le* De sensu *de Charles de Bovelles (1511): conception philosophique des sens et figuration de la pensée* (Geneva: Droz, 2016).

Knod, C. Gustav, *Aus der Bibliothek des Beatus Rhenanus: ein Beitrag zur Geschichte der Humanismus* (Schlettstadt, 1889).

Koyré, Alexandre, *From the Closed World to the Infinite Universe* (New York: Harper Torch, 1958).

Koyré, Alexandre, *Metaphysics and Measurement: Essays in Scientific Revolution* (Cambridge, MA: Harvard University Press, 1968).

Krämer, Fabian, 'Ein papiernes Archiv für alles jemals Geschriebene: Ulisse Aldrovandis Pandechion epistemonicon und die Naturgeschichte der Renaissance', *NTM Zeitschrift für Geschichte der Wissenschaften, Technik und Medizin* 21, no. 1 (2013): 11–36.

Krämer, Fabian, *Ein Zentaur in London: Lektüre und Beobachtung in der frühneuzeitlichen Naturforschung* (Didymos-Verlag, 2014).

Kremer, Richard L., and Michael Shank (eds), Regiomontanus, *Defensio Theonis*, http://regio.dartmouth.edu/, updated 29 March 2004.

Kristeller, Paul Oskar, *Studies in Renaissance Thought and Letters* (Rome: Edizioni di storia e letteratura, 1956).

Kristeller, Paul Oskar, *Iter Italicum: A Finding List of Uncatalogued or Incompletely Catalogued Humanistic Manuscripts of the Renaissance in Italian and Other Libraries* (London: Warburg Institute, 1963).

Kuhn, Thomas S., 'Mathematical vs. Experimental Traditions in the Development of Physical Science', *Journal of Interdisciplinary History* 7, no. 1 (1976): 1–31.

Kusukawa, Sachiko, *The Transformation of Natural Philosophy: The Case of Philip Melanchthon* (Cambridge: Cambridge University Press, 1995).

Ladner, Gerhart B., *The Idea of Reform: Its Impact on Christian Thought and Action in the Age of the Fathers* (New York: Harper Torchbooks, 1959).

Lajarte, Philippe de, *L'Humanisme en France au XVIe siècle* (Paris: Honoré Champion, 2009).

Lawn, Brian, *The Rise and Decline of the Scholastic 'Quaestio Disputata'* (Leiden: Brill, 1993).

Leigh, Matthew, *From Polypragmon to Curiosus: Ancient Concepts of Curious and Meddlesome* (Oxford: Oxford University Press, 2013).

Leitão, Henrique, 'Pedro Nunes against Oronce Fine: Content and Context of a Refutation', in *The Worlds of Oronce Fine: Mathematics, Instruments and Print in Renaissance France*, ed. Alexander Marr (Donington: Shaun Tyas, 2009), 156–71.

Leitão, Henrique, and Antonio Sánchez, 'Zilsel's Thesis, Maritime Culture, and Iberian Science in Early Modern Europe', *Journal of the History of Ideas* 78, no. 2 (2017): 191–210.

Lemay, Richard Joseph, *Abu Ma'shar and Latin Aristotelianism in the Twelfth Century: The Recovery of Aristotle's Natural Philosophy Through Arabic Astrology* (Beirut: American University, 1962).

Lenzi, F., 'I dialoghi morali e religosi di Giulio Landi, Lefèvre d'Étaples ed Erasmo', *Memorie domenicane* 4 (1973): 196–216.

Leonhardt, Jürgen, 'Classics as Textbooks: A Study of the Humanist Lectures on Cicero at the University of Leipzig, ca. 1515', in *Scholarly Knowledge: Textbooks in Early Modern Europe*, ed. Emidio Campi, Simone De Angelis, and Anja-Silvia Goeing (Geneva: Droz, 2008), 89–112.

Letrouit, Jean, 'La prise de notes de cours sur support imprimé dans les collèges parisiens au XVIe siècle', *Revue de la Bibliothèque National de France* 2 (1999): 47–56.

Lewis, Neil, 'Robert Grosseteste's Notes on the Physics', in *Editing Robert Grosseteste: Papers given at the Thirty-Sixth Annual Conference on Editorial Problems, University of Toronto, 3–4 November 2000*, ed. Joseph Ward Goering and Evelyn Anne Mackie (Toronto: University of Toronto Press, 2003), 103–34.

Levitin, Dmitri, *Ancient Wisdom in the Age of the New Science: Histories of Philosophy in England, c. 1640–1700* (Cambridge: Cambridge University Press, 2015).

Lewis, Rhodri, 'A Kind of Sagacity: Francis Bacon, the Ars Memoriae and the Pursuit of Natural Knowledge', *Intellectual History Review* 19, no. 2 (2009): 155–75.

Libera, Alain de, *La querelle des universaux: De Platon à la fin du Moyen Age* (Paris: Édition du Seuil, 1996).

Lindberg, David C., 'On The Applicability of Mathematics to Nature: Rogert Bacon and His Predecessors', *British Journal for the History of Science* 15 (1982): 3–25.

Lines, David A., 'Humanism and the Italian Universities', in *Humanism and Creativity in the Renaissance: Essays in Honor of Ronald G. Witt*, ed. Christopher S. Celenza and Kenneth Gouwens (Leiden: Brill, 2006), 327–46.

Lines, David A., 'Lefèvre and French Aristotelianism on the Eve of the Sixteenth Century', in *Der Aristotelismus in der Frühen Neuzeit: Kontinuität oder Wiederangeignung?*, ed. Günter Frank and Andreas Speer (Weisbaden: Harrassowitz Verlag, 2007), 273–90.

Livesey, Steven J., 'William of Ockham, the Subalternate Sciences, and Aristotle's Theory of Metabasis', *British Journal for the History of Science* 18, no. 2 (1985): 127–45.

Loget, François, 'L'algèbre en France au XVIe siècle: Individus et réseaux', in *Pluralité de l'algèbre à la Renaissance*, ed. Sabine Rommevaux, Maryvonne Spiesser, and Maria Rosa Massa Esteve (Paris: Honoré Champion, 2012), 69–101.

Lohr, Charles H., 'Mathematics and the Divine: Ramon Lull', in *Mathematics and the Divine: A Historical Study*, ed. T. Koetsier and L. Bergmans (Amsterdam: Elsevier, 2005), 211–29.

Long, Pamela O., *Artisan/Practitioners and the Rise of the New Sciences, 1400–1600* (Corvallis, OR: Oregon State University Press, 2011).

Lüthy, Christoph, 'The Fourfold Democritus on the Stage of Early Modern Science', *Isis* 91, no. 3 (2000): 443–79.

Lüthy, Christoph, and Alexis Smets, 'Words, Lines, Diagrams, Images: Towards a History of Scientific Imagery', *Early Science and Medicine* 14, nos 1–3 (2009): 398–439.

Maclean, Ian, 'Diagrams in the Defence of Galen: Medical Uses of Tables, Squares, Dichotomies, Wheels, and Latitudes, 1480–1574', in *Transmitting Knowledge: Words, Images, and Instruments in Early Modern Europe*, ed. Sachiko Kusukawa and Ian Maclean (Oxford: Oxford University Press, 2006), 137–64.

Mahoney, Michael S., 'The Beginnings of Algebraic Thought in the Seventeenth Century', in *Descartes: Philosophy, Mathematics and Physics*, ed. Stephen Gaukroger (Brighton: Harvester Press, 1980), 141–55.

Malet, Antoni, 'Renaissance Notions of Number and Magnitude', *Historia Mathematica* 33 (2006): 63–81.

Mancosu, Paolo, 'Aristotelian Logic and Euclidean Mathematics: Seventeenth-Century Developments of the Quaestio de certitudine mathematicarum', *Studies in History and Philosophy of Science* 23, no. 2 (1992): 241–65.

Mandosio, Jean-Marc, 'Entre mathématiques et physique: note sur les "sciences intermédiaires" à la Renaissance', in *Comprendre et maîtriser la nature au Moyen Âge: Mélanges d'histoire des sciences offerts à Guy Beaujouan* (Geneva: Droz, 1994), 115–38.

Mandosio, Jean-Marc, 'Le *De magia naturali* de Jacques Lefèvre d'Étaples: Magie, alchimie et cabale', in *Les Muses secrètes: kabbale, alchimie et littérature à la Renaissance*, ed. Rosanna Camos Gorris (Geneva: Droz, 2013), 37–79.

Mandosio, Jean-Marc, 'La fabrication d'un faux: l'*Introduction à la rhétorique* pseudo-lullienne (*In rhetoricen isagoge*, Paris, 1515)', *Bibliothèque d'Humanisme et Renaissance* 78 (2016): 311–31.

Mandosio, Jean-Marc, and Marie-Dominique Couzinet, 'Nouveaux éclairages sur les cours de Ramus et de ses collègues au collège de Presles d'après des notes inédites prises par Nancel', in *Ramus et l'Université* (Paris: Rue d'Ulm, 2004), 11–48.

Marenbon, John, *Boethius* (New York: Oxford University Press, 2003).

Margolin, Jean-Claude, 'L'Enseignement des mathématiques en France (1540–70): Charles de Bovelles, Fine, Peletier, Ramus', in *French Renaissance Studies, 1540–70: Humanism and the Encyclopedia*, ed. Peter Sharratt (Edinburgh: Edinburgh University Press, 1976), 109–55.

Margolin, Jean-Claude, 'Le mythe de Prométhee dans la philosophie de la Renaissance', in *Histoire de l'exégèse au XVIe siècle*, ed. Olivier Fatio and Pierre Fraenkel (Geneva: Droz, 1978), 241–69.

Margolin, Jean-Claude, 'Une Géométrie fort singulière: la Géométrie pratique de Charles de Bovelles (Paris, S. de Colines, 1542)', in *Verum et Factum: Beiträge zur Geistesgeschichte und Philosophie der Renaissance zum 60. Geburtstag von Stephan Otto*, ed. Tamara Albertini (Frankfurt am Main: Peter Lang, 1993), 437–51.

Margolin, Jean-Claude, *Lettres et poèmes de Charles de Bovelles* (Paris: Honoré Champion, 2002).

Marr, Alexander, 'A Renaissance Library Rediscovered: The "Repertorium Librorum Mathematica" of Jean I Du Temps', *The Library* 9, no. 4 (2008): 428–70.

Marr, Alexander (ed.), *The Worlds of Oronce Fine: Mathematics, Instruments and Print in Renaissance France* (Donington: Shaun Tyas, 2009).

Marr, Alexander, *Between Raphael and Galileo: Mutio Oddi and the Mathematical Culture of Late Renaissance Italy* (Chicago: University of Chicago Press, 2011).

Marr, Alexander, 'Knowing Images', *Renaissance Quarterly* 69, no. 3 (2016): 1000–13.

Martin, Henri-Jean, et al., *La Naissance du livre moderne: Mise en page et mise en texte du livre français (XIVe–XVIIe siècles)* (Paris: Éditions du Cercle de la Librairie, 2000).

Masi, Michael, 'The Liberal Arts and Gerardus Ruffus' Commentary on the Boethian De Arithmetica', *Sixteenth Century Journal* 10, no. 2 (1979): 23–41.

Masi, Michael, *Boethian Number Theory: A Translation of the De Institutione Arithmetica* (Amsterdam: Rodopi, 1983).

Massaut, Jean-Pierre, *Josse Clichtove, l'humanisme et la réforme du clergé*, 2 vols (Paris: Les Belles Lettres, 1968).

Mathiesen, Thomas J., *Apollo's Lyre: Greek Music and Music Theory in Antiquity and the Middle Ages* (Lincoln, NE: University of Nebraska Press, 2000).

[Mazour-]Matusevich, Yelena, 'Jean Gerson, Nicholas of Cusa and Lefèvre d'Étaples: The Continuity of Ideas', in *Nicholas of Cusa and His Age: Intellect and Spirituality: Essays Dedicated to the Memory of F. Edward Cranz*, ed. Thomas P. McTiche and Charles Trinkaus (Leiden: Brill, 2002), 237–63.

Mazour-Matusevich, Yelena, *Le siècle d'or de la mystique française: étude de la littérature spirituelle de Jean Gerson (1363–1429) à Jacques Lefèvre d'Etaples (1450?–1537)* (Paris: Arche, 2004).

McGinn, Bernard, *The Harvest of Mysticism in Medieval Germany* (New York: Crossroad, 2006).

McGinn, Bernard, *The Varieties of Vernacular Mysticism (1350–1550)* (New York: Crossroad, 2012).

Meier-Oeser, Stephan, *Die Präsenz des Vergessenen: Zur Rezeption der Philosophie des Nicolaus Cusanus vom 15. bis zum 18. Jahrhundert* (Münster: Aschendorff, 1989).

Meserve, Margaret, 'Patronage and Propaganda at the First Paris Press: Guillaume Fichet and the First Edition of Bessarion's Orations against the Turks', *Papers of the Bibliographical Society of America* 97, no. 4 (2003): 521–88.

Methuen, Charlotte, *Kepler's Tübingen: Stimulus to a Theological Mathematics* (Aldershot: Ashgate, 1998).

Mikkeli, Heikki, 'The Aristotelian Classification of Knowledge in the Early Sixteenth Century', in *Renaissance Readings of the 'Corpus Aristotelicum'*, ed. Marianne Pade (Copenhagen: Museum Tusculanum Press, 2001), 103–27.

Milanesi, M., 'Geography and Cosmography in Italy from the XVth to the XVIIth Century', *Memorie della Società Astronomia Italiana* 65 (1995): 443–68.

Miller, Peter N., 'Friendship and Conversation in Seventeenth-Century Venice', *Journal of Modern History* 73, no. 1 (2001): 1–31.

Moé, Émile-A. van, 'Documents nouveaux sur les libraires, parcheminiers et imprimeurs en relation avec l'Université de Paris à la fin du XVe siècle', *Humanisme et Renaissance* 2, no. 1 (1935): 5–25.

Mönch, Walter, *Die italienische Platonrenaissance und ihre Bedeutung für Frankreichs Literatur- und Geistesgeschichte (1450–1550)* (Berlin: Matthiesen Verlags, 1936).

Monfasani, John, *George of Trebizond: A Biography and a Study of His Rhetoric and Logic* (Leiden: Brill, 1976).

Monfasani, John, 'A Tale of Two Books: Bessarion's *In Calumniatorem Platonis* and George of Trebizond's *Comparatio Philosophorum Platonis et Aristotelis*', *Renaissance Studies* 22, no. 1 (2008): 1–15.

Mortimer, Ruth (ed.), *French 16th Century Books*, 2 vols (Cambridge, MA: Belknap Press, 1964).

Mosley, Adam, 'The Cosmographer's Role in the Sixteenth Century: A Preliminary Study', *Archives internationales d'histoire des sciences* 59, no. 2 (2009): 423–39.

Mosley, Adam, 'Early Modern Cosmography: Fine's Sphaera Mundi in Content and Context', in *The Worlds of Oronce Fine: Mathematics, Instruments and Print in Renaissance France*, ed. Alexander Marr (Donington: Shaun Tyas, 2009), 114–37.

Moss, Ann, *Printed Commonplace-Books and the Structuring of Renaissance Thought* (Oxford: Clarendon Press, 1996).

Moss, Ann, *Renaissance Truth and the Latin Language Turn* (Oxford: Oxford University Press, 2003).

Moyer, Ann E., *Musica Scientia: Musical Scholarship in the Italian Renaissance* (Ithaca, NY: Cornell University Press, 1992).

Moyer, Ann E., *The Philosophers' Game: Rithmomachia in Medieval and Renaissance Europe* (Ann Arbor, MI: University of Michigan Press, 2001).

Moyer, Ann E., 'The Quadrivium and the Decline of Boethian Influence', in *A Companion to Boethius in the Middle Ages*, ed. Noel Harold Kaylor and Philip Edward Phillips (Leiden: Brill, 2012), 479–517.

Murdoch, John E., '*Mathesis in philosophiam scholasticam introducta*: The Rise and Development of the Application of Mathematics in Fourteenth Century Philosophy and Theology', in *Arts libéraux et philosophie au Moyen Âge: Actes du quatrième congrès international de philosophie médiévale* (Paris: Vrin, 1969), 215–54.

Murdoch, John E., 'From Social into Intellectual Factors: An Aspect of the Unitary Character of Late Medieval Learning', in *The Cultural Context of Medieval Learning*, ed. John E. Murdoch and Edith D. Sylla (Dortrecht: D. Reidel, 1975), 271–348.

Murdoch, John E., *Album of Science: Antiquity and the Middle Ages* (New York: Charles Scribner's Sons, 1984).

Murdoch, John E., 'The Medieval and Renaissance Tradition of Minima Naturalia', in *Late Medieval and Early Modern Corpuscular Matter Theories*, ed. Christoph H. Lüthy, John E. Murdoch, and William Newman (Leiden: Brill, 2001), 91–132.

Murdoch, John E., and Edith D. Sylla, 'Swineshead (Swyneshed, Suicet, etc.), Richard', in *Dictionary of Scientific Biography*, vol. 13 (New York: Charles Scribner's Sons, 1976), 184–213.

Murdoch, John E., and Edith Dudley Sylla, 'The Science of Motion', in *Science in the Middle Ages*, ed. David C. Lindberg (Chicago: University of Chicago Press, 1978), 206–66.

Nagel, Fritz, 'Scientia experimentalis: Zur Cusanus-Rezeption in England', *Mitteilungen und Forschungsbeiträge der Cusanus-Gesellschaft* 29 (2005): 95–109.

Nauta, Lodi, *In Defense of Common Sense: Lorenzo Valla's Humanist Critique of Scholastic Philosophy* (Cambridge, MA: Harvard University Press, 2009).

Neal, Katherine, 'The Rhetoric of Utility: Avoiding Occult Associations for Mathematics Through Profitability and Pleasure', *History of Science* 37 (1999): 151–78.

Neal, Katherine, *From Discrete to Continuous: The Broadening of Number Concepts in Early Modern England* (Berlin: Springer, 2002).

Nolhac, P. de, 'Recherches sur Fra Giocondo de Vérone', *Courrier de l'art* 8 (1888): 77–9.

Novikoff, Alex J., *The Medieval Culture of Disputation: Pedagogy, Practice, and Performance* (Philadelphia, PA: University of Pennsylvania Press, 2013).

Oberman, Heiko Augustinus, *The Dawn of the Reformation: Essays in Late Medieval and Early Reformation Thought* (Edinburgh: T. & T. Clark, 1986).

Ong, Walter J., *Ramus, Method, and the Decay of Dialogue: From the Art of Discourse to the Art of Reason* (Cambridge, MA: Harvard University Press, 1958).

Oosterhoff, Richard J., 'The Fabrist Origins of Erasmian Science: Mathematical Erudition in Erasmus' Basle', *Journal of Interdisciplinary History of Ideas* 3, no. 6 (2014): 3–37.

Oosterhoff, Richard J., 'Why Marginalia Still Matter: Finding a Voice for Humility in Google Books', *NDIAS Quarterly* (Fall 2014), 6–11.

Oosterhoff, Richard J., 'A Book, a Pen, and the *Sphere*: Reading Sacrobosco in the Renaissance', *History of Universities* 28, no. 2 (2015): 1–54.

Oosterhoff, Richard J., 'Lovers in Paratexts: Oronce Fine's Republic of Mathematics', *Nuncius* 31, no. 3 (2016): 549–83.

Oosterhoff, Richard J., '"Secrets of Industry" for "Common Men": Charles de Bovelles and Early French Readerships of Technical Print', in *Translating Early Modern Science*, ed. Sietske Fransen and Niall Hodson (Leiden: Brill, 2017), 207–29.

Oosterhoff, Richard J., 'Connaissance visuelle comme méthode élémentaire chez Charles de Bovelles', in *Bovelles philosophe et pédagogue*, ed. Anne-Hélène Klinger-Dollé, Emmanuel Faye, and Jocelyne Sfez (Paris: Éditions Beauchesne, forthcoming).

Ovitt, George Jr., *The Restoration of Perfection: Labor and Technology in Medieval Culture* (New Brunswick, NJ: Rutgers University Press, 1987).

Ozment, Steven E., *Homo Spiritualis: A Comparative Study of the Anthropology of Johannes Tauler, Jean Gerson and Martin Luther* (Leiden: Brill, 1969).

Palisca, Claude V., *Humanism in Italian Renaissance Musical Thought* (New Haven, CT: Yale University Press, 1985).

Palisca, Claude V., 'Scientific Empiricism in Musical Thought', in *Studies in the History of Italian Music and Music Theory* (Oxford: Clarendon Press, 1994), 200–35.

Palisca, Claude V., *Music and Ideas in the Sixteenth and Seventeenth Centuries* (Urbana, IL: University of Illinois Press, 2006).

Pantin, Isabelle, 'Les problèmes spécifiques de l'édition des livres scientifiques à la Renaissance: l'exemple de Guillaume Cavellat', in *Le Livre dans l'Europe de la Renaissance* (Paris: Promodis, 1988), 240–52.

Pantin, Isabelle, *La poésie du ciel en France sans la seconde moitié du seizième siècle* (Geneva: Droz, 1995).

Pantin, Isabelle, 'L'illustration des livres d'astronomie à la renaissance: l'évolution d'une discipline à travers ses images', in *Immagini per conoscere: Dal Rinascimento alla Rivoluzione scientifica*, ed. Fabrizio Meroi and Claudio Pogliano (Florence: Leo S. Olschki, 2001), 3–42.

Pantin, Isabelle, 'La représentation des mathématiques chez Jacques Peletier du Mans: Cosmos hiéroglyphique ou ordre rhétorique?', *Rhetorica* 20, no. 4 (2002): 375–89.

Pantin, Isabelle, 'Teaching Mathematics and Astronomy in France: The Collège Royal (1550–1650)', *Science and Education* 15 (2006): 189–207.

Pantin, Isabelle, 'Oronce Fine's Role as Royal Lecturer', in *The Worlds of Oronce Fine: Mathematics, Instruments and Print in Renaissance France*, ed. Alexander Marr (Donington: Shaun Tyas, 2009), 13–30.

Pantin, Isabelle, 'Altior incubuit animus sub imagine mundi: L'inspiration du cosmographe d'après un gravure d'Oronce Finé', in *Les méditations cosmographiques à la Renaissance* (Paris: Presses de l'université Paris-Sorbonne, 2009), 69–90.

Pantin, Isabelle, 'Oronce Finé mathématicien et homme du livre: la pratique éditoriale comme moteur d'évolution', in *Mise en forme des savoirs à la Renaissance: A la croisée des *idées, des techniques et des publics*, ed. Isabelle Pantin and Gérald Péoux (Paris: Armand Colin, 2013), 19–40.

Parent, Annie, *Les métiers du livre à Paris au XVIe siècle* (Geneva: Droz, 1974).

Park, Katharine, 'Albert's Influence on Late Medieval Psychology', in *Albertus Magnus and the Natural Sciences*, ed. James A. Weisheipl (Toronto: Pontifical Institute of Mediaeval Studies, 1980), 501–35.

Perigot, Béatrice, *Dialectique et littérature: les avatars de la dispute entre Moyen Age et Renaissance* (Paris: Honoré Champion, 2005).

Pesic, Peter, 'Hearing the Irrational: Music and the Development of the Modern Concept of Number', *Isis* 101, no. 3 (2010): 501–30.

Pierozzi, L., and Jean-Marc Mandosio, 'L'interprétation alchimique de deux travaux d'Hercule dans le "De magia naturali" de Lefèvre d'Étaples', *Chrysopoeia* 5 (1992–6): 190–264.

Pingree, David, 'Boethius' Geometry and Astronomy', in *Boethius: His Life, Thought and Influence*, ed. Margaret Gibson (Oxford: Oxford University Press, 1981), 155–61.

Poel, Marc van der, *Cornelius Agrippa, the Humanist Theologian and His Declamations* (Leiden: Brill, 1997).

Poovey, Mary, *A History of the Modern Fact: Problems of Knowledge in the Sciences of Wealth and Society* (Chicago: University of Chicago Press, 1998).

Popper, Nicholas, 'The English Polydaedali: How Gabriel Harvey Read Late Tudor London', *Journal of the History of Ideas* 66, no. 3 (2005): 351–81.

Porrer, Sheila M., *Jacques Lefèvre d'Etaples and the Three Maries Debates* (Geneva: Droz, 2009).

Portuondo, María, *Secret Science: Spanish Cosmography and the New World* (Chicago: University of Chicago Press, 2009).

Potter, David, *Renaissance France at War: Armies, Culture and Society* (Woodbridge: Boydell Press, 2008).

Poulle, Emmanuel, *Un constructeur d'instruments astronomiques au XVe siècle: Jean Fusoris* (Paris: Honoré Champion, 1963).

Poulle, Emmanuel, 'Oronce Fine', in *Dictionary of Scientific Biography*, ed. Charles C. Gillespie, vol. 15 (New York: Charles Scribner's Sons, 1978), 153–7.

Prantl, Carl Von, *Geschichte der Logik im Abendlande*, 4 vols (Leipzig: Hirzel, 1855).

Rabouin, David, *Mathesis Universalis: L'idée de 'mathématique universelle' d'Aristote à Descartes* (Paris: Épiméthée, 2009).

Raphael, Renée, *Reading Galileo: Scribal Technologies and the Two New Sciences* (Baltimore, MD: Johns Hopkins University Press, 2017).

Redgrave, Gilbert R., *Erhard Ratdolt and His Work at Venice* (London: Chiswick Press, 1894).

Reid, Jonathan A., *King's Sister – Queen of Dissent: Marguerite of Navarre (1492–1549) and Her Evangelical Network*, 2 vols (Leiden: Brill, 2009).

Reif, Patricia, 'The Textbook Tradition in Natural Philosophy, 1600–1650', *Journal of the History of Ideas* 30, no. 1 (1969): 17–32.

Reiss, Timothy J., *Knowledge, Discovery and Imagination in Early Modern Europe* (Cambridge: Cambridge University Press, 1997).

Remmert, Volker R., *Picturing the Scientific Revolution: Title Engravings in Early Modern Scientific Publications*, trans. Ben Kern (Philadelphia, PA: Saint Joseph's University Press, 2011).

Renaudet, Augustin, *Préréforme et humanisme à Paris pendant les premières guerres d'Italie, 1494–1517* (1916; 2nd edn, Paris: Édouard Champion, 1953).

Renaudet, Augustin, 'L'Humanisme et l'enseignement de l'université de Paris au temps de la renaissance', in *Aspects de l'Universite de Paris* (Paris: Albin Michel, 1949), 135–55.

Renouard, Antoine August, *Annales de l'imprimerie des Estienne; ou, Histoire de la famille des Estienne et de ses editions* (Paris: J. Renouard et cie, 1843).

Renouard, Philippe, *Répertoire des imprimeurs parisiens*, ed. Jeanne Veyrin-Forrer and Brigitte Moreau (Paris: M.J. Minard, 1965).

Renouard, Philippe, *Documents sur les imprimeurs, libraires, cartiers, graveurs, fondeurs de lettres, relieurs, doreurs de livres, faiseurs de fermoirs, enlumineurs, parcheminiers et papetiers ayant exercé à Paris de 1450 à 1600* (Paris: H. Champion, 1901).

Renouard, Philippe, *Bibliographie des impressions et des œuvres de Josse Badius Ascensius, imprimeur et humaniste, 1462–1535*, 3 vols (New York: Burt Franklin, 1980).

Rice, Eugene F., Jr., *The Renaissance Idea of Wisdom* (Cambridge, MA: Harvard University Press, 1958).

Rice, Eugene F., Jr., 'Humanist Aristotelianism in France: Jacques Lefèvre d'Étaples and His Circle', in *Humanism in France*, ed. A. H. T. Levi (Manchester: Manchester University Press, 1970), 132–49.

Rice, Eugene F., Jr., 'The Patrons of French Humanism, 1490–1520', in *Renaissance Studies in Honor of Hans Baron*, ed. Anthony Molho and John A. Tedeschi (Dekalb, IL: Northern Illinois University Press, 1971), 687–702.

Rice, Eugene F., Jr., 'The *De Magia Naturali* of Jacques Lefèvre d'Étaples', in *Philosophy and Humanism: Renaissance Essays in Honor of Paul Oskar Kristeller*, ed. Edward P. Mahoney (Leiden: Brill, 1976), 19–29.

Rice, Eugene F., Jr., 'Jacques Lefevre d'Etaples and the Medieval Christian Mystics', in *Florilegium Historiale: Essays Presented to Wallace K. Ferguson*, ed. J. G. Rowe and W. H. Stockdale (Toronto: University of Toronto Press, 1971), 90–124.

Richardson, Brian, *Print Culture in Renaissance Italy: The Editor and the Vernacular Text, 1470–1600* (Cambridge: Cambridge University Press, 1994).

Richardson, Brian, *Manuscript Culture in Renaissance Italy* (Cambridge: Cambridge University Press, 2009).

Robertson, Kellie, *Nature Speaks: Medieval Literature and Aristotelian Philosophy* (University Park, PA: University of Pennsylvania Press, 2017).

Robichaud, Denis J.-J., 'Angelo Poliziano's Lamia: Neoplatonic Commentaries and the Plotinian Dichotomy between the Philologist and the Philosopher', in *Angelo Poliziano's 'Lamia': Text, Translation, and Introductory Studies*, ed. Christopher S. Celenza (Leiden: Brill, 2010), 131–89.

Rolls, Jonathan James, 'God and the World: Some Interpretations of the "Transcendental" Analogy of Being in Western Theology From the Thirteenth to the Sixteenth Centuries', PhD dissertation, Warburg Institute, University of London, 1999.

Romano, Antonella, *La Contre-réforme Mathématique: Constitution et diffusion d'une culture mathématique jésuite à la Renaissance (1540–1640)* (Rome: École française de Rome, 1999).

Rommevaux, Sabine, Philippe Vendrix, and Vasco Zara (eds.), *Proportions: Science, musique, peinture et architecture* (Turnhout: Brepols, 2011).

Rose, Paul Lawrence, *The Italian Renaissance of Mathematics* (Geneva: Droz, 1975).

Rose, Paul Lawrence, 'Humanist Culture and Renaissance Mathematics: The Italian Libraries of the Quattrocento', *Studies in the Renaissance* 20 (1973): 46–105.

Rossi, Paolo, *Logic and the Art of Memory: The Quest for a Universal Language*, trans. Stephen Clucas (1983; London: Continuum, 2000).

Rouse, Mary A., and Richard H. Rouse, *Authentic Witnesses: Approaches to Medieval Texts and Manuscripts* (Notre Dame, IN: University of Notre Dame Press, 1991).

Rouse, Richard H., and Mary A. Rouse, 'The Book Trade at the University of Paris, ca. 1250–ca. 1350', in *La production du livre universitaire au Moyen Age: exemplar et pecia: actes du symposium tenu au Collegio San Bonaventura de Grottaferrata en mai 1983*, ed. Louis Bataillon, Bertrand G. Guyot, and Richard H. Rouse (Paris: Editions du Centre national de la recherche scientifique, 1988), 41–114.

Ross, Richard, 'The Mathematical Works of Oronce Finé', PhD dissertation, Columbia University, 1971.

Roux, Simone, *La Rive gauche des escholiers (XVe siècle)* (Paris: Éditions Christian, 1992).

Roux, Sophie, 'Forms of Mathematization (14th–17th Centuries)', *Early Science and Medicine* 15 (2010): 319–37.

Rummel, Erika, *Erasmus and His Catholic Critics* (Nieuwkoop: De Graaf, 1989).

Saenger, Paul, and Michael Heinlin, 'Incunable Description and Its Implication for the Analysis of Fifteenth-Century Reading Habits', in *Printing the Written Word: The Social History of Books, circa 1450–1520*, ed. Sandra L. Hindman (Ithaca, NY: Cornell University Press, 1991), 225–58.

Saenger, Paul, 'Reading in the Later Middle Ages', in *A History of Reading in the West*, ed. Guglielmo Cavallo and Roger Chartier, trans. Lydia G. Cochrane (Amherst, MA: University of Massachusetts Press, 1999), 120–48.

Schaffer, Simon, 'Newton on the Beach: The Information Order of *Principia Mathematica*', *History of Science* 47 (2009): 243–76.

Schmidt-Biggemann, Wilhelm, *Topica universalis: eine Modellgeschichte humanistischer und barocker Wissenschaft* (Hamburg: Meiner, 1983).

Schmitt, Charles B., 'Perennial Philosophy: From Agostino Steuco to Leibniz', *Journal of the History of Ideas* 27, no. 4 (1966): 505–32.

Schmitt, Charles B., *Aristotle and the Renaissance* (Cambridge, MA: Harvard University Press, 1983).

Schmitt, Charles B., 'The Rise of the Philosophical Textbook', in *The Cambridge History of Renaissance Philosophy*, ed. Charles B. Schmitt and Quentin Skinner (Cambridge: Cambridge University Press, 1988), 792–804.

Schöner, C., *Mathematik und Astronomie an der Universität Ingolstadt im 15. und 16. Jahrhundert* (Berlin: Dunker & Humblot, 1994).

Schreiber, Fred, *The Estiennes* (New York: E.K. Schreiber, 1982).

Schreiber, Fred, and Jeanne Veyrin-Forrer, *Simon de Colines: An Annotated Catalogue of 230 Examples of His Press, 1520–1546* (London: Oak Knoll Press, 1995).

Scott-Warren, Jason, 'Harvey, Gabriel (1552/3–1631)', *Oxford Dictionary of National Biography* (Oxford: Oxford University Press, 2004; online edn, Jan. 2016).

Serjeantson, Richard, 'The Education of Francis Willughby', in *Virtuoso by Nature: The Scientific Worlds of Francis Willughby FRS (1635–1672)*, ed. Tim Birkhead (Leiden: Brill, 2016), 44–98.

Sfez, Jocelyne, *L'Art des Conjectures de Nicolas de Cues* (Paris: Beauchesne, 2012).

Shalev, Zur, and Charles Burnett (eds), *Ptolemy's Geography in the Renaissance* (London: Warburg Institute, 2011).

Shank, Michael H., *'Unless You Believe, You Shall Not Understand': Logic, University, and Society in Late Medieval Vienna* (Princeton, NJ: Princeton University Press, 1988).

Shea, William R., *The Magic of Numbers and Motion: The Scientific Career of René Descartes* (Canton, MA: Science History Publications, 1991).

Sherman, William H., 'The Begining of "The End": Terminal Paratext and the Birth of Print Culture', in *Renaissance Paratexts*, ed. Helen Smith and Louise Wilson (Cambridge: Cambridge University Press, 2011), 65–87.

Siegel, Steffen, *Tabula: Figuren der Ordnung um 1600* (Berlin: Akademie Verlag, 2009).

Smith, Helen, and Louise Wilson (eds), *Renaissance Paratexts* (Cambridge: Cambridge University Press, 2011).

Smoller, Laura Ackerman, *History, Prophecy, and the Stars: The Christian Astrology of Pierre D'Ailly, 1350–1420* (Princeton, NJ: Princeton University Press, 1994).

Söderlund, Inga Elmqvist, *Taking Possession of Astronomy: Frontispieces and Illustrated Title Pages in 17th-Century Books on Astronomy* (Stockholm: Center for History of Science at the Royal Swedish Academy of Sciences, 2010).

Solère, Jean-Luc, 'L'ordre axiomatique comme modèle d'écriture philosophique dans l'Antiquité et au Moyen Âge', *Revue d'Histoire des Sciences* 56, no. 2 (2003): 323–45.

Somfai, Anna, 'Calcidius's Commentary to Plato's Timaeus and Its Place in the Commentary Tradition: The Concept of Analogia in Text and Diagrams', in *Philosophy, Science and Exegesis in Greek, Arabic and Latin Commentaries*, ed. P. Adamson, H. Baltussen, and M. W. F. Stone, 2 vols (London: Institute of Classical Studies, 2004), 1:203–330.

Spruit, Leen, *Species Intelligibilis: From Perception to Knowledge*, vol. II: *Renaissance Controversies, Later Scholasticism, and the Elimination of the Intelligible Species in Modern Philosophy* (Leiden: Brill, 1995).

Stachowski, Ryszard, *The Mathematical Soul: An Antique Prototype of the Modern Mathematisation of Psychology* (Amsterdam: Rodopi, 1992).

Stadler, Michael, 'Zum Begriff der Mensuratio bei Cusanus. Ein Beitrag zur Ortung der Cusanischen Erkenntnislehre', in *Mensura: Mass, Zahl, Zahlensymbolik im Mittelalter*, ed. Albert Zimmermann (Berlin: de Gruyter, 1983), 118–31.

Stallybrass, Peter, 'Printing and the Manuscript Revolution', in *Explorations in Communication and History*, ed. Barbie Zelizer (London: Routledge, 2008), 111–18.

Stam, Frans Pieter van, 'The Group of Meaux as First Target of Farel and Calvin's Anti-Nicodemism', *Bibliothèque d'Humanisme et Renaissance* 68, no. 2 (2006): 253–75.

Swanson, R. N., *Universities, Academics and the Great Schism* (Cambridge: Cambridge University Press, 1979).

Sylla, Edith Dudley, 'Medieval Concepts of the Latitude of Forms: The Oxford Calculators', *Archives d'histoire doctrinale et littéraire du Moyen Âge* 40 (1973): 223–83.

Sylla, Edith Dudley, 'The Oxford Calculators', in *The Cambridge History of Later Medieval Philosophy*, ed. Norman Kretzmann, Anthony Kenny, and Jan Pinborg (Cambridge: Cambridge University Press, 1982), 540–63.

Sylla, Edith Dudley, 'Compounding Ratios: Bradwardine, Oresme, and the First Edition of Newton's Principia', in *Transformation and Tradition in the Sciences: Essays in Honour of I. Bernard Cohen*, ed. Everett Mendelsohn (Cambridge: Cambridge University Press, 1984), 11–43.

Sylla, Edith Dudley, 'The Oxford Calculators in Context', *Science in Context* 1 (1987): 257–79.

Sylla, Edith Dudley, 'Alvarus Thomas and the Role of Logic and Calculations in Sixteenth Century Natural Philosophy', in *Studies in Medieval Natural Philosophy*, ed. Stefano Caroti (Florence: Leo S. Olschki, 1989), 257–98.

Sylla, Edith Dudley, 'The Origin and Fate of Thomas Bradwardine's *De Proportionibus Velocitatum in Motibus* in Relation to the History of Mathematics', in *Mechanics and Natural Philosophy Before the Scientific Revolution*, ed. Walter Roy Laird and Sophie Roux (Dordrecht: Springer, 2007), 67–119.

Talbot, C. H., 'The Universities and the Mediaeval Library', in *The English Library Before 1700*, ed. F. Wormald and C. E. Wright (London: Athlone Press, 1958), 66–84.

Tanner, Marcus, *The Raven King: Matthias Corvinus and the Fate of His Lost Library* (New Haven, CT: Yale University Press, 2009).

Taylor, Daniel J., *Declinatio: A Study of the Linguistic Theory of Marcus Terentius Varro* (Amsterdam: John Benjamins, 1974).

Taylor, Larissa, *Soldiers of Christ: Preaching in Late Medieval and Reformation France* (Oxford: Oxford University Press, 1992).

Thomas-Stanford, Charles, *Early Editions of Euclid's Elements* (London: Bibliographical Society, 1926).

Thorndike, Lynn, *A History of Magic and Experimental Science*, vol. 4 (New York: Columbia University Press, 1924).

Thorndike, Lynn, *University Records and Life in the Middle Ages* (New York: Columbia University Press, 1944).

Thurot, Charles, *De l'organisation de l'enseignement dans l'Université de Paris au Moyen-Âge* (Deis, 1850).

Traninger, Anita, *Disputation, Deklamation, Dialog: Medien und Gattungen europäischer Wissensverhandlungen zwischen Scholastik und Humanismus* (Stuttgart: Franz Steiner Verlag, 2012).

Tura, Adolfo, *Fra Giocondo et les textes français de géométrie pratique* (Geneva: Droz, 2008).

Ullman, B. L., 'Geometry in the Medieval Quadrivium', in *Studi di bibliografia e di storia in onore di Tammaro de Marinis*, vol. 4 (Verona, 1964), 263–85.

Valleriani, Matteo, *Galileo Engineer* (Dordrecht: Springer, 2010).

Vanautgaerden, Alexandre, *Érasme typographe: humanisme et imprimerie au début du XVIe siècle* (Geneva: Droz, 2012).

Van Engen, John, 'Multiple Options: The World of the Fifteenth-Century Church', *Church History* 77, no. 2 (2008): 257–84.

Van Engen, John, *Sisters and Brothers of the Common Life: The Devotio Moderna and the World of the Later Middle Ages* (Philadelphia, PA: University of Pennsylvania Press, 2008), 14–19.

Vasoli, Cesare, 'Jacques Lefèvre d'Etaples e le origini del "Fabrismo"', *Renascimento* 10 (1959): 238–40.

Vasoli, Cesare, *La dialettica e la retorica dell'Umanesimo: 'Invenzione' e 'Metodo' nella cultura del XV e XVI secolo* (Milan: Feltrinelli, 1968).

Veenstra, Jan R., *Magic and Divination at the Courts of Burgundy and France: Text and Context of Laurens Pignon's 'Contre Les Devineurs' (1411)* (Leiden: Brill, 1997).

Veenstra, Jan R., 'Jacques Lefèvre d'Étaples: Humanism and Hermeticism in the De Magia Naturali', in *Christian Humanism: Essays in Honour of Arjo Vanderjagt*, ed. Arie Johan Vanderjagt, Alasdair A. MacDonald, and Z. R. W. M. von Martels (Leiden: Brill, 2009), 353–62.

Verger, Jacques, 'Le livre dans les universités du Midi de la France à la fin du Moyen Âge', in *Pratiques de la culture écrite en France au XVe siècle*, ed. Monique Ornato and Nicole Pons (Louvain: Fédération internationale des instituts d'études médiévale, 1995), 403–20.

Verger, Jacques, *Men of Learning in Europe at the End of the Middle Ages*, trans. Steven Rendall (Notre Dame, IN: University of Notre Dame Press, 2000).

Verger, Jacques, 'Patterns', in *A History of the University in Europe*, vol. 1: *Universities in the Middle Ages*, ed. H. de Ridder-Symoens (Cambridge: Cambridge University Press, 2003), 35–68.

Verger, Jacques, 'Landmarks for a History of the University of Paris at the Time of Jean Standonck', *History of Universities* 22, no. 2 (2007): 1–13.

Veyrin-Forrer, Jeanne, 'Simon de Colines, imprimeur de Lefèvre d'Etaples', in *Jacques Lefèvre d'Etaples (1450?–1536)*, ed. Jean-François Pernot (Paris: Honoré Champion, 1995), 97–117.

Vicedo, Marga, 'Introduction [to Focus Section]: The Secret Lives of Textbooks', *Isis* 103, no. 1 (2012): 83–7.

Victor, Joseph M., *Charles de Bovelles, 1479–1553: An Intellectual Biography* (Geneva: Droz, 1978).

Victor, Joseph M., 'The Revival of Lullism at Paris, 1499–1516', *Renaissance Quarterly* 28, no. 4 (1975): 504–34.

Victor, Stephen K., *Practical Geometry in the High Middle Ages: Artis Cuiuslibet Consummatio and the Pratike de Geometrie* (Philadelphia, PA: American Philosophical Society, 1979).

Viellard, Jeanne, 'Instruments d'astronomie conservés à la Bibliothèque du Collège de Sorbonne au XVe et XVIe siècles', *Bibliothèque de l'École des Chartres* 131 (1973): 586–93.

Villoslada, Ricardo García, *La universidad de Paris durante los estudios de Francisco de Vitoria (1507–1522)* (Rome: Gregorian University, 1938).

Walker, D. P., 'The Prisca Theologia in France', *Journal of the Warburg and Courtauld Institutes* 17, no. 3/4 (1954): 204–59.

Walter, Joseph, *Catalogue général de la Bibliothèque municipale, première série, les livre imprimés, troisième partie: Incunables & XVIme siècle* (Colmar: Imprimerie Alstasia, 1929).

Walter, Robert, *Trois Profils de Beatus Rhenanus: L'homme, Le Savant, Le Chrétien* (Sélestat: Les amis de la Bibliothèque humaniste de Sélestat, 2002).

Waquet, Françoise, *Parler comme un livre: l'oralité et le savoir, XVIe–XXe siècle* (Paris: Albin Michel, 2003).

Wardhaugh, Benjamin, *Music, Experiment and Mathematics in England, 1653–1705* (Aldershot: Ashgate, 2008).

Wardhaugh, Benjamin, 'Musical Logarithms in the Seventeenth Century: Descartes, Mercator, Newton', *Historia Mathematica* 35, no. 1 (2008): 19–36.

Warwick, Andrew, *Masters of Theory: Cambridge and the Rise of Mathematical Physics* (Chicago: University of Chicago Press, 2003).

Watt, Donald E. R., 'University Clerks and Rolls of Petitions for Benefices', *Speculum* 34, no. 2 (1959): 213–29.

Weier, Reinhold, *Das Thema vom verborgenen Gott von Nikolaus von Kues zu Martin Luther: Dissertation* (Mainz: R. Weier, 1965).

Weijers, Olga, and Louis Holtz (eds), *L'Enseignement des disciplines à la Faculté des arts (Paris et Oxford, XIIIe–XVe siècles)* (Turnhout: Brepols, 1997).

Weijers, Olga, 'The Development of the Disputation Between the Middle Ages and Renaissance', in *Continuities and Disruptions Between the Middle Ages and the Renaissance: Proceedings of the colloquium held at the Warburg Institute, 15–16 June 2007*, ed. Charles Burnett, José Meirinhos, and Jacqueline Hamesse (Louvain-la-Neuve: Brepols, 2008), 139–150.

Weijers, Olga, *In Search of the Truth: A History of Disputation Techniques from Antiquity to Early Modern Times* (Turnhout: Brepols, 2013).

Weijers, Olga, *A Scholar's Paradise: Teaching and Debating in Medieval Paris* (Turnhout: Brepols, 2015).

Weissenborn, Hermann, *Die Übersetzungen des Euklid durch Campano und Zamberti eine mathematisch-historische Studie* (Halle a/S.: Schmidt, 1882).

Wellmon, Chad, *Organizing Enlightenment: Information Overload and the Invention of the Modern Research University* (Baltimore, MD: Johns Hopkins University Press, 2015).

Westman, Robert S., 'The Melanchthon Circle, Rheticus, and the Wittenberg Interpretation of the Copernican Theory', *Isis* 66, no. 2 (1975): 164–93.

Whitney, Elspeth, *Paradise Restored: The Mechanical Arts from Antiquity through the Thirteenth Century*, Transactions of the American Philosophical Society 80 (Philadelphia, PA: American Philosophical Society, 1990).

Wilson, Curtis, *William Heytesbury: Medieval Logic and the Rise of Mathematical Physics* (Madison, WI: University of Wisconsin Press, 1956).

Wissenborn, H., *Acten der Erfurter Universität*, vol. 2 (Halle: O. Hendel, 1881).

Wolfe, Jessica, *Humanism, Machinery, and Renaissance Literature* (Cambridge: Cambridge University Press, 2004).

Yates, Frances A., *The French Academies of the Sixteenth Century* (London: Warburg Institute, 1947).

Yates, Frances A., *The Art of Memory* (Chicago: University of Chicago Press, 1966).

Yeo, Richard, *Notebooks, English Virtuosi, and Early Modern Science* (Chicago: University of Chicago Press, 2014).

Young, Spencer E., 'Faith, Favour, and Fervour: Emotions and Conversion among the Early Dominicans', *Journal of Religious History* 39, no. 4 (2015): 468–83.

Zetterberg, J. Peter, 'The Mistaking of "the Mathematicks" for Magic in Tudor and Stuart England', *Sixteenth Century Journal* 11, no. 1 (1980): 83–97.

Zilsel, Edgar, 'The Sociological Roots of Science', *American Journal of Sociology* 47 (1942): 544–62.

Zorach, Rebecca, 'Meditation, Idolatry, Mathematics: The Printed Image in Europe around 1500', in *The Idol in the Age of Art: Objects, Devotions and the Early Modern World*, ed. Michael Wayne Cole and Rebecca Zorach (Farnham: Ashgate, 2009), 317–42.

Index